Convergent Evolution in Stone-Tool Technology

Vienna Series in Theoretical Biology

Gerd B. Müller, editor-in-chief

Thomas Pradeu and Katrin Schäfer, associate editors

The Evolution of Cognition, edited by Cecilia Heyes and Ludwig Huber, 2000

Origination of Organismal Form, edited by Gerd B. Müller and Stuart A. Newman, 2003

Environment, Development, and Evolution, edited by Brian K. Hall, Roy D. Pearson, and Gerd B. Müller, 2004

Evolution of Communication Systems, edited by D. Kimbrough Oller and Ulrike Griebel, 2004

Modularity: Understanding the Development and Evolution of Natural Complex Systems, edited by Werner Callebaut and Diego Rasskin-Gutman, 2005

Compositional Evolution: The Impact of Sex, Symbiosis, and Modularity on the Gradualist Framework of Evolution, by Richard A. Watson, 2006

Biological Emergences: Evolution by Natural Experiment, by Robert G. B. Reid, 2007

Modeling Biology: Structure, Behaviors, Evolution, edited by Manfred D. Laubichler and Gerd B. Müller, 2007

Evolution of Communicative Flexibility, edited by Kimbrough D. Oller and Ulrike Griebel, 2008

Functions in Biological and Artificial Worlds, edited by Ulrich Krohs and Peter Kroes, 2009

Cognitive Biology, edited by Luca Tommasi, Mary A. Peterson, and Lynn Nadel, 2009

Innovation in Cultural Systems, edited by Michael J. O'Brien and Stephen J. Shennan, 2010

The Major Transitions in Evolution Revisited, edited by Brett Calcott and Kim Sterelny, 2011

Transformations of Lamarckism, edited by Snait B. Gissis and Eva Jablonka, 2011

Convergent Evolution: Limited Forms Most Beautiful, by George McGhee, 2011

From Groups to Individuals, edited by Frédéric Bouchard and Philippe Huneman, 2013

Developing Scaffolds in Evolution, Culture, and Cognition, edited by Linnda R. Caporael, James Griesemer, and William C. Wimsatt, 2014

Multicellularity: Origins and Evolution, edited by Karl J. Niklas and Stuart A. Newman, 2015

Vivarium: Experimental, Quantitative, and Theoretical Biology at Vienna's Biologische Versuchsanstalt, edited by Gerd B. Müller, 2017

Landscapes of Collectivity in the Life Sciences, edited by Snait B. Gissis, Ehud Lamm, and Ayelet Shavit, 2017

Convergent Evolution in Stone-Tool Technology, edited by Michael J. O'Brien, Briggs Buchanan, and Metin I. Eren, 2018

Convergent Evolution in Stone-Tool Technology

edited by Michael J. O'Brien, Briggs Buchanan, and Metin I. Eren

The MIT Press
Cambridge, Massachusetts
London, England

This book was set in Times New Roman by Toppan Best-set Premedia Limited.

Library of Congress Cataloging-in-Publication Data

Names: O'Brien, Michael J. (Michael John), 1950- editor. | Buchanan, Briggs (Briggs Wheeler), editor. | Eren, Metin I., 1982- editor.
Title: Convergent evolution in stone-tool technology / edited by Michael J. O'Brien, Briggs Buchanan, and Metin I. Eren.
Description: Cambridge, MA : The MIT Press, 2018. | Series: Vienna series in theoretical biology | Includes bibliographical references and index.
Identifiers: LCCN 2017038928 | ISBN 9780262037839 (hardcover : alk. paper)
ISBN 9780262552080 (paperback)
Subjects: LCSH: Tools, Prehistoric. | Stone implements. | Paleolithic period.
Classification: LCC GN799.T6 C56 2018 | DDC 930.1/2—dc23 LC record available at https://lccn.loc.gov/2017038928

Contents

Series Foreword

Biology is a leading science in this century. As in all other sciences, progress in biology depends on the interrelations between empirical research, theory building, modeling, and societal context. But whereas molecular and experimental biology have evolved dramatically in recent years, generating a flood of highly detailed data, the integration of these results into useful theoretical frameworks has lagged behind. Driven largely by pragmatic and technical considerations, research in biology continues to be less guided by theory than seems indicated. By promoting the formulation and discussion of new theoretical concepts in the biosciences, this series intends to help fill important gaps in our understanding of some of the major open questions of biology, such as the origin and organization of organismal form, the relationship between development and evolution, and the biological bases of cognition and mind. Theoretical biology has important roots in the experimental tradition of early-twentieth-century Vienna. Paul Weiss and Ludwig von Bertalanffy were among the first to use the term *theoretical biology* in its modern sense. In their understanding the subject was not limited to mathematical formalization, as is often the case today, but extended to the conceptual foundations of biology. It is this commitment to a comprehensive and cross-disciplinary integration of theoretical concepts that the Vienna Series intends to emphasize. Today, theoretical biology has genetic, developmental, and evolutionary components, the central connective themes in modern biology, but it also includes relevant aspects of computational or systems biology and extends to the naturalistic philosophy of sciences. The Vienna Series grew out of theory-oriented workshops organized by the KLI, an international institute for the advanced study of natural complex systems. The KLI fosters research projects, workshops, book projects, and the journal *Biological Theory*, all devoted to aspects of theoretical biology, with an emphasis on—but not restriction to—integrating the developmental, evolutionary, and cognitive sciences. The series editors welcome suggestions for book projects in these domains.

Gerd B. Müller, Thomas Pradeu, Katrin Schäfer

Series Foreword

Preface and Acknowledgments

Stone tools and the debris from stone-tool manufacture are found throughout the archeological record of humans and their ancestors. The first unambiguous hominin-produced tools appeared approximately 2.6 million years ago, although recent studies have shown indirect evidence that hominins began using stone tools nearly 3.4 million years ago. Stone has been used to make tools in nearly all regions of the globe that have been inhabited. Given the nearly ubiquitous use of stone tools by hominins, their study is an important line of inquiry for shedding light on questions of evolution and behavior. Researchers have long studied stone tools and have investigated a wide range of topics, including the evolution of technology, prehistoric economy, hominin global dispersals, and ancient engineering, but the topic of evolutionary convergence remains an understudied yet potentially important avenue of research.

Convergence is the phenomenon by which evolutionary processes result in the same, or similar, forms in independent lineages as a result of functional or developmental constraints. In studies of stone tools, identifying cases of convergence is of particular importance because similarities in form and function are often used to suggest historical connections among prehistoric groups. Identifying cases of convergence would refute hypotheses that otherwise would suggest some degree of physical or cultural connection among toolmakers. The reason that convergence remains understudied has to do in large part with the unsupported assumption that there are "endless" stone-tool production techniques and forms.

Considerable reason exists to doubt this widely held belief because the manufacture of stone tools is a reductive process, whereby stone flakes are removed from larger cores to make smaller tool forms. As stone is reduced, the number of possible outcomes in terms of form becomes increasingly constrained. Widespread convergence of lithic technologies is also possible as a result of the fracturing properties of stone, which are governed by a specific set of physical constraints.

The vast array of stone types that are appropriate for stone-tool manufacture—flint, obsidian, basalt, and quartzite, for example—are controlled by the same basic set of fracture properties, which increases the possibility of convergence. Moreover, prehistoric people, albeit in different times and places, would have faced similar adaptive challenges that in turn would ostensibly have directed tool forms toward similar optimal designs, or "adaptive peaks." Considering all these factors, it should be no surprise that several recent studies are empirically consistent with the hypothesis that convergence in lithic technology is not rare.

To explore these issues, we organized the 33rd Altenberg Workshop in Theoretical Biology at the Konrad Lorenz Institute (KLI) in Klosterneuburg, Austria, from June 16 to 19, 2016. We invited a variety of scholars from different disciplines—archaeology as well as paleontology, philosophy, and paleoanthropology—to present papers and initiate in-depth conversation. As is usual for KLI workshops, the organizers were asked to edit a book for submission to the Vienna Series in Theoretical Biology, which is published by MIT Press. We made it clear to participants that the volume would not consist simply of conference proceedings; rather, it would further develop the novel ideas and concepts generated at the meeting.

In general terms, our intent is to generate new conceptual advances that, because of their explicit interdisciplinary nature, will be attractive to a wide range of experts in the human sciences and neighboring disciplines. Specifically, this book aims to address the lack of research devoted to the principle of evolutionary convergence in stone tools. Through specific empirical case studies, the book chapters address the following questions:

- Why does convergence occur in the stone-tool record?
- Did particular stone-tool technologies originate once, or did prehistoric people converge on various innovations several times independently?
- How often does convergence occur in the stone-tool record, and in what forms?
- Are there particular environmental or behavioral situations in which we can predict convergence in stone tools?
- How is the process of evolutionary convergence in stone tools similar to, or different than, that seen in biological species?

By all measures, the KLI workshop was a success—a point hopefully underscored by the content of the chapters included here. We are extremely grateful to the KLI for funding the workshop, with a generous subsidy from the University of Missouri. As usual, our hosts—Gerd Müller, Isabella Sarto-Jackson, and Eva Lackner—went out of their way to make the event memorable. We also thank the fellows of the KLI, who added substantially to the discussions between sessions. Professor Müller, who is editor-in-chief for MIT Press' *Vienna Series in Theoretical Biology*, guided us through the proposal process with the press. Finally, we thank our colleagues at MIT Press, especially Anne-Marie Bono, Katherine Almeida, and Bob Prior, executive editor of MIT Press. This is the fourth book that MJO has published with Bob's unflagging support, and he acknowledges his deep gratitude.

I INTRODUCTION

1 Issues in Archaeological Studies of Convergence

Michael J. O'Brien, Briggs Buchanan, and Metin I. Eren

As we note in the preface, the nearly ubiquitous use of stone tools by hominins over the past several million years, coupled with the fact that those tools and the by-products of their manufacture are highly durable, make them obvious places to look for significant insights into human evolution and behavior. Researchers have long studied stone tools to examine a wide range of topics, including the evolution of technology, prehistoric economy, hominin global dispersals, and ancient engineering, but the topic of evolutionary convergence has until recently been an understudied yet potentially important avenue of research. Indeed, perhaps "critical" is a better adjective than "important." Without the ability to differentiate between cases of convergence (organisms land on similar solutions to common problems) and cases of divergence (organisms exhibit similar solutions as a result of common ancestry), evolutionary explanations end up being just-so stories.

Our modest goal here is not to delve too deeply into all the myriad issues surrounding convergence or attempt to list the numerous archaeological case studies that have focused on convergence in stone tools. We leave those topics to our colleagues, whose excellent work appears in the following chapters. Rather, our goal is to show how archaeologists have long wrestled with how to distinguish convergence from divergence and to demonstrate the impact that confusing the two processes can have on archaeological explanations.

Convergence in Archaeology: Getting Past a Stalemate

As the chapters in this volume make clear, archaeology has made remarkable strides in being able to distinguish between convergence and divergence. In our opinion, these strides were possible only after there was a wholesale change in how archaeology was viewed—a change that was at once ontological, epistemological, and methodological (O'Brien, Lyman, and Schiffer 2005). We might also add "painful" to the mix, because all true shifts in paradigm—and that is what this was, in the pure sense of how Kuhn (1962) used the term—are highly disruptive. The change from what has been termed a "culture-historical" perspective (Lyman, O'Brien, and Dunnell 1997) to an evolutionary perspective began in the early 1980s and picked up speed throughout the decade, to the point where today, for example, the term "evolutionary archaeology" isn't even used anymore because evolution is so ingrained in the discipline.

This statement, however, comes with a caveat. Although evolution has a deep history in archaeology the manner in which the term is used today in the field, parallel to its use in biology, has little or nothing in common with how archaeologists used the term throughout much of the twentieth century. Whereas biological evolution is based on descent with modification and emphasizes processes such as mutation, drift, and selection, the brand of evolution made popular by culture historians relied primarily on unilinear, progressive developmental sequences popular in the nineteenth-century work of E. B. Tylor (1871), Lewis Henry Morgan (1877), and others. These sequences, or "progressions," supposedly were driven by processes such as diffusion and population movement (e.g., Colton 1942; Ford 1952, 1969; Gladwin 1936) or even, in some quarters, a "psychic unity of mankind" (Koepping 1984). This reliance on unilinear developmental schemes, built on a view of cultural evolution as being synonymous with *change*, ensured a lack of analytical clarity between convergence and divergence.

The Stalemate

To archaeologists working throughout most of the twentieth century, culture was an evolving entity, but any similarity between biological and cultural evolution was strictly metaphorical (Kroeber 1923). The former was seen as being inextricably linked to genetic transmission, whereas the latter was not (Brew 1946). Thus, any attempt to link Darwin's mechanism for change—natural selection—to the evolution of culture was nothing more than misapplied biology. Julian Steward (1941: 367) pointed this out rather forcefully: "It is apparent … that strict adherence to a method drawn from biology inevitably fails to take into account the distinctively cultural and unbiological fact of blends and crosses between essentially unlike types. … A taxonomic scheme cannot indicate this fact without becoming mainly a list of exceptions. It must pigeon-hole, … [which] inevitably distorts true cultural relationships."

Somewhat ironically, despite their distaste for biological evolution as a framework for understanding the evolution of culture, many culture historians showed not just a knowledge of, but an appreciation for, the distinction between convergence and divergence, the former creating analogous similarity and the latter homologous similarity. They knew, as did their biologist colleagues, that analogous characters are those that two or more organisms possess that, while they might serve similar purposes, did not evolve because of common ancestry. For example, birds and bats both have wings, and those characters share properties in common, yet the organisms are placed in two widely separate taxonomic groups because birds and bats are only distantly related. Those two large groups diverged from a common vertebrate ancestor long before either one of them developed wings. Thus wings are of no utility in reconstructing lineages, because they evolved independently in the two lineages after they diverged. The character of having wings is held in common by birds and bats, but the state of the character—the details of its osteological composition and anatomical structure—differs between the two groups.

Darwin, unlike his predecessors, offered a new and logical causal explanation as to *why* there would be formal similarities between organs and organisms. He argued that "by unity of type is meant that fundamental agreement in structure, which we see in organic beings of the same class, and which is quite independent of their habits of life. On my theory, unity of type is explained by unity of descent" (Darwin 1859: 206). In those two short sentences, Darwin clearly distinguished between analogous and homologous characters and provided the first explicitly scientific and theoretical explanation for the existence of homologs: homologous similarity is historical because it results from *heritable continuity* (Lyman and O'Brien 1998). With such a definition, a biologist could answer the questions, Why are the wings of swallows and crows similar, yet the two organisms are different species? and Why are the wings of brown bats and fruit bats similar, yet bat wings and bird wings are similar but structurally different?

On the cultural side, culture historians might have appreciated the distinction between homologs and analogs, but they were at a standstill as to *how* to make the distinction analytically. It's one thing to recognize the distinction; it's another to be able to sort one kind of feature from the other. Writing in the 1930s, A. L. Kroeber (1931: 152–153) had this to say on the subject:

> There are cases in which it is not a simple matter to decide whether the totality of traits points to a true relationship or to secondary convergence. … Yet few biologists would doubt that sufficiently intensive analysis of structure will ultimately solve such problems of descent. … There seems no reason why on the whole the same cautious optimism should not prevail in the field of culture; why homologies should not be positively distinguishable from analogies when analysis of the whole of the phenomena in question has become truly intensive. That such analysis has often been lacking but judgments have nevertheless been rendered, does not invalidate the positive reliability of the method.

Although Kroeber was clear that there are two forms of similarity—one homologous and the other analogous—he was less than clear as to how the two could actually be distinguished. He suggested that identifying "similarities [that] are specific and structural and not merely superficial … has long been the accepted method in evolutionary and systematic biology" (Kroeber 1931: 151), but he offered no advice on how to separate what is "specific and structural" from what is "merely superficial" beyond undertaking a "sufficiently intensive analysis of structure." To Kroeber—and he was not alone—formal similarities between cultural phenomena signified some kind of ethnic relation, which was a predictable result of using ethnologically documented mechanisms such as diffusion and enculturation to account for typological similarities in the archaeological record (Lyman et al. 1997).

No one realized it at the time, but this was tautological and put the cart before the horse. Thus Gordon Willey's (1953: 363) statement that "typological similarity is an indicator of cultural relatedness (and this is surely axiomatic to archeology), [and thus] such relatedness carries with it implications of a common or similar history" caused little or no concern within the discipline. It might have caused considerable concern because the axiom falls prey to a caution raised by paleontologist George Gaylord Simpson (1961), using monozygotic twins as an example: they are twins not because they are similar; rather, they are similar because they are twins and thus share a common history.

We should not be too surprised that most culture historians were far more interested in divergence and homology than they were in convergence and analogy. There were key exceptions—Julian Steward (1938), for example—but, for the most part, it was precisely because of their interest in cultural evolution (which, again, was synonymous in most cultural-historical usage with *change*) that culture historians put so much effort into constructing units that allowed them to keep things in chronological and spatial order (Lyman et al. 1997). These units almost by definition were based on suspected homology (divergence), not on analogy (convergence). There was often an automatic assumption that, for example, similar pottery designs are homologous because the probability of duplication by chance is astronomically low. Similarly, as several authors in this volume point out (see chapters 8 and 13), the presence of Levallois-like flaking industries in widely separated regions has often been invoked as a case of homology by way of diffusion and population movement. We now know this is not necessarily the case.

By the 1960s, the number of articles and monographs touting diffusion and migration as explanatory devices had reached the point that John Rowe (1966) felt compelled to write a withering critique, commenting that most diffusionist accounts were nothing more than off-the-wall speculation:

> We are now being subjected in archaeological meetings to ever more strident claims that Mesoamerican culture was derived from China or southeast Asia, early Ecuadorian culture from Japan, Woodland culture from Siberia, Peruvian culture from Mesoamerica, and so forth. In the science-fiction world of the diffusionists, a dozen similarities of detail prove cultural contact, and time, distance, and the difficulties of navigation are assumed to be irrelevant. (Rowe 1966: 334)

As O'Brien and Lyman (1998) point out, Rowe was general enough in his comments that the uninitiated wouldn't have known to whom he was referring in his biting essay—he cited no specific works—but it was clear enough to those he lampooned.

Certainly the most celebrated case for migration, and one that Rowe made frequent reference to in his critique, began life in the late 1950s when an Ecuadorian archaeologist Emilio Estrada, and two American colleagues working in Ecuador, Betty Meggers and Clifford Evans, raised the possibility of transoceanic contact between Japan and coastal Ecuador around 5,000 years ago. Soon, the possibility became firm belief (Estrada, Meggers, and Evans 1962; Evans, Meggers, and Estrada 1959; Meggers, Evans, and Estrada 1965). The basis for their claim lay in similarities between some of the pottery they were excavating on the coast of Ecuador, part of what Estrada and his colleagues called the Valdivia phase, and Middle Jomon pottery they had seen in the collections of amateurs and local museums on Kyushu, the southernmost island in the Japanese chain. Bolstering their claim, as they viewed it, was the apparent contemporaneity of the Japanese and Ecuadorian pottery as determined through radiocarbon dating. If the pottery on the Ecuadorian coast was derived from Japan, how did it get there? The investigators had an answer for this: Japanese fishermen were blown off course, and Pacific currents carried them to the Ecuadorian coast. It was there that they taught local fishermen the art of pottery making.

In his monograph on diffusion across the Americas, James Ford (1969: 154) was enthusiastic about Meggers et al.'s views on the similarities between Japanese and coastal Ecuadorian pottery:

> The reader is now faced with the classic dilemma of American archeology: either both complexes were independent inventions of ceramics, or one derived from the other. Those who choose the first conclusion should stop reading here and head for the roulette wheel and dice table. Obviously they have a superior faith in, and perhaps mastery of, the laws of probability and coincidence than does the writer. (Ford 1969: 154)

Rowe (1966: 336) was much less enthusiastic: "No sound basis has ever been established for the romantic theory that occasional castaways will be listened to, like the Connecticut Yankee at King Arthur's court, rather than knocked on the head or put to work cleaning fish."

Two archaeologists who broke with the culture historians—and who, because of their de-emphasis on diffusion and emphasis on culture process, became known as founders of archaeological "processualism"—showed in the late 1960s that methodologically things had not improved since Kroeber (1931) laid out the homology-analogy conundrum. Lewis Binford (1968: 8) identified the lack of a method to distinguish between homologous and analogous cultural similarities as still being "a basic, unsolved problem" in archaeology. His analytical interest was on tool function, not on design, and he needed a means of distinguishing between convergence and divergence, as was made evident in his debates with François Bordes over the nature of Mousterian tool kits from the Dordogne (e.g., Binford 1973). Similarly, David Clarke (1968: 211) wrote:

> One of the fundamental problems that the archaeologist repeatedly encounters is the assessment of whether a set of archaeological entities are connected by a direct cultural relationship linking their generators or whether any affinity between the set is based on more general grounds. This problem usually takes the form of an estimation of the degree of affinity or similarity between the entities and then an argument as to whether these may represent a genetic and phyletic lineage or merely a phenetic and non-descent connected affinity.

Clarke then basically reiterated the criteria long used by anthropologists and archaeologists for assessing affinity: the more similar two phenomena are, the more characteristics they share, and the more correlations between "idiosyncratic attributes" they share, the stronger the hypothesis of "phyletic relationship" (Clarke 1968: 211). This really wasn't much different than Kroeber's (1931: 151) distinction between similarities that are "specific and structural" and those that are "merely superficial," and it missed the point made by Simpson about similarity and monozygotic twins.

Solving the Conundrum

A way forward, initially suggested by Robert Dunnell (1980), was to view the archaeological record in more or less the same way that a paleobiologist would view a fossil bed: as a population of "things" that represent the hard parts of past phenotypes. Any evolutionary study, whether it's archaeological biological, or paleobiological, encompasses "description[s] of the historical patterns of differential trait representation and arguments as to how evolutionary [processes] acted to create those patterns" (Jones, Leonard, and Abbott 1995: 29). Both steps employ concepts embedded within evolutionary theory, such as (a) lineage, or a line of development owing its existence to heritability; (b) natural selection, which is a mechanism of change; (c) a transmission mechanism, which itself is a source of new variants; (d) invention, another source of new variants; and (e) heritability, which denotes continuity such that similarity is homologous (Lyman and O'Brien 1998; O'Brien and Lyman 2000). Heritability ensures that we are examining change within a lineage (or a set of related lineages—what biologists refer to as a "clade") rather than merely convergence.

Objects in the archaeological record, because they were parts of past phenotypes, were shaped by the same evolutionary processes as were the somatic (bodily) features of their makers and users (Leonard and Jones 1987). That artifacts were once part of phenotypes is nonproblematic to most biologists (e.g., Bonner 1988; Dawkins 1990; Turner 2000, 2012), who routinely view such things as a bird's nest, a beaver's dam, or a chimpanzee's twig tools as phenotypic traits, and it certainly is not problematic to paleobiologists, who have to rely on the hard parts of phenotypes (shells, for example) to study the evolution of extinct organisms and their lineages. Archaeologists should have no trouble accepting that the behaviors that lead to creation of a ceramic vessel or a stone tool are phenotypic. Accepting the results of behaviors as phenotypic, then, requires only another small step. Once one makes that step, one can begin discussing such things as selection and drift in terms of how they shaped the variation that shows up in the archaeological record—variation that provides the phylogenetic clues that one looks for to reconstruct evolutionary history (O'Brien and Holland 1995).

As numerous chapters in this volume demonstrate, these phylogenetic clues are precisely what tell us about whether tools, their by-products of manufacture, and even the behaviors that tool makers used are analogs—incidences of convergence—or homologs—incidences of divergence from a common ancestor. One method that is becoming more common in archaeology is one that has been used in biology since the 1960s: phylogenetic analysis, commonly referred to as "cladistics." Another is stone-tool replication, which is a form of reverse engineering. Because both methods play such prominent roles in the chapters in this volume, we briefly discuss them below.

Phylogenetic Analysis

Phylogenetic reconstruction is based on a model of descent with modification in which new taxa arise from the bifurcation of existing ones. It defines ancestor-descendant relationships in terms of relative recency of common ancestry. This means that two groups are deemed to be more closely related to one another than either is to a third group if they share a common ancestor that is not also shared by the third group (figure 1.1). The evidence for exclusive common ancestry lies in derived (inherited) characters or, more precisely, character states, with "state" defined as the value that a character possesses in a group. For example, the character "number of digits" could have several possible states, such as "three" and "four." Two groups are inferred to share a common ancestor to the exclusion of a third group if they exhibit derived characters that are not also exhibited by a third group (figure 1.2). We refer to those as *shared derived* characters—for example, groups A and B in figure 1.2 share characters 3 and 4 in common—as opposed to *shared ancestral states,* which the three groups hold in common—for example, groups A, B, and C all share characters 1 and 2. Both kinds of characters (or states) are *homologous* because they are inherited from a common ancestor. Characters that arise independently in more than one lineage—for example, character 7 in figure 1.2—are *analogous.* They can result from a number of processes, including convergence, parallelism, and horizontal transmission (see chapter 2, this volume). In cladistics, analogous traits are often referred to as *homoplasies.*

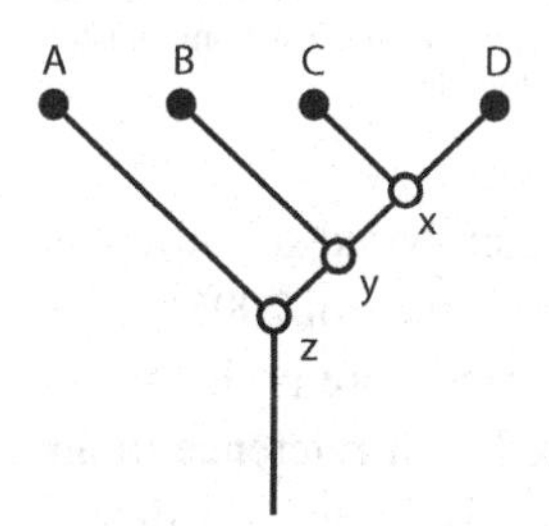

Figure 1.1
A phylogenetic tree showing the historical relationship of four groups (A–D) and three ancestors (x–z). Based on a certain character-state distribution (see figure 1.2 for an example), groups C and D are more similar to one another than either is to any other group. Also, groups B, C, and D are more similar to each other than any of the three is to group A. Related groups and their ancestors form ever-more-inclusive groups, or clades: C + D + x is one clade; B + C + D + y is a second; and A + B + C + D + z is a third.

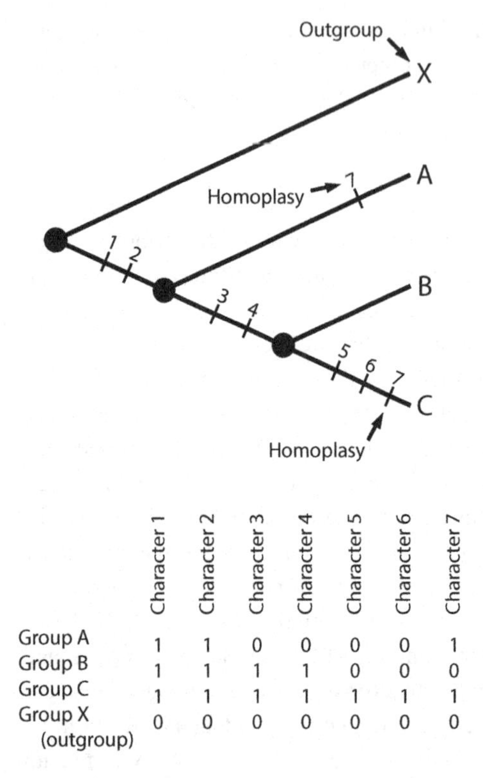

	Character 1	Character 2	Character 3	Character 4	Character 5	Character 6	Character 7
Group A	1	1	0	0	0	0	1
Group B	1	1	1	1	0	0	0
Group C	1	1	1	1	1	1	1
Group X (outgroup)	0	0	0	0	0	0	0

Figure 1.2
An example of a tree of evolutionary relationships generated by means of cladistics, together with the character-state data matrix from which it was derived. The tree indicates that groups B and C, plus their common ancestor, form a bigger group, or clade, that, based on the shared possession of derived character states for characters 3 and 4, excludes group A. It also indicates that groups A, B, and C form a larger group (clade) based on the shared possession of derived character states for characters 1 and 2. Group C is the most derived group, having derived states for characters 5, 6, and 7, in addition to the other derived characters. Character 7 is homoplastic in that it occurs in groups A and C but not in B. In other words, character 7 is analogous.

Various methods have been used for phylogenetic inference, each based on different models and each having its own strengths and weaknesses (Goloboff and Pol 2005). Two commonly used methods, maximum likelihood and Bayesian inference, are probabilistically based, where the criterion for constructing trees is calculated with reference to an explicit evolutionary model from which the data are assumed to be distributed identically (Kolaczkowski and Thornton 2004). Cultural phylogenies that are based on language evolution have relied largely on probabilistic methods (e.g., Currie and Mace 2011; Currie, Greenhill, Gray, Hasegawa, and Mace 2010; Gray, Drummond, and Greenhill 2009). Those not based on language evolution—archaeological phylogenies, for example, which are more prospective—tend to rely on maximum parsimony, which is based on a model that seeks to identify the smallest number of evolutionary steps required to arrange the taxonomic units under study (e.g., Buchanan and Collard 2008 García Rivero and O'Brien 2014; Jordan and Mace 2006; O'Brien et al. 2014a; O'Brien, Darwent, and Lyman 2001; O'Brien, Buchanan, and Eren 2016; Tehrani and Collard 2002; chapters 11 and 12, this volume).

In the briefest of terms (see Buchanan and Collard 2008 for details), cladistic analysis proceeds via four steps. First, a character-state matrix is generated, as shown in figure 1.2. This shows the states of the characters exhibited by each group. Second, the direction of evolutionary change among the states of each character is established. Several methods have been developed to facilitate this, including communality, ontogenetic analysis, and stratigraphic-sequence analysis (see O'Brien and Lyman 2003). Currently, the favored method is outgroup analysis (figure 1.2), which entails examining a close relative of the study group (chapters 11 and 12, this volume). When a character occurs in two states among the study group, but only one of the states is found in the outgroup, the principle of parsimony is invoked, and the state found only in the study group is deemed to be evolutionarily novel with respect to the outgroup state.

After the probable direction of change for the character states has been determined, the third step is to construct a branching diagram of relationships for each character. This is done by joining the two most derived groups by two intersecting lines and then successively connecting each of the other groups according to how derived they are (figure 1.2). Ideally, the distribution of character states among the groups will be such that all character trees imply relationships among the groups that are congruent with one another. This happens only in text books. Rather, a number of the character trees will suggest relationships that are incompatible, meaning that homoplasies are muddying the waters (figure 1.2). This problem is overcome through the fourth step, generating an ensemble tree that is consistent with the largest number of characters and therefore requires the smallest number of homoplasies to account for the distribution of character states among the taxa (chapters 11 and 12, this volume). We refer to such a tree as the "most parsimonious" solution (figure 1.2).

Numerous techniques exist for measuring the goodness of fit between a dataset and a given tree. The most commonly used are the consistency index and the retention index (chapters 11 and 12, this volume). The consistency index measures the relative amount of homoplasy in a dataset but is dependent on the number of groups being analyzed. Thus, the expected consistency index for a given tree must be assessed relative to the number of groups. The retention index measures the number of similarities in a dataset that are retained as homologies in relation to a given tree. It is insensitive to both the presence of derived character states that are present in only a single group and the number of characters or groups employed. Thus, it can be compared among studies. Both indices range from zero, which indicates a lack of fit between the cladogram and the dataset, to 1.0, which represents a perfect fit (see O'Brien and Lyman 2003 for a nontechnical discussion of how the indices are calculated).

It is difficult to overemphasize that phylogenetic trees, whether they comprise organisms or cultural products, by-products, and behaviors, are hypothetical statements of relatedness, "given the model and parameters used" (Archibald, Mort, and Crawford 2003: 189), not irrefutable statements of precise phylogenetic relationships (O'Brien, Collard, Buchanan, and Boulanger 2013). Critics of cultural phylogenetics have overlooked this point, in part because they view cultural transmission as destroying evidence of evolutionary processes (e.g., Moore 2001; Terrell, Kelly, and Rainbird 2001; see chapter 14, this volume), but also in part because cultural phylogenists have failed to make clear the distinction between methods of phylogenetic inference—"tree-building" methods—and phylogenetic comparative methods, which rely on the trees to understand patterns of descent in order to examine the distribution of adaptive (functional) features—those that arose in two or more lines of descent not united through an immediate ancestor. Together, the methods are based on the "logical proposition that given data about the present distribution of traits across taxa and knowledge about the historical relationships between these taxa, it is possible to infer what the traits were like in the past and how they have changed to give rise to their present distribution" (Currie and Mace 2011: 1110).

The modern comparative method is designed to escape what Francis Galton pointed out in 1889: comparative studies of adaptation (convergence) are irrelevant if we cannot rule out the possibility of a common origin of the adaptive features under examination (Naroll 1970). To escape Galton's problem requires a working knowledge of the phylogeny of taxa included in an analysis. As Joseph Felsenstein (1985: 14) put it, "phylogenies are fundamental to comparative biology; there is no doing it without taking them into account." The same applies to comparative studies of cultural phenomena, as many of the chapters in this volume make clear.

Replication

With respect to stone tools, we can define replication as the act of creating or using nonartifactual flaked-stone specimens for one or more of three purposes: as a framework for generating hypotheses about stone-tool manufacture; to test a specific hypothesis about certain parameters of stone-tool technology; and as a means of validating methods, such as using experimentally knapped tools to assess quantitative methods that will be used to study archaeological tools and their by-products of manufacture (Eren et al. 2016). Like in any scientific study, hypotheses and their predictions determine the variables required for an experiment. Experimental variables include such things as the sample size of participants or specimens, the measurement and test protocols, whether the experiment is a blind test, and the chosen quantitative methods and statistical analyses. Additional experimental variables specific to replication include reduction strategy, skill level of the knapper, material type, and the number and types of knapping tools available (Eren et al. 2016). In some instances, it might not matter whether the raw material is a basalt from Africa or a chert from Texas, whether the knapper is skilled or not, or whether soft- or hard-hammer percussion is used. In other instances, it might. The design of an experiment and the variables that go into it must be considered carefully to understand what matters and what does not in the context of a specific question, and what could thus validate or confound the results of an experiment (Eren et al. 2016).

Archaeologists who use replication experiments face several challenges (Kelly 1994). Researchers must always be wary of the "flintknapper's fundamental conceit" (Thomas 1986): exploiting their intuitive knowledge of stone-tool replication as an authoritative trump card to overrule colleagues who are not flintknappers and thus influencing them into believing that because knappers know how to make stone tools, they automatically understand such things as prehistoric foraging behavior, evolution, adaptation, and the like. Alternatively, and perhaps in reaction to this latter behavior, there are archaeologists who dismiss the usefulness of any stone-tool replication experiment. Although they are opposed, both viewpoints stem from a poor articulation of the principle of uniformitarianism. The first, "intuitive" view exaggerates the principle of uniformitarianism to such an extent that a scientific framework no longer becomes necessary to test hypotheses because the knapper simply "knows" the past because he or she is "reproducing" it. The second, "reactionary" view ignores the fact that stone breaks the same way today as it did in the past and possesses the same physical properties as it did in the past—sharp cutting edge, durability, morphology, and so on—readily facilitating some level of uniformitarian link that is exploitable scientifically (Eren et al. 2016).

As the chapters in this volume amply demonstrate, if one accepts that rocks in the past fractured similarly to rocks in the present, then it goes without saying that particular hypotheses and predictions about such things as stone-tool efficiency, morphology, and function are validly examined through stone-tool replication experiments conducted within a scientific framework of modeling and testing. That's easy to say, of course, but, as the saying goes, the devil is in the detail. Here, the detail is being able to distinguish between convergence and divergence, which depends on using appropriate methods. As we'll see in the following section, archaeologists who ignore such methods do so at their own peril. At that point, not even deep knowledge of the archaeological record and expertise in knapping can save a hypothesis.

Across Atlantic Ice?

In 2012, Dennis Stanford and Bruce Bradley published *Across Atlantic Ice: The Origin of America's Clovis Culture* (Stanford and Bradley 2012), which was the latest iteration of their proposal that North America was first colonized by people from Europe rather than from East Asia, as most archaeologists accept. They argued that groups from southern France and the Iberian Peninsula used watercraft to make their way across the North Atlantic and into North America during the Last Glacial Maximum, some 20,000–24,000 years ago. This 6,000-kilometer journey was facilitated, in their view, by a continuous ice shelf that provided fresh water and a stable food supply.

In its initial formulation, the hypothesis was based primarily on similarities between the stone tools and production techniques of Solutrean people from Western Europe, which date about 23,500–18,000 calibrated radiocarbon years before present (cal BP; Straus 2005), and those of North American Clovis people, which date about 13,300–12,800 cal BP (Collard, Buchanan, Hamilton, and O'Brien 2010). Flaws in the argument were quickly pointed out. Straus (2000), for example, noted that the existence of a several-thousand-year gap between Solutrean and Clovis made an ancestor-descendant relationship highly improbable. He argued that similarities in tool design and production were therefore much more likely to be the result of convergence. Stanford and Bradley subsequently revised their hypothesis in an effort to deal with the chronological gap (Bradley and Stanford 2004; Stanford and Bradley 2002). Instead of highlighting similarities between the Solutrean and Clovis, they pointed out supposed similarities among Solutrean, Clovis, *and* pre-Clovis tool types and production techniques, although they still emphasized Solutrean-Clovis similarities. It was that version of the hypothesis that was given extended treatment in *Across Atlantic Ice*. Unfortunately for their hypothesis, however, if the pre-Clovis dates and artifact contexts Stanford and Bradley cite in defense of their argument are correct—and the evidence overwhelmingly indicates they are not (O'Brien et al. 2014b, 2014c)—then they actually *predate* the Solutrean, not the other way round. If we are to believe that the technological similarities Stanford and Bradley observe between North America and Western Europe are historically related, we are forced to conclude, from their own chronological data, that they appeared *first* in North America and then were transferred to Europe. Of course, this is implausible.

Although there are numerous other reasons for rejecting the hypothesis (Boulanger and Eren 2015; Eren, Patten, O'Brien, and Meltzer 2013; Eren, Patten, O'Brien, and Meltzer 2014; Eren, Boulanger, and O'Brien 2015; O'Brien et al. 2014b, 2014c; Straus 2000; Straus, Meltzer, and Goebel 2005), including the absence of large chunks of Solutrean culture, such as rock art, from North American contexts—an instance, Straus et al. (2005) wryly suggested, of "cultural amnesia"—we focus here on only two topics, phylogenetic analysis and hypothesis-driven replication. Stanford and Bradley used neither. With respect to phylogeny, Stanford and Bradley (2012) argued that the similarities among Solutrean, pre-Clovis, and Clovis tool types and production techniques are evidence of ancestral-descendant relationships, such that Solutrean was the ancestor of pre-Clovis, and pre-Clovis was the ancestor of Clovis. They argued that the direct, lineal relationship was supported by the results of a phenetic analysis they carried out. Note they used the term *phenetic* and not *phylogenetic*. This is because they did not perform a phylogenetic analysis and indeed *did* undertake a phenetic analysis. Unfortunately, this is precisely the kind of analysis one does *not* want to pin one's hopes on if ancestry is of interest. The reason is that phenetics—often referred to as *numerical taxonomy* (Sneath and Sokal 1973)—informs us only about the overall similarity of assemblages and not about historical relatedness. Whereas in phylogenetic analysis the evidence for exclusive common ancestry is the presence of evolutionarily novel, or derived, character states, phenetics places objects in groups according to the degree to which they are alike or not alike, with no distinction made among the kinds of character states used, meaning that it treats analogs and homologs the same. It neither establishes the existence of historical relationships nor demonstrates that the likeness indicates that sets of phenomena are related.

Another reason for being skeptical about the ancestral-descendant relationships proposed by Stanford and Bradley has to do with the key trait that they argue supports the hypothesis that the Solutrean was ancestral to Clovis. The trait is *overshot flaking*, in which long flakes are struck from prepared edges of a biface and travel from one edge across the face and remove a portion of the opposite margin. Controlled, intentional overshot flaking is a difficult knapping technique that few modern knappers have mastered (Eren et al. 2013, 2014), but Stanford and Bradley (2012: 28) were "completely convinced" that the technique was intentionally used by Solutrean and Clovis peoples because of its presumed advantages, most prominent of which is that overshot flaking is an "incredibly efficient" or "highly effective" strategy for rapidly thinning stone bifaces (Bradley and Stanford 2004: 461; Bradley and Stanford 2006: 708–710; Stanford and Bradley 2012).

Stanford and Bradley (2012: 28, 157) make two other claims: first, that there is "clear archaeological evidence of widespread use" of overshot flaking by Solutrean and Clovis knappers, and second, that the "level of correspondence between the two technologies is amazing," such that "even the details of flaking are virtually identical." Based on these claims, they then argue that because the intentional use of a complex, difficult, even "counterintuitive" (Stanford and Bradley 2012: 28) strategy is unlikely to occur by chance, its presence in two separate groups "suggests that it is unlikely to have been independently invented." Thus, the occurrence of supposedly intentional overshot flaking on both sides of the Atlantic in Late Pleistocene times is said to "demonstrate historical connections between [the] technologies" (2012: 138). As we'll see below, this hypothesis didn't fare much better than the others that Stanford and Bradley put forward.

Eren et al. (2013, 2014) used replication experiments and quantitative analysis of the archaeological record to evaluate the claim. They found that overshot flaking is most parsimoniously explained as a technological by-product rather than a complex knapping strategy. Specifically, they found that overshot flaking is not more efficient at thinning a biface than non-overshot flaking, and that there is no frequent occurrence of overshot evidence at Clovis sites and *no* published data on the frequency or regularity of overshot flaking at Solutrean sites. Further, they showed that even if one accepted the unpublished data provided by Stanford and Bradley (2012) for the amount of overshot flaking at Solutrean sites, the data are statistically different from what Stanford and Bradley reported for Clovis sites. The final nail was provided by the almost complete lack of evidence—Stanford and Bradley reported but a single flake—for overshot flaking in their 14 purported pre-Clovis assemblages. Further, bifacial points from purported pre-Clovis sites fail to exhibit overshot scars, and, in fact, rarely are scars present that travel past the biface medial axis (Eren et al. 2013, 2014). In the final analysis, when considered in sum, the experimental and archaeological results are consistent only with the proposition that Solutrean and later Clovis knappers independently invented a basic, straightforward, efficient technique for thinning bifaces that occasionally happened to produce the analogous detritus of overshot flakes.

Conclusion

The "Across Atlantic Ice" hypothesis tells us a lot about the potential dangers involved in taking less than a careful and detailed methodological approach to distinguishing between analogy and homology in the archaeological record. Think about this: how many times throughout the history of archaeology have homologous relations been posited on a whole lot less evidence than what has been brought to bear in the Solutrean-Clovis debate? Or in Ford's (1966, 1969) elaborate hyperdiffusionist reconstructions of the prehistory of the Americas that relied on scores of archaeological sequences, including Meggers et al.'s (1965) Ecuadorian sequence? The number is probably countless. In addition, consider that Ford saw more pottery than any ten archaeologists combined, and, similarly, that the two proponents of the Solutrean-origin hypothesis, Bruce Bradley and Dennis Stanford, see more Clovis-age tools in a year than most of us will see in a lifetime. On top of that, Bradley is an expert flintknapper who knows Clovis stone-tool technology inside and out. If they can be wrong in assessing homologous relations, then that should give us some reason to pause. We might ask ourselves if it wouldn't be better to rely on some of the quantitative methods discussed here rather than on experience and intuition to explain the archaeological record (O'Brien 2010).

Our colleagues whose work appears in the following chapters discuss in detail some of the various innovative methods they have applied to the thorny problem of distinguishing between convergence and divergence. Their contributions make clear that we have moved away from the intuitive nature of archaeological inquiry that characterized the discipline throughout much of the twentieth century, although there is still considerable work to be done. Most of this work can be focused on method instead of on theory because there now exists in archaeology a firm theoretical foundation built on Darwinian evolutionary principles and processes, including transmission, descent with modification, invention/mutation, selection, and drift. Importantly, this body of theory complements biological evolutionary theory as opposed to borrowing it wholesale and praying that it contains something of value (O'Brien and Shennan 2010; Shennan 2008). We hope that readers perusing the chapters in this volume agree that there is considerable value in such an approach.

References

Archibald, J. K., Mort, M. E., & Crawford, D. J. (2003). Bayesian Inference of Phylogeny: A Non-technical Primer. *Taxon*, *52*, 187–191.

Binford, L. R. (1968). Archeological Perspectives. In S. R. Binford & L. R. Binford (Eds.), *New Perspectives in Archeology* (pp. 5–32). New York: Aldine.

Binford, L. R. (1973). Interassemblage Variability—The Mousterian and the "Functional Argument." In C. Renfrew (Ed.), *The Explanation of Culture Change* (pp. 227–254). London: Duckworth.

Bonner, J. T. (1988). *The Evolution of Complexity*. Princeton, N.J.: Princeton University Press.

Boulanger, M. T., & Eren, M. I. (2015). On the Inferred Age and Origin of Lithic Bi-points from the Eastern Seaboard and Their Relevance to the Pleistocene Peopling of North America. *American Antiquity*, *80*, 134–145.

Bradley, B., & Stanford, D. (2004). The North Atlantic Ice-Edge Corridor: A Possible Paleolithic Route to the New World. *World Archaeology*, *36*, 459–478.

Bradley, B., & Stanford, D. (2006). The Solutrean-Clovis Connection: Reply to Straus, Meltzer, and Goebel. *World Archaeology*, *38*, 704–714.

Brew, J. O. (1946). *The Archaeology of Alkali Ridge, Southeastern Utah*. Peabody Museum of Archaeology and Ethnology Papers, vol. 21. Cambridge, Mass.: Harvard University.

Buchanan, B., & Collard, M. (2008). Phenetics, Cladistics, and the Search for the Alaskan Ancestors of the Paleo-indians: A Reassessment of Relationships among the Clovis, Nenana, and Denali Archaeological Complexes. *Journal of Archaeological Science*, *35*, 1683–1694.

Clarke, D. L. (1968). *Analytical Archaeology*. London: Methuen.

Collard, M., Buchanan, B., Hamilton, M. J., & O'Brien, M. J. (2010). Spatiotemporal Dynamics of the Clovis-Folsom Transition. *Journal of Archaeological Science*, *37*, 2513–2519.

Colton, H. S. (1942). Archaeology and the Reconstruction of History. *American Antiquity*, *8*, 33–40.

Currie, T. E., Greenhill, S. J., Gray, R. D., Hasegawa, T., & Mace, R. (2010). Rise and Fall of Political Complexity in Island South-East Asia and the Pacific. *Nature*, *467*, 801–804.

Currie, T. E., & Mace, R. (2011). Mode and Tempo in the Evolution of Socio-political Organization: Reconciling "Darwinian" and "Spencerian" Approaches in Anthropology. *Philosophical Transactions of the Royal Society of London. Series B, Biological Sciences*, *366*, 1108–1117.

Darwin, C. (1859). *On the Origin of Species by Means of Natural Selection; or the Preservation of Favoured Races in the Struggle for Life*. London: Murray.

Dawkins, R. (1990). *The Extended Phenotype: The Long Reach of the Gene* (2nd ed.). Oxford: Oxford University Press.

Dunnell, R. C. (1980). Evolutionary Theory and Archaeology. In M. B. Schiffer (Ed.), *Advances in Archaeological Method and Theory* (Vol. 3, pp. 35–99). New York: Academic Press.

Eren, M. I., Patten, R. J., O'Brien, M. J., & Meltzer, D. J. (2013). Refuting the Technological Cornerstone of the Ice-Age Atlantic Crossing Hypothesis. *Journal of Archaeological Science*, *40*, 2934–2941.

Eren, M. I., Patten, R. J., O'Brien, M. J., & Meltzer, D. J. (2014). More on the Rumor of "Intentional Overshot Flaking" and the Purported Ice-Age Atlantic Crossing. *Lithic Technology*, *39*, 55–63.

Eren, M. I., Boulanger, M. T., & O'Brien, M. J. (2015). The Cinmar Discovery and the Proposed Pre-Late Glacial Maximum Occupation of North America. *Journal of Archaeological Science: Reports*, *2*, 708–713.

Eren, M. I., Lycett, S. J., Patten, R. J., Buchanan, B., Pargeter, J., & O'Brien, M. J. (2016). Test, Model, and Method Validation: The Role of Experimental Stone Artifact Replication in Hypothesis-Driven Archaeology. *Ethnoarchaeology, 8*, 103–136.

Estrada, E., Meggers, B. J., & Evans, C. (1962). Possible Transpacific Contact on the Coast of Ecuador. *Science, 135*, 371–372.

Evans, C., Meggers, B. J., & Estrada, E. (1959). *Cultura Valdivia*. Publication no. 6. Museo Victor Estrada: Guayaquil, Ecuador.

Felsenstein, J. (1985). Phylogenies and the Comparative Method. *American Naturalist, 125*, 1–15.

Ford, J. A. (1952). Measurements of Some Prehistoric Design Elements in the Southeastern States. *American Museum of Natural History. Anthropological Papers, 44*(3), 313–384.

Ford, J. A. (1966). Early Formative Cultures in Georgia and Florida. *American Antiquity, 31*, 781–798.

Ford, J. A. (1969). *A Comparison of Formative Cultures in the Americas: Diffusion or the Psychic Unity of Man?* Smithsonian Contributions to Anthropology (Vol. 11). Washington, D.C.: Smithsonian Institution.

García Rivero, D., & O'Brien, M. J. (2014). Phylogenetic Analysis Shows That Neolithic Slate Plaques from the Southwestern Iberian Peninsula Are Not Genealogical Recording Systems. *PLoS One, 9*(2), e88296.

Gladwin, H. S. (1936). Editorials: Methodology in the Southwest. *American Antiquity, 1*, 256–259.

Goloboff, P., & Pol, D. (2005). Parsimony and Bayesian Phylogenetics. In V. A. Albert (Ed.), *Parsimony, Phylogeny, and Genomics* (pp. 148–159). New York: Oxford University Press.

Gray, R. D., Drummond, A. J., & Greenhill, S. J. (2009). Language Phylogenies Reveal Expansion Pulses and Pauses in Pacific Settlement. *Science, 323*, 479–483.

Jones, G. T., Leonard, R. D., & Abbott, A. L. (1995). The Structure of Selectionist Explanations in Archaeology. In P. A. Teltser (Ed.), *Evolutionary Archaeology: Methodological Issues* (pp. 13–32). Tucson: University of Arizona Press.

Jordan, P., & Mace, T. (2006). Tracking Culture-Historical Lineages: Can "Descent with Modification" Be Linked to "Association by Descent"? In C. P. Lipo, M. J. O'Brien, M. Collard, & S. J. Shennan (Eds.), *Mapping Our Ancestors: Phylogenetic Approaches in Anthropology and Prehistory* (pp. 149–167). New York: Aldine.

Kelly, R. L. (1994). Some Thoughts on Future Directions in the Study of Stone Tool Organization. In P. Carr (Ed.), *The Organization of North American Prehistoric Chipped Stone Tool Technology* (pp. 132–136). No. 7. International Monographs in Prehistory: Ann Arbor, Mich.

Koepping, K.-P. (1984). *Adolf Bastian and the Psychic Unity of Mankind: The Foundations of Anthropology in Nineteenth-Century Germany*. Brisbane: University of Queensland Press.

Kolaczkowski, B., & Thornton, J. W. (2004). Performance of Maximum Parsimony and Likelihood Phylogenetics When Evolution Is Heterogeneous. *Nature, 431*, 980–984.

Kroeber, A. L. (1923). *Anthropology*. New York: Harcourt, Brace.

Kroeber, A. L. (1931). Historical Reconstruction of Culture Growths and Organic Evolution. *American Anthropologist, 33*, 149–156.

Kuhn, T. S. (1962). *The Structure of Scientific Revolutions*. Chicago: University of Chicago Press.

Leonard, R. D., & Jones, G. T. (1987). Elements of an Inclusive Evolutionary Model for Archaeology. *Journal of Anthropological Archaeology, 6*, 199–219.

Lyman, R. L., & O'Brien, M. J. (1998). The Goals of Evolutionary Archaeology: History and Explanation. *Current Anthropology, 39*, 615–652.

Lyman, R. L., O'Brien, M. J., & Dunnell, R. C. (1997). *The Rise and Fall of Culture History*. New York: Plenum.

Meggers, B. J., Evans, C., & Estrada, E. (1965). *Early Formative Period of Coastal Ecuador: The Valdivia and Machalilla Phases*. Smithsonian Contributions to Anthropology (Vol. 11). Washington, D.C.: Smithsonian Institution.

Moore, J. H. (2001). Ethnogenetic Patterns in Native North America. In J. E. Terrell (Ed.), *Archaeology, Language and History: Essays on Culture and Ethnicity* (pp. 30–56). Westport, Conn.: Bergin and Garvey.

Morgan, L. H. (1877). *Ancient Society*. New York: Holt.

Naroll, R. (1970). Galton's Problem. In R. Naroll & R. Cohen (Eds.), *A Handbook of Method in Cultural Anthropology* (pp. 974–989). New York: Columbia University Press.

O'Brien, M. J. (2010). The Future of Paleolithic Studies: A View from the New World. In S. J. Lycett & P. R. Chauhan (Eds.), *New Perspectives on Old Stones: Analytical Approaches to Paleolithic Technologies* (pp. 311–334). New York: Springer.

O'Brien, M. J., Boulanger, M. T., Buchanan, B., Collard, M., Lyman, R. L., & Darwent, J. (2014a). Innovation and Cultural Transmission in the American Paleolithic: Phylogenetic Analysis of Eastern Paleoindian Projectile-Point Classes. *Journal of Anthropological Archaeology*, *34*, 100–119.

O'Brien, M. J., Boulanger, M. T., Collard, M., Buchanan, B., Tarle, L., Straus, L. G., et al. (2014b). On Thin Ice: Problems with Stanford and Bradley's Proposed Solutrean Colonisation of North America. *Antiquity*, *88*, 606–613.

O'Brien, M. J., Boulanger, M. T., Collard, M., Buchanan, B., Tarle, L., Straus, L. G., et al. (2014c). Solutreanism. *Antiquity*, *88*, 622–624.

O'Brien, M. J., Buchanan, B., & Eren, M. I. (2016). Clovis Colonization of Eastern North America: A Phylogenetic Approach. *Science and Technology of Archaeological Research*, *2*, 67–89.

O'Brien, M. J., Collard, M., Buchanan, B., & Boulanger, M. T. (2013). Trees, Thickets, or Something in Between? Recent Theoretical and Empirical Work in Cultural Phylogeny. *Israel Journal of Ecology & Evolution*, *59*(2), 45–61.

O'Brien, M. J., Darwent, J., & Lyman, R. L. (2001). Cladistics Is Useful for Reconstructing Archaeological Phylogenies: Palaeoindian Points from the Southeastern United States. *Journal of Archaeological Science*, *28*, 1115–1136.

O'Brien, M. J., & Holland, T. D. (1995). Behavioral Archaeology and the Extended Phenotype. In J. M. Skibo, W. H. Walker, & A. E. Nielsen (Eds.), *Expanding Archaeology* (pp. 143–161). Salt Lake City: University of Utah Press.

O'Brien, M. J., & Lyman, R. L. (1998). *James A. Ford and the Growth of Americanist Archaeology*. Columbia: University of Missouri Press.

O'Brien, M. J., & Lyman, R. L. (2000). *Applying Evolutionary Archaeology: A Systematic Approach*. Plenum, New York: Kluwer Academic.

O'Brien, M. J., & Lyman, R. L. (2003). *Cladistics and Archaeology*. Salt Lake City: University of Utah Press.

O'Brien, M. J., Lyman, R. L., & Schiffer, M. B. (2005). *Archaeology as a Process: Processualism and Its Offspring*. Salt Lake City: University of Utah Press.

O'Brien, M. J., & Shennan, S. J. (2010). Issues in Anthropological Studies of Innovation. In M. J. O'Brien & S. J. Shennan (Eds.), *Innovation in Cultural Systems: Contributions from Evolutionary Anthropology* (pp. 3–17). Cambridge, Mass.: MIT Press.

Rowe, J. H. (1966). Diffusionism and Archaeology. *American Antiquity*, *31*, 334–337.

Shennan, S. J. (2008). Evolution in Archaeology. *Annual Review of Anthropology*, *37*, 75–91.

Simpson, G. G. (1961). *Principles of Animal Taxonomy*. New York: Columbia University Press.

Sneath, P. H. A., & Sokal, R. R. (1973). *Numerical Taxonomy*. San Francisco: Freeman.

Stanford, D., & Bradley, B. (2002). Ocean Trails and Prairie Paths? Thoughts about Clovis Origins. In N. G. Jablonski (Ed.), *The First Americans: The Pleistocene Colonization of the New World* (pp. 255–271). Memoirs, vol. 27. San Francisco: California Academy of Sciences.

Stanford, D., & Bradley, B. (2012). *Across Atlantic Ice: The Origin of America's Clovis Culture*. Berkeley: University of California Press.

Steward, J. H. (1938). *Basin-Plateau Aboriginal Sociopolitical Groups*. Bulletin no. 120. Washington, D.C.: Bureau of American Ethnology.

Steward, J. H. (1941). Review of "Prehistoric Culture Units and Their Relationships in Northern Arizona" (H. S. Colton). *American Antiquity*, *6*, 366–367.

Straus, L. G. (2000). Solutrean Settlement of North America? A Review of Reality. *American Antiquity*, *65*, 219–226.

Straus, L. G. (2005). The Upper Paleolithic of Cantabrian Spain. *Evolutionary Anthropology*, *14*, 145–158.

Straus, L. G., Meltzer, D. J., & Goebel, T. (2005). Ice Age Atlantis? Exploring the Solutrean-Clovis "Connection." *World Archaeology*, *37*, 507–532.

Tehrani, J. J., & Collard, M. (2002). Investigating Cultural Evolution through Biological Phylogenetic Analyses of Turkmen Textiles. *Journal of Anthropological Archaeology*, *21*, 443–463.

Terrell, J. E., Kelly, K. M., & Rainbird, P. (2001). Foregone Conclusions? In Search of "Papuans" and "Austronesians." *Current Anthropology*, *42*, 97–124.

Thomas, D. H. (1986). Points on Points: A Reply to Flenniken and Raymond. *American Antiquity*, *51*, 619–627.

Turner, J. S. (2000). *The Extended Organism: The Physiology of Animal-Built Structures*. Cambridge, Mass.: Harvard University Press.

Turner, J. S. (2012). Evolutionary Architecture? Some Perspectives from Biological Design. *Architectural Design*, *82*, 28–33.

Tylor, E. B. (1871). *Primitive Culture*. London: Murray.

Willey, G. R. (1953). Archaeological Theories and Interpretation: New World. In A. L. Kroeber (Ed.), *Anthropology Today* (pp. 361–385). Chicago: University of Chicago Press.

II RECOGNIZING CONVERGENCE AND CONSTRAINTS

2 Limits on the Possible Forms of Stone Tools: A Perspective from Convergent Biological Evolution

George R. McGhee

In all cases of two very distinct species furnished with apparently the same anomalous organ, it should be observed that, although the general appearance and function of the organ may be the same, yet some fundamental difference can generally be detected. I am inclined to believe that in nearly the same way as two men have sometimes independently hit on the very same invention, so natural selection, working for the good of each being and taking advantage of analogous variations, has sometimes modified in very nearly the same manner two parts in two organic beings, which owe but little of their structure in common to inheritance from the same ancestor.

—Charles Darwin (1859: 193–194)

In 1859 Darwin faced the dilemma of dealing with convergence in biological evolution—the appearance of morphological traits and instinctive behaviors that are "almost identically the same in animals so remote in the scale of nature, that we cannot account for their similarity by inheritance from a common parent, and must therefore believe that they have been acquired by independent acts of natural selection" (Darwin 1859: 235–236). In his list of difficulties for the theory of natural selection to explain (chapter VI), he included several cases of apparent convergent evolution that particularly vexed him: the independent evolution of specialized electric organs in distantly related fish, similar luminous organs in distantly related insects, and sterile workers in distantly related eusocial hive insects such as ants, bees, and termites. The independent evolution of multiple groups of bizarre carnivorous plants that actively trap, kill, and digest animals so fascinated him that he devoted an entire book to their study (Darwin 1875).

In an attempt to explain to his readers how natural selection might produce convergent evolution, Darwin called on the human analogy that two people sometimes independently hit on the same invention—a phenomenon he was well aware of, as the theory of natural selection itself had been independently formulated by both Darwin and Wallace, just as the logic of the calculus had been independently formulated by both Newton and Leibniz. As pointed out by Lycett (2011: 157), when Darwin in 1859 used the analogy of two people independently creating the same invention, he was de facto giving an "example of convergence within the realm of human technology." Thus I find it fascinating that in the twenty-first century archaeologists and anthropologists, facing the problem of convergence in human technological evolution, are turning to convergent biological evolution for insight into this phenomenon—a reversal of the dilemma faced by Darwin in 1859.

The Phenomenon of Convergent Biological Evolution

As in the biological sciences, a phylogenetic analysis of stone-tool evolution is essential for determining whether a new and similar stone-tool trait found in separate human populations is shared by those humans simply because they inherited it from a common ancestral population that first created that type of stone tool—that is, it is a *synapomorphic*, or *shared* derived (see chapter 1, this volume) trait—or whether that tool arose independently in nonrelated human populations and is thus a *convergent* (analogous) trait. The same techniques used in the phylogenetic analysis of biological forms can be used in the analysis of stone-tool forms, and this important analytic step has already been taken in evolutionary archaeology (e.g., Buchanan and Collard 2008; Lipo, O'Brien, Collard, and Shennan 2006; Lycett, 2009, 2011; O'Brien and Lyman 2003; O'Brien, Darwent, and Lyman 2001; O'Brien et al. 2008; Shott 2008; chapter 1, this volume).

Given a phylogenetic analysis (figure 2.1a), convergence can be shown to arise in three ways (figures 2.1b–2.1d). In figure 2.1a, the phylogenetic relationship of six hypothetical species is shown. Species 1, 2, and 3 all possess synapomorphy S and belong to clade S. Likewise, species 4, 5, and 6 all belong to clade T, as all possess synapomorphy T. Although clades S and T have diverged during their evolution, they nevertheless evolved from a common ancestor that evolved the derived trait R, a trait that all six species still possess by inheritance. Thus trait R is a synapomorphy for the larger clade R, which contains both clade S and clade T.

Convergent evolution is illustrated in figure 2.1b, in which a new trait, Z, independently evolves in species 3 and species 6 from *different* preexisting traits, trait A in the case of species 3 and trait B in the case of species 6. We can prove that trait Z evolved *independently* in these two species lineages, and that trait Z is not a synapomorphy species 3 and species 6 inherited from a common ancestor, because we have previously conducted a phylogenetic analysis for the entire group of species (figure 2.1a) and know that species 3 and species 6 belong to two entirely different clades, S and T.

Parallel evolution is illustrated in figure 2.1c, in which a new trait, Z, evolves independently in species 3 and species 6 from the *same* preexisting trait, R. Most, though not all, phylogenetic systematists consider that "parallelism is a special case of convergence," in which the same trait has independently evolved "from the same ancestral character in different taxa" (Lecointre and Le Guyader 2006: 541). Logically, convergent evolution, in which similar traits evolve independently in different lineages, can be subdivided into two subphenomena, *nonparallel* convergent evolution (figure 2.1b) and *parallel* convergent evolution (figure 2.1c), but most systematists simply use the terms *conver-*

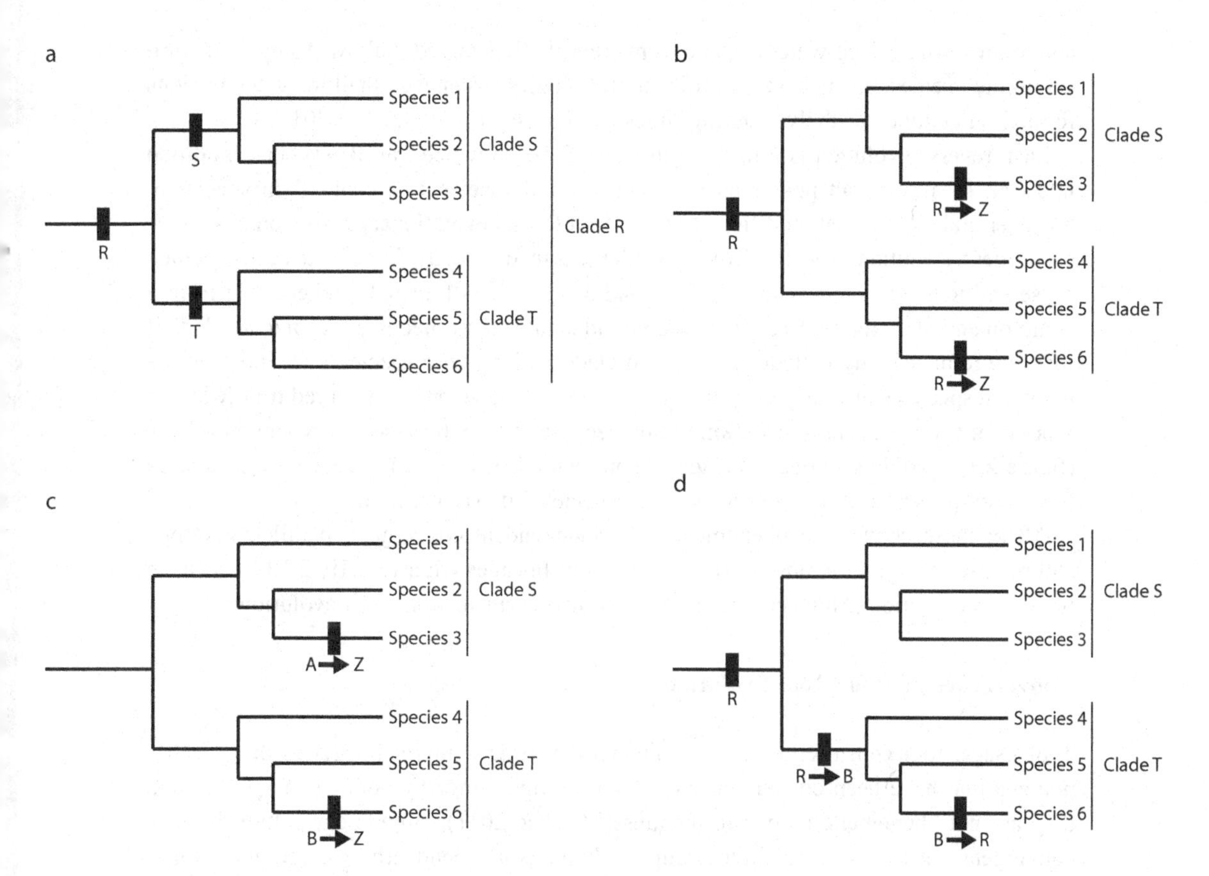

Figure 2.1
Hypothetical cladograms showing relations among six species: (A) two traits, S and T, both derived from R, create two clades (S and T); (B) convergent evolution of trait Z in species 3 and 6; (C) parallel evolution of trait Z in species 3 and 6; (D) reverse evolution of trait R in species 6 within clade T, whereas all the species in clade S simply inherited trait R from a common ancestor (from McGhee 2011).

gent and *parallel*. In actual practice, the difference between the two is often difficult to establish. For example, it was long thought that the eye evolved by convergence in at least 49 independent animal lineages (McGhee 2011), but we now know that all of these eye types are produced by modifying the *same* conserved regulatory gene present in the animal genome (the *Pax-6* gene). Thus the eye is now seen as a product of parallelism.

Iterative evolution is the special case of parallel evolution where the same morphology is evolved repeatedly in time from the same ancestral trait source. Iterative evolution can also be convergent, as in the case where multiple *different* ancestral lineages repeatedly evolve the same morphology, repeatedly in parallel within each lineage but convergently between the different lineages. Iterative evolution was particularly common in the ancient ammonites, where deep-water oceanic forms repeatedly evolved shallow-water shelf-form ammonite species, all with very similar morphologies, whenever shallow-water habitats formed on continental shelves during times of high sea level (McGhee 2012).

Last, reverse evolution is illustrated in figure 2.1d, in which trait R has been modified into trait B, a new trait possessed by species 4 and species 5 in clade T. In species 6, however, trait B has been modified back into trait R—an evolutionary reversion or *reverse* convergent evolution. Species 6 did not inherit trait R directly from its ancestor (which possessed trait B), whereas species 1, 2, and 3 all possess trait R by inheritance from a common ancestor (figure 2.1d). Thus we would create an erroneous, *polyphyletic*, clade if we were to mistakenly include species 6 in clade S along with species 1, 2, and 3 on the basis that species 6 also possesses trait R. Species 6 independently acquired trait R by the process of reverse evolution; we know this because species 6 possesses synapomorphy T (figure 2.1a) and thus belongs in clade T along with species 4 and 5, even though species 6 does not possess trait B, as species 4 and species 5 do (figure 2.1d).

All of these convergent phenomena—the independent convergent, parallel, iterative, and reverse evolution of similar traits in different lineages (figures 2.1b–2.1d)—occur in biological evolution (McGhee 2011). Do they also occur in stone-tool evolution?

Convergence in Stone-Tool Evolution?

Tool usage, tool construction, and architectural and agricultural behaviors are all phenomena that have been convergently evolved multiple times by nonhuman species, and they are not phenomena unique to humans (McGhee 2011). Human agriculture itself is convergent—at least four separate groups of humans independently evolved agricultural technologies around the world about 10,000 years ago (Gupta 2004). Thus it is not only reasonable, it is to be expected that human stone-tool technologies could also evolve through convergence.

I am not an archaeologist, but a brief survey of the literature yields at least two examples of what appear to me to be parallel stone-tool evolution. First, Lycett (2009) has argued that Levallois technology independently evolved in human populations in the Karoo region of South Africa from an ancestral Acheulean bifacial technology. Second, Adler et al. (2014) have argued that Levallois technology independently evolved in human populations in the Armenian region of Eurasia, also from an ancestral Acheulian bifacial technology. Adler et al. (2014: 1609) explicitly state that this technological "transition occurred independently within geographically dispersed … hominin populations with a shared technological ancestry." Thus, if the new Levallois stone technology evolved twice independently from the *same* preexisting Acheulean bifacial technology, then these are examples of parallel evolution (figure 2.1c). Interestingly, Lycett (2009) points out that in 1936 Louis Leakey himself considered the origin of Levalloisian-like stone-tool technology in South Africa to have been a possible case of parallel evolution.

But is it? As I pointed out earlier, the difference between convergent and parallel evolution is often difficult to establish. Thus, if one could argue that some significant differences existed in the ancestral Acheulean bifacial technology present in South Africa from that present in Armenia—say, Acheulean technology type A in South Africa and Acheulean technology type B in Armenian—then one could correspondingly argue that these are two cases of convergent evolution, given that the new Levallois technology evolved twice independently from two *different* preexisting Acheulean technologies, type A and type B (figure 2.1b).

One example of what appears to be a clear-cut case of convergent stone-tool evolution is given by Eren, Patten, O'Brien, and Meltzer (2013), who argue for the independent evolution of ultrashot flaking techniques within the Western European Solutrean knapping technology and the North American Clovis knapping technology (see chapter 1, this volume). Because the same flaking technique arose independently from two demonstrably different preexisting stone-tool technologies, Solutrean and Clovis, then this is an example of convergent evolution (figure 2.1b).

Last, one fascinating case of iterative stone-tool evolution is given by Shea (2006), who argues that similar Levantine-Mousterian technologies were independently evolved in the eastern Mediterranean Levant region by two separate populations of *Homo neanderthalensis* at two different times during the Middle Paleolithic. Given that similar stone-tool technologies were repeatedly evolved in the Levant by invasive populations of *H. neanderthalensis* from western Eurasia, this phenomenon fits the criterion of "iteration of a similar morphology from a single source" (McGhee 2012: 36), the special case of parallel evolution occurring repeatedly in time from the same ancestral trait source.

What makes this example particularly interesting is that Shea (2006) also argues that the similar Levantine-Mousterian technologies were also iteratively evolved in the Levant region by two separate invasive populations of *H. sapiens* from northern Africa at different times (different from the *H. neanderthalensis* invasions) during the Middle Paleolithic. Shea (2006: 189) argues that these two separate cycles of iterative evolution of stone-tool technologies by two separate species of *Homo* "arise from *convergence* in hominin behavioral evolution, probably in the context of competition for the same ecological niche" (emphasis mine).

Thus we have an example—from stone-tool evolution—of yet another biological phenomenon: iterative evolution that is both parallel and convergent. Within each hominin species (the same ancestral source), the repeated evolution of the same technology at different times in the Levant region is a case of iterative evolution occurring in parallel. Between the two hominin species (two different ancestral sources), the repeated evolution of the same stone-tool technology in the Levant region is a case of iterative evolution that is convergent.

I know of no examples of reverse evolution in stone-tool technologies. In such a scenario, a new technology (say, technology B in figure 2.1d) evolved from an ancestral technology (say, technology R in figure 2.1d), but then the new technology, B, was abandoned by later human populations in favor of the older technology, R (figure 2.1d). Are there any examples of such a technological reversion in human history, or is reverse evolution a biological phenomenon that is not found in human technological evolution?

Limits in Biological Evolution: Theoretical Possibilities

Why does convergent biological evolution occur? Convergence arises because the evolutionary pathways available to life are not endless but instead are *limited.* If the number of possible evolutionary pathways were infinite, then each species on Earth would be morphologically different from every other species, and each would have its own unique ecological role, or niche. Such an Earth does not exist. Instead, repeated evolutionary convergence on similar morphologies, niches, molecules, and even mental states is the norm for life (McGhee 2011).

The analytical techniques of theoretical morphology allow us to take a spatial approach to understanding convergent evolution. Any given biological form, *f*, may be described by a set of measurements taken from that form—how tall it is, how wide, how long, and so on. Each type of measurement can be considered as a dimension of form. The total set of the possible dimensions can be used to construct a hyperdimensional morphospace of possible form coordinates (figure 2.2). Each point within this theoretical morphospace represents a specific combination of form measurements that will produce the form coordinate for a hypothetical form *f*. Convergence occurs when forms originally present in different regions of the morphospace evolve in such a way that they move to the same spatial region.

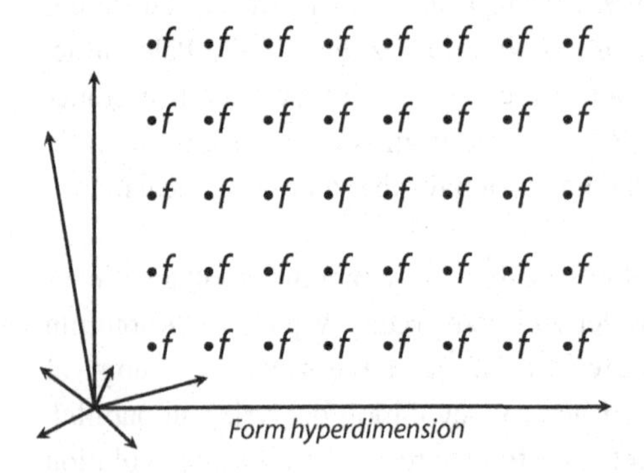

Figure 2.2
A theoretical morphospace of possible form. Each dimension of the space represents a morphological trait that can be measured on a given biological form, *f*. All possible coordinate combinations (points) within the theoretical morphospace represent the set of all possible forms. Although only eight dimensions are shown, the dimensionality of an actual hyperspace of form will be much larger (from McGhee 2011).

Using figure 2.2, we can begin to consider the effects of evolutionary constraint in producing convergent evolution. First, we know that some types of forms function in nature; these are the myriad forms of life that surround us. The opposite of this observation is the concept that there exist forms that do not function in nature and that if an organism were to produce one of these forms, it would be lethal. Mutations that produce nonfunctional, lethal forms are well known in biology. We can thus conceptually divide the spectrum of form within the morphospace into spatial regions that contain nonfunctional forms and functional forms. The boundary between the spatial regions is itself a spatial representation of the concept of functional constraint. That is, if the evolving form is to remain functional, it must stay within the functional region of forms in the morphospace. The boundaries of these regions are not mere conceptual abstractions; they can actually be mapped in theoretical morphospace, as I discuss in the next section.

Second, we know that organisms develop biological form from an original cell. We also know that the possible types of form that can be developed from a given cell are limited: they depend on the DNA coding within the cell and the interaction of the different molecules and tissue geometries produced as the cell grows. These observations lead to the concept of developmental constraint, meaning that the different types of forms that different organisms can develop are limited.

Now let us consider the spatial representation of these two types of evolutionary constraint within our morphospace (figure 2.3). The functional-constraint boundary in figure 2.3 is shown by the solid line, and the developmental constraint boundary is shown by the dotted line. That is, an evolving form must remain within the functional and

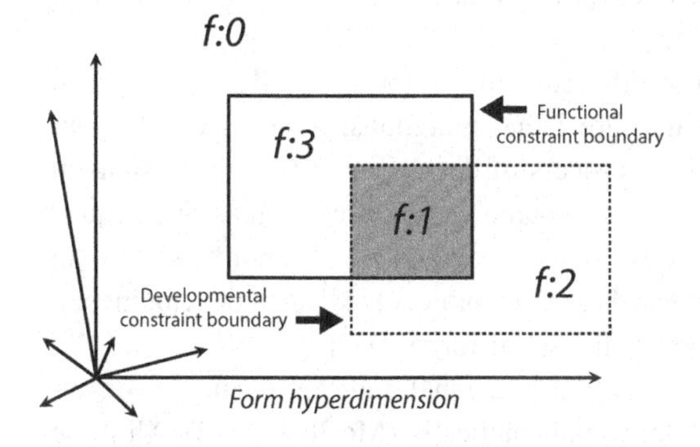

Figure 2.3
A spatial representation of functional and developmental constraint in theoretical morphospace. Boundaries of the solid-line rectangle delimit the functional-constraint boundary: forms located within this rectangle are functional, and forms outside the rectangle are nonfunctional. Boundaries of the dotted-line rectangle delimit the developmental-constraint boundary on possible form: forms within the dotted-line rectangle are developmentally possible, whereas forms outside the rectangle are developmentally impossible. Forms *f:0* are thus both nonfunctional and developmentally impossible; forms *f:1* (gray-shaded region) are both functional and developmentally possible; forms *f:2* are nonfunctional but developmentally possible; and forms *f:3* are functional but developmentally impossible (from McGhee 2011).

developmentally possible regions of forms in the morphospace of all possible forms (figure 2.2). However, as can be seen in figure 2.3, the functional-constraint boundary within the morphospace does not have to spatially coincide with the developmental-constraint boundary.

The mapping of the functional and developmental limit boundaries in figure 2.3 allows us to create a Venn diagram of four distinct sets of theoretical forms within our morphospace (McGhee 2011), described below.

1. Biological form set {*f:0*}: these are the forms, *f:0*, that are nonfunctional and cannot be developed by life on Earth.

2. Biological form set {*f:1*}: these are the forms, *f:1*, that are both functional and that can be developed by life on Earth (the gray-shaded region in figure 2.3).

3. Biological form set {*f:2*}: these are the forms, *f:2*, that can be developed but that are nonfunctional and thus lethal for life on Earth.

4. Biological form set {*f:3*}: these are the forms, *f:3*, that are functional but that cannot be developed by life on Earth.

Now let us consider these four regions of possible form from the point of view of existent life. The myriad forms of life that surround us clearly are both functional and developable. Thus every living thing on Earth, from butterflies to bacteria, belongs to the form set {*f:1*}. The phenomenon of convergent evolution immediately reveals to us that the size of form set {*f:1*} is not infinite, in that life has been constrained to repeatedly re-evolve the same forms within this set over and over. The hypothetical universe in which every species has its own unique functional morphology, and is morphologically different from every other species, does not exist.

Biologists have long studied the many different mutant forms of life—two-headed snakes, three-legged frogs, and so on—that are lethal mutations, which are important in that they give us valuable clues about the process of development (see discussions in Alberch [1989] and Blumberg [2009]). Developmental abnormalities such as two-headed snakes and frogs with three hind legs instead of two are real and are also nonfunctional, in that they do not survive in the wild (as opposed to the laboratory). All such developmental "freaks of nature" (Blumberg 2009) belong to the set of forms {*f:2*}.

Visualizing the types of forms that are both nonfunctional and that cannot be developed is a bit more difficult, but it can be done mathematically (McGhee 2012). All these hypothetical form coordinates within the morphospace belong to the form set {*f:0*}. They do not really matter for our discussion here, but they do need to be listed so we will have considered the complete spectrum of all possible existent, nonexistent, and impossible biological forms. So, thus far we have three sets of form with respect to Earth life: one functional set of form, {*f:1*}, one nonfunctional set of forms, {*f:2*}, and one impossible set of forms {*f:0*}.

The fourth set of forms is crucial to our understanding of the implications of convergent evolution for extraterrestrial life. These are the possible forms of life that are functional, that work just fine in nature, but nevertheless that cannot be developed by life on Earth. These possible, but nonexistent, forms on Earth, would belong to the final set of forms, {*f:3*}. This logical possibility leads immediately to the question: does the form set {*f:3*} actually exist in the universe? We know that it does not on Earth, but do living forms exist elsewhere on alien worlds that belong to the form set {*f:3*}?

At first, this question may seem to be so abstract as to be of no real importance, but it is of critical importance in our consideration of predictability in evolution. We know that much of the convergent evolution of life on Earth is driven by developmental constraint, yet we do not know the answer to two of the most fundamental questions concerning developmental constraint itself: "How did development originate?" and "How did the developmental repertoire evolve?" (Müller 2007: 944). Only when we have the answer to those two questions can we make predictions about the possible evolution of alien developmental repertoires and whether those repertoires might be similar to those found in Earth life. The existence of the form set {*f:3*} would imply that there exists somewhere in the universe an alien set of functional biological forms that nevertheless cannot be developed by Earth life and thus would *not be convergent* on any of the forms of life seen on Earth.

A similar logical issue can be raised concerning the possible subdivision of form set {*f:1*} into two subsets (figure 2.4).

5. Biological form subset {*f:4*}: these are the forms, *f:4* (black region in figure 2.4), within form set {*f:1*} that are functional, developmentally possible, and that *actually have been evolved* by life on Earth.

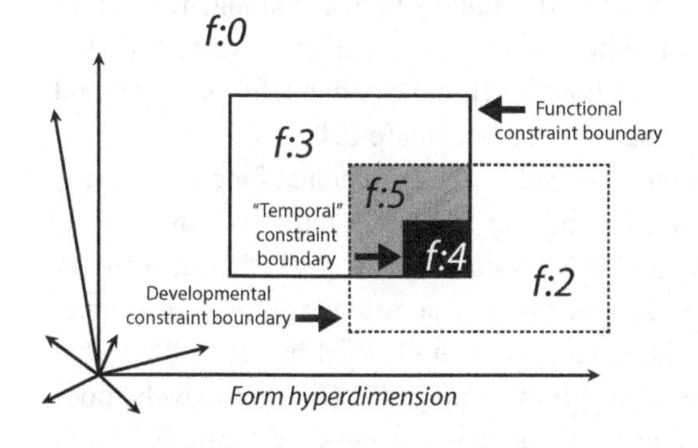

Figure 2.4
A spatial representation of "temporal" constraint within theoretical morphospace. The {*f:1*} form-set region shown in figure 2.3 has been subdivided into two subsets, {*f:4*} (black region) and {*f:5*}. Forms within both subsets are functional and developmentally possible, but only those forms in subset {*f:4*} have actually been evolved—the process of evolution is postulated to have "not yet had enough time" to discover the forms in subset {*f:5*}.

6. Biological form subset {*f:5*}: these are the *potential* forms, *f:5*, within form set {*f:1*} that are functional and developmentally possible but that nonetheless have not yet been evolved by life on Earth.

The boundary between these two form subset regions within form set {*f:1*} is labeled the "temporal" constraint boundary in figure 2.4. The concept here is the possibility that existent organisms simply have *not had enough time* to evolve morphologies with the set {*f:1*} forms found in the empty region of morphospace (subset {*f:5*} region in figure 2.4).

This concept is an old one, going back to Raup's (1966) classic theoretical morphospace study of computer-simulated mollusc-shell form (for an extensive discussion of constraint concepts from a theoretical morphologic perspective, see McGhee 2012). The most striking feature of Raup's theoretical morphospace was the fact that the overwhelming majority of the morphospace was empty—that there exist numerous geometrically possible shell forms that no mollusc has ever evolved. From the point of view of natural-selection theory, an obvious causal explanation for empty morphospace would be that the forms present in the empty region of morphospace are nonfunctional—that is, they belong to form set {*f:2*} (figure 2.3). However, an alternative point of view could just as easily maintain that such hypothetical nonexistent morphologies might function perfectly well in nature but that the process of evolution simply has not yet produced them.

The question of temporal constraint has been revived recently by Powell and Mariscal (2014), but in a different context. Gould (1989) argued that all evolutionary trends are chains of historically contingent events. Given that the actual functional biological forms that have been "discovered" by evolution are a function of past contingent events, there could exist numerous other functional forms that have not yet been discovered purely as a function of history. This argument has been labeled "Gould's radical contingency thesis" by Powell and Mariscal (2014: 119–120), who further argue that it predicts "smaller boundaries around the space of forms that are functional, developmentally possible, and actual," as illustrated for the boundaries of subset {*f:4*} in figure 2.4.

Logically, the question is not one of whether the subset {*f:4*} of actual functional forms exists—it does, as all the living forms on Earth belong to it—but rather whether subset {*f:5*} exists. If subset {*f:5*} does not exist, then neither does subset {*f:4*}; that is, if subset {*f:5*} = ∅, then subset {*f:4*} = set {*f:1*}. To rephrase the question: are there functional and developmentally possible forms that have not yet been evolved by Earth life? That is, does {*f:4*} exist as a separate possible form subset from {*f:5*}? Or, alternatively, does {*f:4*} = {*f:1*}; that is, have all functional and developmentally possible forms for Earth life in fact been discovered by Earth life?

An obvious question for future research concerns the existence or nonexistence of form subset {*f:5*}. Is temporal constraint real? That is, can we prove that perfectly functional, developmentally possible biological forms exist that nevertheless have not yet been evolved by life on Earth? In order to answer that question, we must be able to create these nonexistent forms and to analyze their functionality. The analytic techniques of theoretical morphology allow the researcher to do exactly that (McGhee 2012).

Analyzing Limits in Biological Evolution

Below I illustrate the actual mapping of limits in evolution in theoretical morphospaces with one example from the evolution of cephalopod form (Lophotrochozoa: Eutrochozoa: Mollusca) and one example from the evolution of brachiopod form (Lophotrochozoa: Lophophorata). More-detailed discussion of these and other analytic examples of limits mapping in theoretical morphospaces can be found in McGhee (2012, 2015).

In figure 2.5, a computer-simulated spectrum of possible cephalopod shell forms is illustrated, where the dimensions of the morphospace are the geometric parameters *W*, which is the whorl-expansion rate of the shell, and *D*, which is the displacement of the aperture of the shell from its coiling axis. The heavy black line zig-zagging across the center of the morphospace is a hydrodynamic-limit line: all shell forms in the morphospace

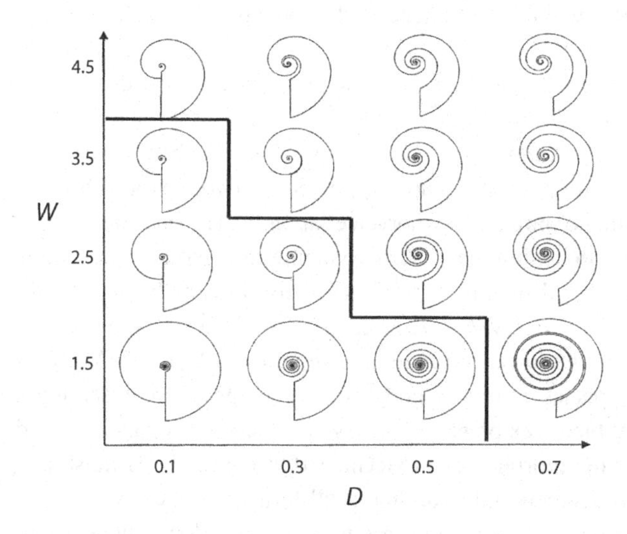

Figure 2.5
A theoretical morphospace of hypothetical cephalopod shell form. Actual fossil ammonoid shell forms are found only in the lower-left corner of the morphospace, under the limit-line extending diagonally across the center of the morphospace (after McGhee 2015).

above this limit, in the upper-right region, have swimming-efficiency coefficients (SEC) that are less than SEC = 40, whereas all shell forms below this limit, in the lower-left region of the morphospace, have swimming-efficiency coefficients that are greater than or equal to SEC = 40.

Numerous analyses of the shell geometries of over a thousand fossil species of ancient ammonoid cephalopods have shown that evolved shells are found only in the lower-left region of the morphospace (Chamberlain 1976, 1981; McGowan 2004; Raup 1967; Saunders, Work, and Nikolaeva 2004; also see McGhee 2012). The shells illustrated in the upper-right region of the morphospace are all geometrically possible forms, yet they were never evolved by actual ammonoids. Why not?

The hydrodynamic-limit line shown in figure 2.5 is an example of mapping an actual functional-constraint boundary in theoretical morphospace. The ancient ammonoid cephalopods evolved streamlined shells that minimized drag resistance while swimming and possessed shells with SECs ranging from 70 down to the minimum limit of 40 (figure 2.5). Note that the shells in the lower-left region of the morphospace, below the SEC = 40 line, also have geometries in which the whorls of the shell overlap one another to a greater or lesser degree, as shell geometries with substantial whorl overlap minimize turbulence and drag. Note also that shell geometries in the upper-right region of the morphospace include forms in which the whorls not only do not overlap, they do not even touch! These geometries have very poor swimming efficiencies; for example, the shell shown at $W = 2.5$ and $D = 0.7$ has an SEC = 8.

In figure 2.6, a computer-simulated spectrum of possible biconvex brachiopod shell forms is illustrated, where the dimensions of the morphospace are the logarithms of the whorl expansion rates, W, of the dorsal and ventral valves of the shell. The heavy black line enclosing the central region of the morphospace maps the position of two separate limits on brachiopod form: an internal-volume limit and a whorl-overlap limit. The internal-volume limit is a functional constraint: the minimum internal-volume-to-external-surface-area ratio (V/A) for a functional brachiopod shell is V/A = 7. All shell forms in the upper-right region of the morphospace have ratios of $V/A < 7$, whereas all shell forms below and to the left of the central limit line, in the lower-left region, have ratios of $V/A \geq 7$. Analyses of the shell geometries of 324 species of brachiopods, with each species representing a different genus, have shown that biconvex brachiopods have evolved shells that are found only in the lower-left region of the morphospace (McGhee 1999). The shells illustrated in the upper-right region are all geometrically possible shell forms, yet they were never evolved by the biconvex brachiopods and, as in the case with the cephalopod nonexistent shell forms, we can ask why not.

Biconvex brachiopods filter feed by pumping water through the interior of their shells, where the lophophore tentacular feeding organ is located. The larger the lophophore, the more water the brachiopod can filter for food. The lophophore, however, must be contained within the protective shell of the animal, and it costs metabolic energy to grow

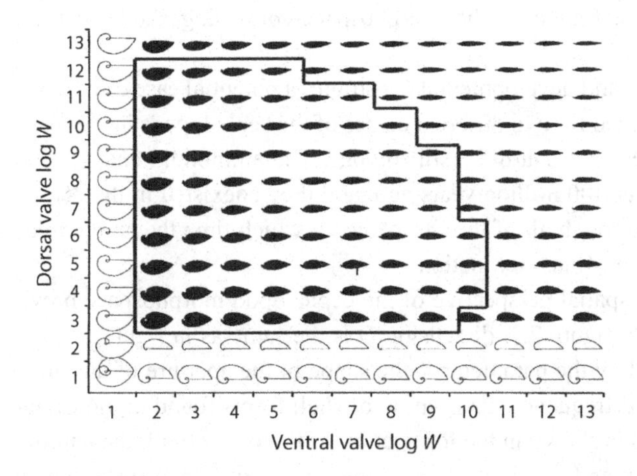

Figure 2.6
A theoretical morphospace of hypothetical brachiopod shell form. Actual brachiopod shell forms are found only within the limit-line enclosed region in the center of the morphospace (after McGhee 2015).

the shell. Thus, the best shell geometry for a filter-feeding biconvex brachiopod is one that maximizes internal volume (for the largest lophophore) but one that at the same time minimizes external surface area (for the smallest metabolic cost of growth). The one three-dimensional geometry that has the smallest surface area, and the largest volume relative to its surface area, is the sphere. The most abundant shell geometries found in actual biconvex brachiopod shells have ratios of V/A = 10, and the frequency of actual brachiopod shell geometries drops sharply with decreasing values of V/A until the minimum limit of V/A = 7 is reached (figure 2.6).

The second limit, the whorl-overlap limit, is a developmental constraint. In order to articulate two brachiopod valves into a functional bivalve shell, whorl overlap must be absent in both valves. In typical biconvex brachiopod shell geometries, whorl overlap occurs in the dorsal valve if log $W < 3$ and in the ventral valve if log $W < 2$. Note that, in figure 2.6, one vertical column of ventral valves with log $W = 1$, and two horizontal rows of dorsal valves with log $W \leq 2$, are shown with whorl overlap. These valves are not articulated with an opposite valve. The reason becomes apparent when the two simulations given at coordinates ventral valve log $W = 1$ and dorsal valve log $W = 1$ and log $W = 2$ are examined. In these simulations we can see that the whorls of the two valves would have to interpenetrate one another in the posterior part of the shell, where the valves articulate. That is, portions of the two separate valves would have to occupy the same space at the same time in the articulation region, and that is geometrically impossible. Thus, the vertical limit line at ventral valve log $W < 2$ and the horizontal limit line at dorsal valve log $W < 3$ separate the region of morphospace-containing shells that are geometrically possible from the region of the morphospace-containing shells that are geometrically impossible.

Last, in addition to functional and developmental constraint, a potential case of "temporal" constraint might also be argued for two different clades of the cephalopods in the fossil record: the ammonoids (extinct) and nautilids (still living). The ammonoid and nautilid cephalopod clades diverged over 400 million years ago, and they coexisted in the oceans for most of the Paleozoic and through all of the Mesozoic, at which time the ammonoids went extinct in the end-Cretaceous mass extinction.

Figure 2.7 shows a different spatial perspective of the cephalopod morphospace previously examined in figure 2.5. In figure 2.7, dimension D is the same as in figure 2.5, but dimension W has been replaced by the parameter S, the shape of the aperture of the shell. Within this space, the contoured frequency distribution of shell forms found in the Cretaceous ammonoids and nautilids is shown in the top section of figure 2.7, with the nautilid frequency distribution on the left and the much larger ammonoid frequency distribution on the right. Note that there is very little overlap in shell forms between the two clades—only in the lower-left corner, in the region of $S = 0.50$ to 0.75 and $D = 0.01$ to 0.14.

In the bottom section of figure 2.7, the contoured frequency distribution of shell forms found in Cenozoic nautilids is shown. There, the ammonoids are extinct and no longer present in the morphospace. Ward (1980) pointed out that post-extinction nautilids shifted their shell-form distribution in the Cenozoic down toward the region of morphospace previously occupied by the now-extinct ammonoids. Specifically, the peak of the nautilid frequency distribution (the most abundant shell form within the clade) shifted from $S = 1.15$ to $S = 0.9$, a shift shown by the vertical vectors on the left in the bottom section of figure 2.7. Ward (1980: 32) argued that this shift evidenced a form of ecological release: that the extinction of the ammonoids "may have opened up new opportunities for nautilid evolution during the Tertiary, because Tertiary nautilids are dominated by moderately compressed, hydrodynamically efficient shell shapes which were rarely present among Jurassic and Cretaceous nautilids" but that were common among the ammonoids (figure 2.7).

However, the very large region of morphospace that was once occupied by ammonoids (top section of figure 2.7) still remains empty today. The fact that the Cenozoic nautilids never invaded the large empty region of morphospace is indicated by the horizontal vectors and question mark in the bottom section of figure 2.7. Shell forms in this region clearly are functional, as they were successfully used by ammonoids for the greater part of the Paleozoic and Mesozoic. Thus, functional constraint is not a limiting factor for the potential re-evolution of these shell morphologies.

Could this phenomenon be a possible case of temporal constraint—that is, did the nautilids simply not have enough time to re-evolve the numerous functional shell forms previously evolved by the ammonoids? Herein lies the difficulty in trying to demonstrate

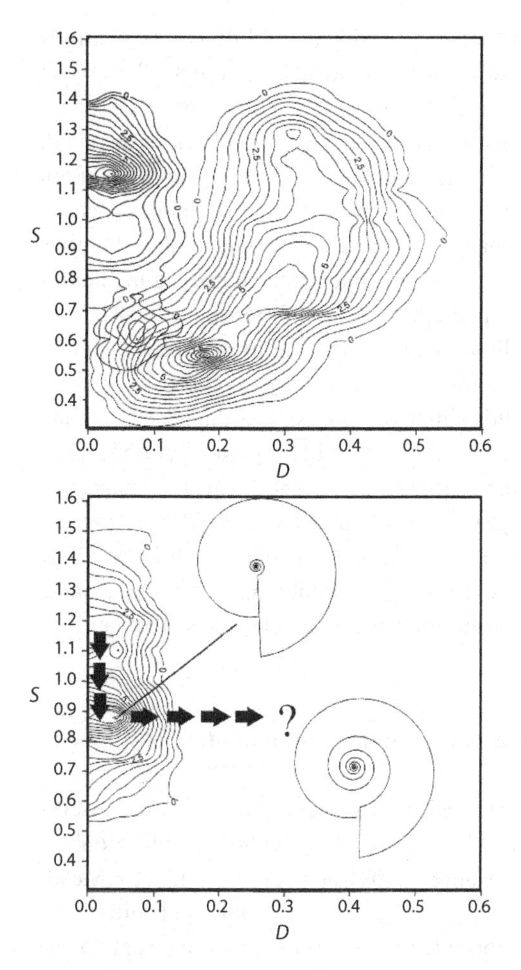

Figure 2.7
Contoured frequency distribution of shell forms found in Cretaceous ammonoids and nautilids (top) and Cenozoic nautilids (bottom). In the top section, the nautilid frequency distribution is on the left, and ammonoid shell forms are the much larger frequency distribution on the right. The fact that the Cenozoic nautilids never invaded the large empty region of the morphospace previously occupied by ammonoids is indicated by the horizontal vectors and question mark in the bottom section (from McGhee 2012; data from Ward 1980).

"temporal" constraint. First, the nautilids have had 65 million years to evolve the forms in the empty region of the morphospace, yet they have not. Second, the Cenozoic nautilids evolved some morphologies that are indeed more ammonoid-like (Ward 1980), yet the amount of ammonoid morphospace invaded by the nautilids is minor (figure 2.7). Thus, in 65 million years, the nautilids successfully invaded only a tiny region of ammonoid morphospace—a fact that can be used to argue that some other constraint other than insufficient time is at work.

On probabilistic grounds, one can argue that it is more likely the inability of the nautilids to evolve ammonoid morphologies was a result of developmental constraint. The extinct ammonoids had long, eel-like bodies encased in narrow shells with numerous whorls. Often, their eel-like bodies extended in a narrow tube a full revolution back in the shell, as we can see in their fossil shells, where the living chamber is typically over 360° of arc in length (Saunders et al. 2004). In contrast, the nautilids have short, thick bodies encased in broad shells with few whorls. The living chamber of the shell containing their bulbous bodies is typically very short, less than 180° of arc. The phylogenetic legacy of the bodies and organ systems of these two groups of cephalopods is very different. It is also very ancient. You have to go back over 400 million years to find the most recent common ancestor of these two groups of superficially similar swimming cephalopods.

The probability is thus high that the nautilids simply do not possess the genetic coding necessary to produce the shell morphologies found in the empty region of morphospace (figure 2.7). These shell geometries are fundamentally at variance with the shape and organ distribution of the nautilid body itself, and the 400-million-year-old phylogenetic legacy of the nautilids would have to undergo a radical reorganization in order to allow these animals to use the ammonite shell forms. These shell morphologies are unattainable by the nautilids as a result of constraints imposed by the specific biology of the nautilids themselves.

Discussion: Analyzing Limits and Optimization in Stone-Tool Evolution

The concept of theoretical morphospace originated in evolutionary biology (see discussion in McGhee 1999), but it has subsequently caught the attention of philosophers (Maclaurin 2003), cultural anthropologists (Hauser 2009), and even architects (Steadman 2014), who are seeking to explore the spectrum of both possible and impossible cultures and building morphologies. Why not extend the morphospace concept to the analysis of stone-tool morphologies, as O'Brien and his colleagues have recently suggested (e.g., O'Brien and Bentley 2011; O'Brien et al. 2016)? Curiously, I had already taken a step in this direction back in 1999 when I was trying to demonstrate that the construction of a theoretical morphospace does not necessarily have to involve computer simulations or complex mathematics. To do so, I created a theoretical morphospace of triangular form (figure 2.8) using nothing more complex than a ruler, a pencil, a piece of paper, and a simple geometric model (isosceles triangle). I then noted that although the triangles in themselves have no biological meaning, they could easily be used by a botanist to examine leaf shape in plants, by a vertebrate paleontologist to examine tooth shape in dinosaurs, or even by an anthropologist to examine arrowhead shape in Neolithic human cultures. The last example is obviously nonbiological, in that a stone arrowhead is not, nor ever has been, alive, yet the techniques of theoretical morphology might be

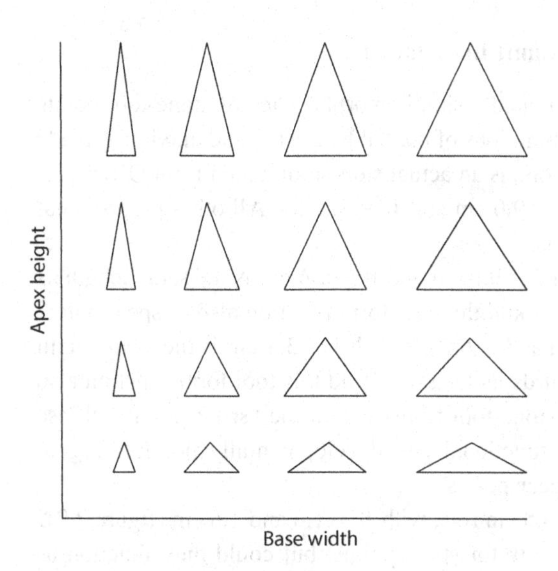

Figure 2.8
A theoretical morphospace of isosceles triangular form. Apex height and base width increase in equal increments for each simulated triangle (from McGhee 1999).

fruitfully used by archaeologists to simulate arrowhead shapes that have never been created and to ask why not—a question that would lead to interesting questions concerning aerodynamics and structural limitations in using stone as an engineering medium (McGhee 1999).

Mapping Limits in Stone-Tool Evolution

Clearly the concepts of functional and developmental constraint, and perhaps the idea of temporal constraint, can be directly applied to stone-tool formation. That is, it should be possible to divide a stone-tool morphospace into the six theoretically possible regions that are used in analyzing biological evolution (figures 2.3 and 2.4). The archaeological literature is full of studies that analyze why certain stone-tool forms function better than others, but what about nonfunctional possibilities? Likewise, the literature contains numerous studies that analyze why certain stone-tool forms are easier to make than others (e.g., Eren, Lycett, Roos, and Sampson 2011), but do we pay enough attention to the spectrum of functional tool forms that cannot be made of stone and the transition of those forms to other engineering media (e.g., different metal types)? The developmental-limit boundaries for potential stone-tool evolution can be found best by the analysis of both developable and nondevelopable tool form.

Mapping Functional Regions within Limit Boundaries

To explore some of these questions, I created a simple morphospace of stone-tool points (figure 2.9). Of all the proportional combinations of medial length (*L*) and maximum blade width (*W*) illustrated in figure 2.9, only one is an actual stone tool, taken from O'Brien et al. (2001)—the spear point located at *L* = 9.0 cm and *W* = 3.3 cm. All other proportional combinations of *L* and *W* are simulations.

The hypothetical stone tool located at *L* = 9.0 cm and *W* = 1.6 cm is the same length as the real one, but is about half its width. Would this tool form function also as spear point? Likewise, the hypothetical point located at *L* = 6.3 cm and *W* = 3.3 cm is the same width as the real one but is a little over two-thirds its length. Would this tool form function also as a spear point? Could all four of the stone-tool forms within the "spear points" ellipse illustrated in figure 2.9 constitute the functional and developmentally possible region within the morphospace for potential spear points?

And what about the points in the *L* = 4.5 cm row, with *W* = 1.0 and 1.6 cm (figure 2.9)? They are perhaps too short and too narrow for spear points, but could they function as arrowheads? And what about the points located in the *W* = 1.0 cm column at lengths *L* = 6.3 and 9.0 cm (figure 2.9)? They are perhaps too long for arrowheads but also too thin for spear points. Could they function as awls instead? Thus the stone-tool morphospace might contain not only hunting forms such as spear points and arrowheads, but also stone-tool forms used in clothing manufacture.

What about the forms in figure 2.9 that lie outside the limits of the spear points, arrowheads, and awls? Are these forms not possible for either functional or developmental reasons? Are the forms located within the "too wide?" ellipse (figure 2.10), nonfunctional for either arrowheads or spear points, because their excessive width would cause too much drag friction in the atmosphere? Are the forms located within the "too large?" ellipse (figure 2.10), nonfunctional as spear points because they are too massive and heavy and are also the wrong shape even to be used as handaxes? Alternatively, in figure 2.10, could it be that the very long and thin forms located in the *W* = 1.0 cm column with lengths *L* = 13.5 and 18.0 cm be developmentally impossible? That is, are they too fragile to make from stone, but perhaps not from metal?

Mapping Degrees of Optimization within Functional Regions

Last, it should be possible to analyze the degree of functional optimization in the spectrum of existent stone-tool forms (figure 2.9) using the same adaptive-landscape analytical techniques that are used in evolutionary biology. In figure 2.11, a detailed contoured mapping of the frequency distribution of actual ammonoid shell forms is given in the cephalopod theoretical morphospace shown in figure 2.5. In the bottom part of figure 2.11, a contoured mapping of the distribution of swimming-efficiency coefficients for the spectrum of different shell forms is given in the morphospace—in essence, an adaptive landscape of SEC

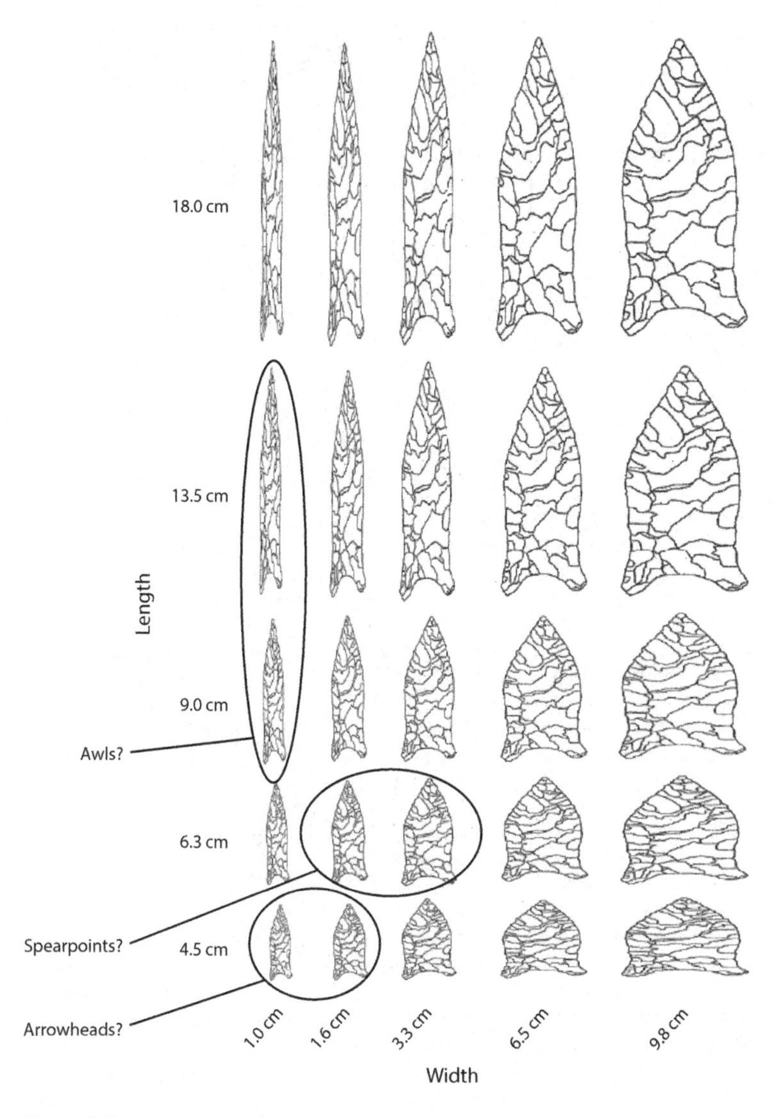

Figure 2.9
A theoretical morphospace of hypothetical stone-tool form. Dimensions of the space are *L* (medial length; see O'Brien et al. 2001, figure 8) and *W* (maximum blade width). Within the morphospace, a hypothetical mapping of the functional and developmentally possible regions for spear points, arrowheads, and awls is shown.

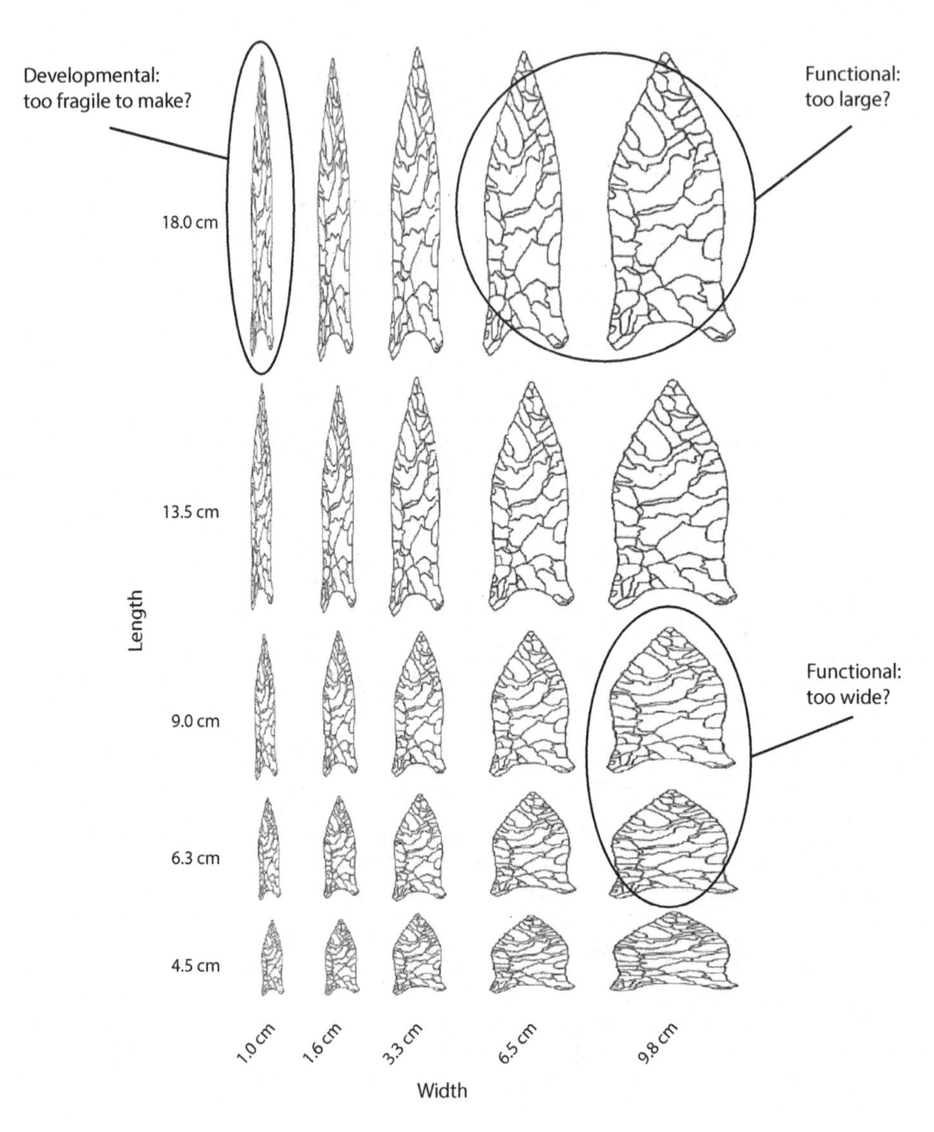

Figure 2.10
A hypothetical mapping of nonfunctional and developmentally impossible stone-tool-form regions in the morphospace.

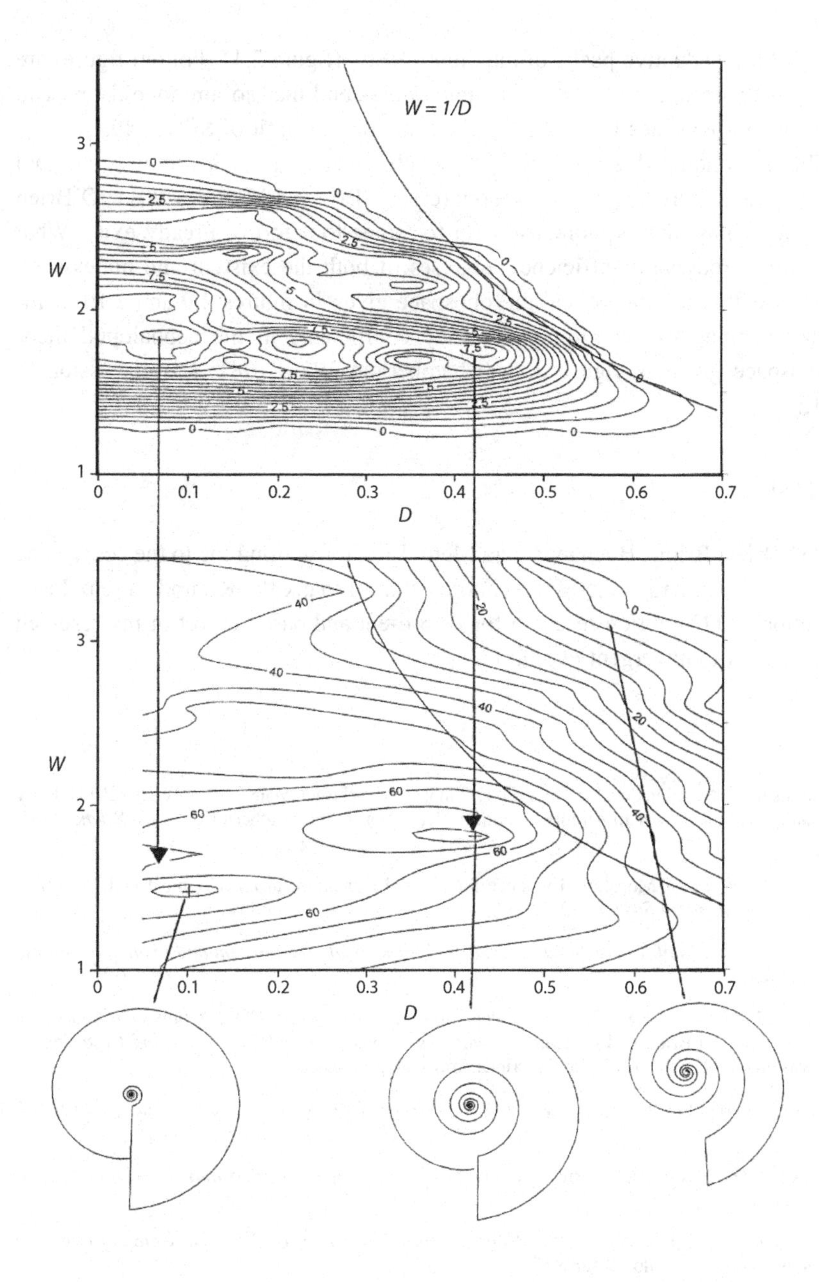

Figure 2.11
The contoured frequency distribution of 597 species of Paleozoic ammonoids (top) in the cephalopod theoretical morphospace illustrated in figure 2.5, compared with contoured distribution of swimming-efficiency coefficients within the theoretical morphospace (bottom). Note that both adaptive peaks of maximum swimming-efficiency coefficients are shown to have been occupied by Paleozoic ammonoids (ammonoid morphometric data from Saunders et al. 2004).

values. Note that both adaptive peaks of maximum SECs (figure 2.11, bottom figure) are shown to have been occupied by Paleozoic ammonoids and that no ammonoids exist in the region of the morphospace beyond the minimum contour limit of SEC = 40.

It is possible to conduct this same type of adaptive-landscape analysis of stone-tool forms, as archaeologists are beginning to show (e.g., O'Brien and Bentley 2011; O'Brien et al. 2016). Mountains of morphometric data for stone-tool forms already exist. What we now need are aerodynamic-efficiency analyses of both the existent and nonexistent stone-tool forms within the theoretical morphospace given in figures 2.9 and 2.10, comparable to the mapping of swimming-efficiency coefficients in the cephalopod theoretical morphospace given in figure 2.11 (Chamberlain 1976, 1981; see discussion in McGhee 2012).

Acknowledgments

I thank Mike O'Brien, Briggs Buchanan, and Metin Eren for inviting me to the conference on convergent evolution and stone-tool technology, and I thank the Konrad Lorenz Institute for Evolution and Cognition research for its present and past support of my research on the phenomenon of convergent evolution.

References

Adler, D. S., Wilkinson, K. N., Blockley, S., Mark, D. F., Pinhasi, S., Schmidt-Magee, B. A., et al. (2014). Early Levallois Technology and the Lower to Middle Paleolithic Transition in the Southern Caucasus. *Science, 345*, 1609–1613.

Alberch, P. (1989). The Logic of Monsters: Evidence for Internal Constraint in Development and Evolution. *Geobios (Lyon, France). Mémoire Spécial, 12*, 21–57.

Blumberg, M. S. (2009). *Freaks of Nature: What Anomalies Tell Us about Development and Evolution.* Oxford: Oxford University Press.

Buchanan, B., & Collard, M. (2008). Testing Models of Early Paleoindian Colonization and Adaptation Using Cladistics. In M. J. O'Brien (Ed.), *Cultural Transmission and Archaeology: Issues and Case Studies* (pp. 359–376). Washington, D.C.: Society for American Archaeology Press.

Chamberlain, J. A. (1976). Flow Patterns and Drag Coefficients of Cephalopod Shells. *Palaeontology, 19*, 539–563.

Chamberlain, J. A. (1981). Hydromechanical Design of Fossil Cephalopods. *Systematics Association Special Volume, 18*, 289–336.

Darwin, C. (1859). *On the Origin of Species by Means of Natural Selection; or the Preservation of Favoured Races in the Struggle for Life*. London: Murray.

Darwin, C. (1875). *Insectivorous Plants*. London: Murray.

Eren, M. I., Lycett, S. J., Roos, C. I., & Sampson, C. G. (2011). Toolstone Constraints on Knapping Skill: Levallois Reduction with Two Different Raw Materials. *Journal of Archaeological Science, 38*, 2731–2739.

Eren, M. I., Patten, R. J., O'Brien, M. J., & Meltzer, D. J. (2013). Refuting the Technological Cornerstone of the Ice-Age Atlantic Crossing Hypothesis. *Journal of Archaeological Science*, *40*, 2934–2941.

Gould, S. J. (1989). *Wonderful Life: The Burgess Shale and the Nature of History*. New York: Norton.

Gupta, A. K. (2004). Origin of Agriculture and Domestication of Plants and Animals Linked to Early Holocene Climate Amelioration. *Current Science*, *87*, 54–59.

Hauser, M. D. (2009). The Possibility of Impossible Cultures. *Nature*, *460*, 190–196.

Lecointre, G., & Le Guyader, H. (2006). *The Tree of Life: A Phylogenetic Classification*. Cambridge, MA: Belknap.

Lipo, C. P., O'Brien, M. J., Collard, M., & Shennan, S. (Eds.). (2006). *Mapping Our Ancestors: Phylogenetic Approaches in Anthropology and Prehistory*. New Brunswick, N.J.: Aldine.

Lycett, S. J. (2009). Are Victoria West Cores "Proto-Levallois"? A Phylogenetic Assessment. *Journal of Human Evolution*, *56*, 175–191.

Lycett, S. J. (2011). "Most Beautiful and Most Wonderful": Those Endless Stone Tool Forms. *Journal of Evolutionary Psychology (Budapest)*, *9*, 143–171.

Maclaurin, J. (2003). The Good, the Bad and the Impossible. *Biology & Philosophy*, *18*, 463–476.

McGhee, G. R. (1999). *Theoretical Morphology: The Concept and Its Applications*. New York: Columbia University Press.

McGhee, G. R. (2011). *Convergent Evolution: Limited Forms Most Beautiful*. Cambridge, Mass.: MIT Press.

McGhee, G. R. (2012). *The Geometry of Evolution: Adaptive Landscapes and Theoretical Morphospaces*. Cambridge: Cambridge University Press.

McGhee, G. R. (2015). Limits in the Evolution of Biological Form: A Theoretical Morphologic Perspective. *Interface Focus*, *5*, 1–6. doi: 10.1098/rsfs.2015.0034.

McGowan, A. J. (2004). The Effect of the Permo-Triassic Bottleneck on Triassic Ammonoid Morphological Evolution. *Paleobiology*, *30*, 369–395.

Müller, G. B. (2007). Evo-Devo: Extending the Evolutionary Synthesis. *Nature Reviews. Genetics*, *8*, 943–949.

O'Brien, M. J., & Bentley, R. A. (2011). Stimulated Variation and Cascades: Two Processes in the Evolution of Complex Technological Systems. *Journal of Archaeological Method and Theory*, *18*, 309–335.

O'Brien, M. J., & Lyman, R. L. (2003). *Cladistics and Archaeology*. Salt Lake City: University of Utah Press.

O'Brien, M. J., Darwent, J., & Lyman, R. L. (2001). Cladistics Is Useful for Reconstructing Archaeological Phylogenies: Palaeoindian Points from the Southeastern United States. *Journal of Archaeological Science*, *28*, 1115–1136.

O'Brien, M. J., Lyman, R. L., Collard, M., Holden, C. J., Gray, R. D., & Shennan, S. J. (2008). Transmission, Phylogenetics, and the Evolution of Cultural Diversity. In M. J. O'Brien (Ed.), *Cultural Transmission and Archaeology: Issues and Case Studies* (pp. 39–58). Washington, D.C.: Society for American Archaeology Press.

O'Brien, M. J., Boulanger, M. T., Buchanan, B., Bentley, R. A., Lyman, R. L., Lipo, C. P., et al. (2016). Design Space and Cultural Transmission: Case Studies from Paleoindian Eastern North America. *Journal of Archaeological Method and Theory*, *23*, 692–740.

Powell, R., & Mariscal, C. (2014). There Is Grandeur in This View of Life: The Bio-philosophical Implications of Convergent Evolution. *Acta Biotheoretica*, *62*, 115–121.

Raup, D. M. (1966). Geometric Analysis of Shell Coiling: General Problems. *Journal of Paleontology*, *40*, 1178–1190.

Raup, D. M. (1967). Geometric Analysis of Shell Coiling: Coiling in Ammonoids. *Journal of Paleontology, 41*, 43–65.

Saunders, W. B., Work, D. M., & Nikolaeva, S. V. (2004). The Evolutionary History of Shell Geometry in Paleozoic Ammonoids. *Paleobiology, 30*, 19–46.

Shea, J. J. (2006). The Middle Paleolithic of the Levant: Recursion and Convergence. In E. Hovers & S. L. Kuhn (Eds.), *Transitions before the Transition: Evolution and Stability in the Middle Paleolithic and Middle Stone Age* (pp. 189–211). New York: Springer.

Shott, M. (2008). Darwinian Evolutionary Theory and Lithic Analysis. In M. J. O'Brien (Ed.), *Cultural Transmission and Archaeology: Issues and Case Studies* (pp. 146–157). Washington, D.C.: Society for American Archaeology Press.

Steadman, P. (2014). *Building Types and Built Forms.* Leicestershire, England: Matador.

Ward, P. (1980). Comparative Shell Shape Distributions in Jurassic—Cretaceous Ammonites and Jurassic—Tertiary Nautilids. *Paleobiology, 6*, 32–43.

3 The Transparency of Imitation versus Emulation in the Middle Paleolithic

R. Alexander Bentley

Within the issue of convergent evolution is the question of why change in stone-tool evolution can be remarkably slow. Although stone-tool evolution was faster among Paleoindian groups of North America, it was still stable enough, or convergent enough, that Clovis projectile points would bear a superficial resemblance to Solutrean technology of Late Paleolithic Europe (chapter 1, this volume). The shift toward more cutting edge per mass in Pleistocene stone tools, for example, occurred over hundreds of thousands of years. Such stasis was occasionally punctuated by rapid change, such that by about 40,000 years ago, the Mousterian industry evolved relatively quickly into Châtelperronian assemblages, just as anatomically modern humans were spreading into Europe (chapter 10, this volume).

Outside such major hominin dispersals, why so little change over such a long time? For some researchers, the slowness of change in the Middle Paleolithic reflects reduced cognitive capacity (Bruner and Lozano 2014). Kuhn and Zwyns (chapter 8, this volume) argue that later Paleolithic flexibility and creativity in technology reflects a cognitive change among hominins. Cognitive theories for Plio-Pleistocene technological evolution often invoke change in capacity for working memory as a prerequisite for hominin toolmaking. Chimpanzees, for example, have sufficient "procedural memory" to make and use a tool—an elongated rod used to reach for a grape—that they'd been taught four years before but had not practiced since (Vale et al. 2016). A general definition of working memory includes multiple subcomponents, specifically a "phonological loop," which allows for recursive language understanding; visual and spatial memory; and a central executive component, which pulls everything together using the "episodic buffer" as memory storage (Baddeley 2001; Wynn and Coolidge 2010). If memory had to evolve physiologically before the corresponding stone-tool technology could evolve, then change would indeed be slow.

Stone tools and other material culture, however, are themselves a form of memory. The coevolution may have allowed for "a complex brain [that] produces a complex culture which, through feedback, selects for a more complex brain, generating a loop towards increasing complexity" (Bruner and Lozano 2014: 273). The relationship among stone-tool evolution, cognitive memory, and information embodied in material culture depends on the balance of *imitation* versus *emulation.* The difference lies between finding one's way to achieving the same goal (emulation) versus copying the means of getting there (imitation) (chapter 9, this volume). Imitation implies that the *châine operatoire*, or "cultural recipe," for making certain tools is the unit of transmission, such that most variation arises through copying errors (chapter 5, this volume). Emulation, on the other hand, allows the form of the artifact itself to serve as the unit of transmission, such that error happens through reproduction based on that mental model of the artifact. Under emulation, new variation is introduced by imperfect perception, model memorization or recall; stone tools therefore evolve through minute changes in their morphological characters (chapter 5, this volume).

Wilkins (chapter 9, this volume) sees population size, environmental variability, and resource competition as key factors in whether imitation versus emulation was the more advantageous learning mechanism. Imitation is explicitly invoked in models of Upper Paleolithic cumulative culture (Powell, Shennan, and Thomas 2009), for example. This demographic view of technological innovation (Muthukrishna and Henrich 2016) provides an alternative hypothesis for the sporadic creativity in the Middle Paleolithic record—for example, marked bones, deliberate use of pigments, ornamental perforated animal teeth, and rock art (Bednarik 1992)—in that they occurred in populations too small to sustain the behaviors as traditions over subsequent generations.

In terms of stone-tool production and *châine operatoire*, it's useful to consider a continuum from pure emulation at one end to pure imitation at the other. For relatively simple tasks, humans can achieve similar quality through imitation or emulation (Caldwell and Millen 2009), and the balance between these learning modes is at least partly affected by culture (Mesoudi, Chang, Murray, and Lu 2015). If an object reveals its own assembly instructions—through flake scars, for example—then one can learn to make the object through emulation. Complex tasks, however, more likely require learning through imitation of successive steps from an expert.

In either case, we can imagine situations where imitation or emulation is less transparent. For example, in the case of cave art, the artist(s) might have lived in a past generation and cannot be directly imitated. In terms of emulation, complex multicomponent tools or refined art may take a lot of practice to emulate because the reverse engineering is difficult to figure out. This idea of learning transparency introduces a second continuum to the learning process, ranging from opaque to transparent. With imitation-emulation as the horizontal axis, we can represent transparency as a vertical axis, such that the combination forms a two-dimensional heuristic "map" (figure 3.1).

This space, which illustrates four different "quadrants" of learning, carries implications for culture evolution. In the upper left, the extreme would be individual learning by emulation, presumably trial and error, where the goal is quite transparent. The upper left therefore accommodates Boyd and Richerson's (1985) *guided variation*, where individual

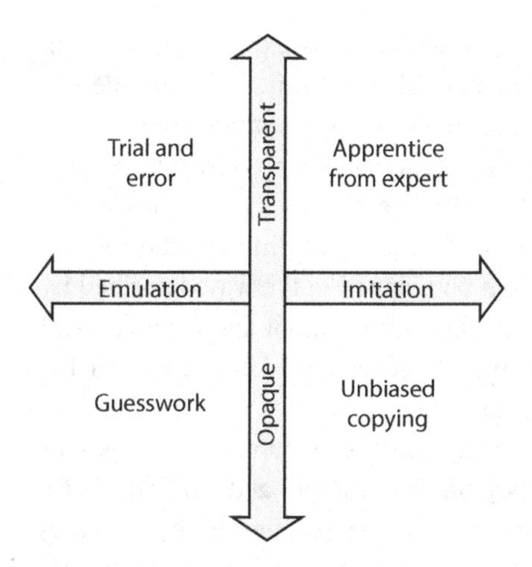

Figure 3.1
A heuristic map for understanding different domains of human decision making, based on whether a decision is made individually or socially on the horizontal axis and the transparency of payoffs that inform a decision on the vertical axis (after Bentley et al. 2014).

learning is inherited and modified over successive generations. In the upper right, the extreme would be transparent imitation of experts through apprenticeship, for example. The expert in each generation is likely determined by superior skill or knowledge. In the so-called Tasmania model, Henrich (2004) provided a parameter for the transparency of this learning from experts, represented inversely by an error rate in learning from the most-skilled individual in each generation.

Whereas the upper two quadrants represent well-explored areas of cultural evolution and social learning (e.g., Hoppitt and Laland 2013), the bottom two quadrants, where learning is less transparent, provide the space for exploring different modes of cultural evolution (Bentley, O'Brien, and Brock 2014). The lower right, where expertise is not transparent, can be characterized by "neutral" models, where individuals simply copy one another at random (Acerbi and Bentley 2014; Kandler and Shennan 2013, 2015; Premo 2014). In the lower left, the extreme is random choice, where an individual might try a sequence of actions randomly. The lower left could include, for example, early stone tools that are barely distinguishable from naturally fractured stones.

Several colleagues and I have used this four-quadrant heuristic to show how decision making can be mapped using data on the relative frequency of different choices in a well-resolved sequence of discrete time slices (Bentley, Earls, and O'Brien 2011; Bentley et al. 2014; Bentley, Brock, Caiado, and O'Brien 2016; Brock, Bentley, O'Brien, and Caiado 2014; Caiado, Brock, Bentley, and O'Brien 2016). In a data-rich situation, different expected data patterns for each quadrant—frequency distributions, time series behavior, and the like—can be compared to the empirical record, which can also be compared to explicit expectations of the parameterized map, where each axis is explicitly defined in terms of a numerical transparency variable and a numerical social-influence variable.

Unfortunately, it is difficult to distinguish one from the other without meticulous time-stratified observational data of one individual influencing another (Hobaiter, Poisot, Zuberbühler, Hoppitt, and Gruber 2014). Middle Paleolithic archeological remains would rarely exhibit the completeness in terms of the relative frequency of different choices and temporal resolution to generate data series that could be mapped in the way described by Brock et al. (2014) for simulated data. Nevertheless, the four-quadrant approach can still be used qualitatively, or semiquantitatively, in terms of archeological inferences of the learning process and expectations for change through time.

Archaeological evidence may yet reveal, however, some signature of imitation or emulation in Middle Paleolithic technology (Corbey, Jagich, Vaesen, and Collard 2016). The variation present in lithic debris may be indicative. Wilkins (chapter 9, this volume) uses variation to infer emulation in middle Pleistocene blade production at Kathu Pan, South Africa, dating some 500,000 years ago. For Wilkins, emulation predicts diversity in reduction strategies that lead to similar end products. There is a limit, however, to what reduction-sequence debris can reveal about imitation versus emulation. Using experimental archeology, Eren et al. (chapter 4, this volume) show that Levallois cores, Upper and Lower Paleolithic blade cores, and Acheulean hand-axe production techniques produce flake-debris patterns that are difficult to distinguish statistically.

This implies uncertainty in where Kathu Pan sits on the emulation-imitation axis, as well as difficulty in inferring the reduction sequence from the lithic remains themselves. Given that *Homo heidelbergensis* would have encountered the same difficulty reconstructing the reduction sequence from the remains alone, the transparency of emulation at Kathu Pan would be lower than the transparency of imitation or learning directly from an expert. Hence, we can at least draw a diagonal region on the map, moving up from left to right, labeled Kathu Pan (figure 3.2).

Such inferences ideally would draw upon the wider cultural complex, including perishable objects of wood, plant fibers, and animal products. Consider the assemblage of wooden spears, stone tools and debris, and faunal remains at Schöningen, Germany (Thieme 1997), associated with early Neanderthals or *H. heidelbergensis* about 300,000 years ago (Richter and Krbetschek 2015). Where would learning to use these spears fit on the heuristic map? Although it is debated whether the spears were thrown as opposed to just thrusted (Schoch, Bigga, Böhner, Richter, and Terberger 2015)—the asymmetric Neanderthal humerus morphology, for instance, could reflect either throwing or scraping hides, while the expertise of Neanderthals in hunting (e.g., Marean and Kim 1998) suggests they did throw these wooden spears up to about 35 meters, according to experiments (Schoch et al. 2015). Also, following Leiberman's argument (chapter 6, this volume) that throwing was among the primary adaptations of the *Homo* lineage, learning to throw these spears could have been achieved through individual practice and emulation. We therefore

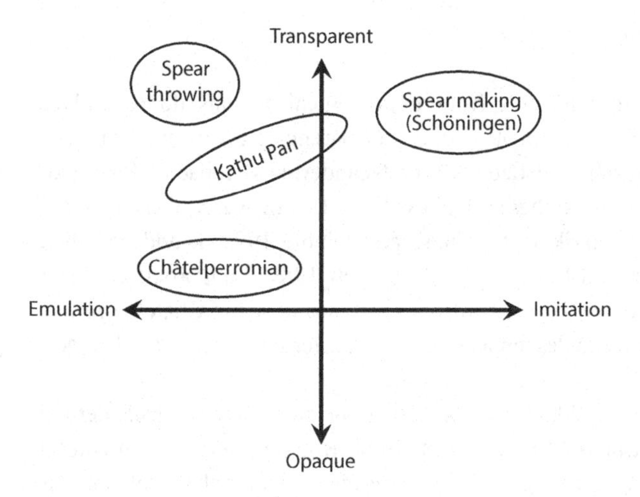

Figure 3.2
Examples discussed in the text as they are located on the map in figure 3.1 of the different domains of human decision making.

might place Neanderthal spear throwing near the highly transparent emulation quadrant in the upper left of the map (figure 3.2).

Where might the *making* of these spears plot on the map? They were produced by first stripping the bark off spruce or pine, which was then cut, scraped, and smoothed into the spear, whose tip, which was located away from the central axis of the trunk, was later resharpened (Schoch et al. 2015). Their context is instructive; from the same horizon at Schöningen as the spears, the stone remains include 1,500 artifacts that are technologically late Lower Paleolithic (Serangeli and Conard 2015). Patterns of wear and residues are indicative of hafting (Rots, Hardy, Serangeli, and Conard 2015), which may or may not be the oldest evidence of the technique (Rots and Plisson 2014; Wilkins, Schoville, Brown, and Chazan 2012), but nonetheless also suggest more imitative learning than emulation. We can therefore map the making of Schöningen spears in the upper-right quadrant of the map (figure 3.2), indicative of transparent imitation.

The upper right could be consistent with learning through apprenticeship, for which Neanderthals might have had sufficient working memory (Wynn and Coolidge 2004). Neanderthals presumably had the phonological loop, based on the Kebara 2 hyoid from Israel, ca. 63,000 years ago (Arensburg et al. 1989; D'Anastasio et al. 2013; Lieberman 1993), as well as the *FOXP2* gene (Krause et al. 2007). Integration between the phonological loop and visuo-spatial memory is suggested by the probable musical instrument at Divje Babe I cave, Slovenia—a hollow cave-bear femur broken at both ends, with four holes in a straight line—dating between 60,000 and 50,000 years ago (Turk and Dimkaroski 2011).

Example: The Châtelperronian

These heuristic examples aside, it is difficult, from archeological remains alone, to place Neanderthal learning on the imitation-emulation and transparency axes of our map. The question also runs into the longstanding debate over Neanderthals' capacity for art or ritual versus overinterpretation of the archeological evidence (Akazawa, Muhesen, Dodo, Kondo, and Mizoguchi 1995; Bednarik 1992; Chase and Dibble 1992; Davidson 1992; d'Errico, Zilhao, Julien, Baffier, and Pelegrin 1998; Gargett 1989; Langbroek 2014). A good place to address this is the last 20 years of research into how the Châtelperronian industry—the early Upper Paleolithic descendent of the Mousterian—relates to Neanderthals in their last phases ca. 40,000 years ago in Europe.

Two sites have become prime candidates for a secure association between Neanderthal remains and Châtelperronian artifacts (d'Errico et al. 1998; Roussel, Soressi, and Hublin 2016). One is Saint-Césaire, France, where an almost complete Neanderthal skeleton was found with Châtelperronian artifacts that date to about 36,000 years ago (Hublin, Spoor, Braun, Zonneveld, and Condemi 1996). The other is Grotte du Renne at Arcy-sur-Cure, France, dated at about 34,000 years ago, where a temporal bone in context with Châtelperronian assemblage was determined to be from a Neanderthal juvenile. The Châtelperronian ornaments from Grotte du Renne included personal ornaments such as pierced and grooved animal teeth, ivory beads and rings, colorants, and bone awls (Caron, d'Errico, Del Moral, Santos, and Zilhão 2011). Nearby, in the Châtelperronian levels of the Grotte du Bison, Arcy-sur-Cure, Mousterian tools were found in context with a Neanderthal maxilla (Bodu et al. 2014).

The hominid remains at Grotte du Renne have now all but been proven to be Neanderthal (Bailey and Hublin 2006; Bailey, Weaver, and Hublin 2009; Hublin et al. 1996). The first determination was based on principal-components analysis of seven measurements, including radii of curvature of the anterior, posterior, and lateral semicircular canals, evaluated by Mahalanobis distances from group centroids defined by Neanderthal, *H. erectus*, modern *H. sapiens,* and Upper Paleolithic modern human samples (Hublin et al. 1996). New evidence from Grotte du Renne was obtained by mass spectrometry of the collagen protein from additional bone fragments found in Châtelperronian levels. Welker et al. (2016) identified 28 *Homo* fragments, among 196 bone fragments screened overall, through peptide and amino-acid characterization. In these newly identified hominin remains, the collagen protein, COL10α10, contained asparagine in amino acid position 128, which is identified as Neanderthal through comparison with protein-sequences from modern humans, a Denisovian, and three Neanderthals (Welker et al. 2016). The mtDNA sequences from the remains also indicate Neanderthal, and enrichment in collagen ^{15}N suggests the infant had not yet been weaned (Welker et al. 2016). As baby teeth were found in the same excavation square, Welker et al. (2016) concluded that the remains were from an infant Neanderthal.

With Neanderthal skeletal remains and Châtelperronian artifacts found in the same contexts, post-depositional mixing has been suggested—that is, the personal adornments were pushed down from overlying Aurignacian levels. The evidence indicates this is unlikely, as there were no reported Aurignacian stone tools in the Châtelperronian levels at Grotte du Renne, for instance (Caron et al. 2011).

This helps resolve the dispute over the integrity of the Châtelperronian levels at Grotte du Renne (Higham et al. 2010; Hublin et al. 2012). Originally, d'Errico et al. (1998) maintained that the Châtelperronian artifacts show stylistic continuity with the Mousterian, and that in terms of both stratigraphy and radiocarbon dates, the oldest Châtelperronian (about 45,000 years ago) predates the oldest Aurignacian (about 40,000 years ago) and thus was not derived from the Aurignacian. Currently, Châtelperronian is considered a transitional industry between Upper and Middle Paleolithic "because of its chronological position, and the association of Neanderthal remains with blades, bone tools and personal ornaments" (Roussel, Soressi, and Hublin 2016). At the open-air site of Ormesson in the Paris Basin, about 300 kilometers northeast of Arcy, Châtelperronian and Middle Solutrean settlements lie immediately below Gravettian remains of a bison hunt (Bodu et al. 2014). In Level 4 at Ormesson, Bodu et al. (2014) see evidence that the bone pieces, fireplaces, and nodules of red coloring materials were used by the last Mousterians.

If Neanderthals did produce the ornaments in the Châtelperronian assemblage (Roussel et al. 2016), can we locate their learning process on our four-quadrant map? The first question is whether Neanderthals imitated or emulated Aurignacian material culture. Originally, d'Errico et al. (1998) argued that Neanderthals at Arcy-sur-Cure made simple bone tools, decorated themselves with perforated animal teeth, and put red ochre on their living floors (d'Errico et al. 1998). Maintaining that the symbolic ornaments at Grotte du Renne were Neanderthal imitations of Aurignacian culture, d'Errico et al. (1998) effectively argued that the process was emulation, not imitation, due to differences between Châtelperronian objects versus Aurignacian artifacts. They also pointed to techniques of grooving the root of the tooth ornaments at Grotte du Renne, which had been found at other Châtelperronian sites such as Quinçay but not observed at Aurignacian sites.

Given the differences between Châtelperronian and Aurignacian assemblages, it seems doubtful that Neanderthals learned the Châtelperronian from imitation of Aurignacians, as opposed to emulation at a distance (socially, spatially, temporally). At Quinçay Cave, Roussel et al. (2016) found that the retouched bladelets in the Châtelperronian layers were distinct enough from their Proto-Aurignacian counterparts to rule out technological convergence. Instead, the "idea of retouched bladelets" may have diffused from the northern Proto-Aurignacian to the Quinçay Châtelperronian (Roussel et al. 2016). Although ancient DNA indicates that modern humans and Neanderthals interbred on occasion (Fu et al. 2015; Kuhlwilm et al. 2016; Stringer 2014, 2016), it is doubtful that Neanderthal-human interactions would have been prolonged enough for apprenticeship to have arisen. Indeed, given evidence that the early *H. sapiens* clade is at least 300,000 years old (Hublin et al. 2017), the cognitive capacities of Neanderthal and *H. sapiens* lineages probably evolved independently and in parallel over at least 400,000 years (Stringer and Galway-Witham 2017). Roussel et al. (2016) specifically favor emulation, in terms of transmission of the shape of desired end products without the transmission of their manufacturing process.

Indeed, social interaction is not necessary for emulation to move a technology forward. A good case in point is cave art, which may be observed and emulated centuries after painting. The unexpectedly old uranium-thorium dates for Upper Paleolithic cave art (Pike et al. 2012), such as a claviform symbol at Altamira Cave dated at 35,600 years old, or at Tito Bustilo Cave, Spain, where six-foot-long paintings of horses overlay paintings that are about 29,000 years old. Pike et al. (2012) suggest the 37,000-year old hand stencils at El Castillo Cave in Spain might have been Neanderthal hands (Appenzeller 2013). Analyzing the technology and morphology of red pigments in the disks from El Castillo Cave by a range of spectrographic methods, d'Errico et al. (2016) identify clear differences in technology and pigment composition. Different individuals made, and emulated, these disks at different times.

Hence, Neanderthal art and ornaments would seem to reflect emulation of varying degrees of transparency (figure 3.2). Anatomical evidence suggests that Neanderthals and their ancestors were exceptional emulators, excellent at carrying out tasks by mental representation (Langbroek 2014). At Sima de los Huesos, Spain, specimens of European *H. heidelbergensis*, the probable ancestors of Neanderthals, have wear on incisors and canines to indicate they relied extensively on their teeth as a "third hand" for tasks such as cutting or holding (Bruner and Lozano 2014; Lozano, Bermúdez de Castro, Carbonell, and Arsuaga 2008). Langbroek (2014) argues that, by comparison, *H. sapiens*, with better developed intraparietal sulcus and less frequent dental evidence for the "third hand," is more reliant on imitation—eye-hand coordination and direct vision—in toolmaking. *H. sapiens* is the imitative species *par excellence* (Horner and Whiten 2005).

Implications for Convergent and Other Evolution

If archaeologists were omniscient observers, they could look in detail at the social network of communication (Centola and Baronchelli 2015; Hobaiter et al. 2014; Hoppitt and Laland 2013), which, unfortunately, is impossible. Our four-quadrant map (figure 3.1), though, at least draws attention to important factors in technological evolution-imitation versus emulation and transparency of model (object or expert)—and allows us to generate some expectations for what we might be seeing.

In the lower left, where learning is both individual and opaque, we expect the stone-tool record to be variable and inconsistent. In the lower right, with its indiscriminate copying of others without focus on any one in particular, we expect to find forms and reduction sequences that potentially could be quite similar within groups but with significant differences between groups (Lipo, Madsen, and Dunnell 2015; Premo 2014). We also expect continual change in the most frequently learned characteristics through time (Acerbi and Bentley 2014). In the lower right, the frequency distribution of different traits should be highly right-skewed; that is, non-Gaussian (Bentley et al. 2014).

In the upper-right quadrant, with its apprenticeship, the frequency distribution of different traits should also be right-skewed, but in this case without much turnover in what ranks as the most frequent form. In other words, stasis is expected, which brings us back to the start of this chapter and its emphasis on very slow change in Middle Paleolithic tools. In the upper right, population size affects the efficiency with which agents learn and retain new and better behavioral strategies (Henrich 2010; Shennan 2011). A linear correlation between these variables is predicted for small-scale, adaptive societies (Henrich 2010; Kline and Boyd 2010).

In the upper-left quadrant, the frequency of certain trait types will tend to be normally (Gaussian) distributed as a result of cost/benefit constraints underlying them, with the maximal behavior becoming the most popular option and remaining so until circumstances change or a better solution becomes available. This, too, is consistent with much of the Middle Paleolithic lithic evidence. A way to distinguish the upper left from the upper right may be in frequency distribution among the different choices—Gaussian for upper left and right-skewed for upper right (Bentley et al. 2014).

If we can manage to plot different case examples on the map, what does this imply for their evolution? Here we can make use of the fitness-landscape function in the spirit of Sewell Wright (1932), as O'Brien and colleagues point out (O'Brien and Bentley 2011; O'Brien et al. 2016). As McGhee notes (chapter 2, this volume), the analogy for lithic studies would be design space, or "technospace" in Charbonneau's (chapter 5, this volume) terms: a volume of all possible designs that helps us map out functional constraints or developmental constraints, including what is functional and what is not functional. McGhee considers frequency distributions as contours in morphospace plots: discrete character morphospaces represented as n-dimensional spaces (or as networks or phylogenetic trees) that can identify homology/convergence at the morphological level. In the technospace model, imitation prevails, and convergence occurs in terms of manufacturing techniques rather than of morphological forms.

Caiado et al. (2016) explored how to quantify fitness on the four-quadrant map by employing a hill-climbing algorithm leading to the expected values of the optimal decisions as peaks on the fitness landscape. When imitation was significant, there were multiple equilibria at each point on the fitness landscape. The landscape was rugged even for three choices. If there is utility to doing as others do, as in our consideration of imitation, then the maximum fitness can be highly sensitive to initial conditions or choice frequencies (Caiado et al. 2016). This path-dependence is much different from a "Mount Fuji"–type fitness landscape, the peak of which has a clear solution toward which all uphill paths converge.

Conclusion

Stone-tool evolution in the Middle Paleolithic, which changed very slowly and apparently tracked payoffs in the local environment, appears to reflect transparent emulation, represented by the upper-right quadrant of the map. Neanderthal learning was apparently especially adapted for emulation—for example, of Aurignacian tools, art, and personal adornments. This appears to be the best explanation for the Châtelperronian, which might be plotted on the middle left of the map as emulation with moderate transparency. The onset of the Châtelperronian was a relatively abrupt change in Paleolithic terms. Had Neanderthals survived longer, their improving intergenerational learning might have evolved toward apprenticeship—transparent imitation in the upper-right quadrant—where we would expect slower change. In any case, the four-quadrant map provides insight, or at least new hypotheses, into what happens when abrupt technological change punctuates periods of much slower technological change.

References

Acerbi, A., & Bentley, R. A. (2014). Biases in Cultural Transmission Shape the Turnover of Popular Traits. *Evolution and Human Behavior, 35*, 228–236.

Akazawa, T., Muhesen, M., Dodo, Y., Kondo, O., & Mizoguchi, Y. (1995). Neanderthal Infant Burial. *Nature, 377*, 585–586.

Appenzeller, T. (2013). Neanderthal Culture: Old Masters. *Nature, 497*, 302–304.

Arensburg, B., Tillier, A.-M., Vandermeersch, B., Duday, H., Schepartz, L. A., & Rak, Y. (1989). A Middle Palaeolithic Hyoid Bone. *Nature, 338*, 758–760.

Baddeley, A. D. (2001). Is Working Memory Still Working? *American Psychologist, 56*, 851–864.

Bailey, S. E., & Hublin, J.-J. (2006). Dental Remains from the Grotte du Renne at Arcy-sur-Cure (Yonne). *Journal of Human Evolution, 50*, 485–508.

Bailey, S. E., Weaver, T. D., & Hublin, J.-J. (2009). Who Made the Aurignacian and Other Early Upper Paleolithic Industries? *Journal of Human Evolution, 57*, 11–26.

Bednarik, R. G. (1992). Paleoart and Archaeological Myths. *Cambridge Archaeological Journal, 2*, 27–57.

Bentley, R. A., Earls, M., & O'Brien, M. J. (2011). *I'll Have What She's Having: Mapping Social Behavior*. Cambridge, Mass.: MIT Press.

Bentley, R. A., O'Brien, M. J., & Brock, W. A. (2014). Mapping Collective Behavior in the Big-Data Era. *Behavioral and Brain Sciences, 37*, 63–119.

Bentley, R. A., Brock, W. A., Caiado, C. C. S., & O'Brien, M. J. (2016). Evaluating Reproductive Decisions as Discrete Choices under Social Influence. *Philosophical Transactions of the Royal Society of London. Series B, Biological Sciences, 371*, 20150154.

Bodu, P., Salomon, H., Leroyer, M., Naton, H.-G., Lacarriere, J., & Dessoles, M. (2014). An Open-Air Site from the Recent Middle Palaeolithic in the Paris Basin (France): Les Bossats at Ormesson (Seine-et-Marne). *Quaternary International, 331*, 39–59.

Boyd, R., & Richerson, P. J. (1985). *Culture and the Evolutionary Process*. Chicago: University of Chicago Press.

Brock, W. A., Bentley, R. A., O'Brien, M. J., & Caiado, C. C. S. (2014). Estimating a Path through a Map of Decision Making. *PLoS One*, *9*(11), e111022.

Bruner, E., & Lozano, M. (2014). Extended Mind and Visuo-spatial Integration: Three Hands for the Neandertal Lineage. *Journal of Anthropological Sciences*, *92*, 273–280.

Caiado, C. C. S., Brock, W. A., Bentley, R. A., & O'Brien, M. J. (2016). Fitness Landscapes among Many Options under Social Influence. *Journal of Theoretical Biology*, *405*, 5–16.

Caldwell, C. A., & Millen, A. E. (2009). Social Learning Mechanisms and Cumulative Cultural Evolution: Is Imitation Necessary? *Psychological Science*, *20*, 1478–1483.

Caron, F., d'Errico, F., Del Moral, P., Santos, F., & Zilhão, J. (2011). The Reality of Neandertal Symbolic Behavior at the Grotte du Renne, Arcy-sur-Cure, France. *PLoS One*, *6*(6), e21545.

Centola, D., & Baronchelli, A. (2015). The Spontaneous Emergence of Conventions: An Experimental Study of Cultural Evolution. *Proceedings of the National Academy of Sciences of the United States of America*, *112*, 1989–1994.

Chase, P. G., & Dibble, H. L. (1992). Scientific Archaeology and the Origins of Symbolism: A Reply to Bednarik. *Cambridge Archaeological Journal*, *2*, 43–51.

Corbey, R., Jagich, A., Vaesen, K., & Collard, M. (2016). The Acheulean Handaxe: More Like a Bird's Song than a Beatles' Tune? *Evolutionary Anthropology*, *25*, 6–19.

D'Anastasio, R., Wroe, S., Tuniz, C., Mancini, L., Cesana, D. T., Dreossi, D., et al. (2013). Micro-Biomechanics of the Kebara 2 Hyoid and Its Implications for Speech in Neanderthals. *PLoS One*, *8*(12), e82261.

Davidson, I. (1992). There's No Art—To Find the Mind's Construction—In Offence (Reply to Bednarik). *Cambridge Archaeological Journal*, *2*, 52–57.

D'Errico, F., Zilhao, J., Julien, M., Baffier, D., & Pelegrin, J. (1998). Neanderthal Acculturation in Western Europe? *Current Anthropology*, *39*, S1–S44.

D'Errico, F., Bouillot, L. D., García-Diez, M., Martí, A. P., Pimentele, D. G., & Zilhão, J. (2016). The Technology of the Earliest European Cave Paintings: El Castillo Cave, Spain. *Journal of Archaeological Science*, *70*, 48–65.

Fu, Q., Hajdinjak, M., Moldovan, O. T., Constantin, S., Mallick, S., Skoglund, P., et al. (2015). An Early Modern Human from Romania with a Recent Neanderthal Ancestor. *Nature*, *524*, 216–219.

Gargett, R. H. (1989). Grave Shortcomings: The Evidence for Neanderthal Burial. *Current Anthropology*, *30*, 157–190.

Henrich, J. (2004). Demography and Cultural Evolution. *American Antiquity*, *69*, 197–214.

Henrich, J. (2010). The Evolution of Innovation-Enhancing Institutions. In M. J. O'Brien & S. J. Shennan (Eds.), *Innovation in Cultural Systems: Contributions from Evolutionary Anthropology* (pp. 99–120). Cambridge, Mass.: MIT Press.

Higham, T., Jacobi, R., Julien, M., David, F., Basell, L., Wood, R., et al. (2010). Chronology of the Grotte du Renne (France) and Implications for the Context of Ornaments and Human Remains within the Châtelperronian. *Proceedings of the National Academy of Sciences of the United States of America*, *107*, 20234–20239.

Hobaiter, C., Poisot, T., Zuberbühler, K., Hoppitt, W., & Gruber, T. (2014). Social Network Analysis Shows Direct Evidence for Social Transmission of Tool Use in Wild Chimpanzees. *PLoS Biology*, *12*(9), e1001960.

Hoppitt, W., & Laland, K. N. (2013). *Social Learning: An Introduction to Mechanisms, Methods, and Models*. Princeton, N.J.: Princeton University Press.

Horner, V., & Whiten, A. (2005). Causal Knowledge and Imitation/Emulation Switching in Chimpanzees (*Pan troglodytes*) and Children (*Homo sapiens*). *Animal Cognition*, *8*, 164–181.

Hublin, J.-J., Ben-Ncer, A., Bailey, S. E., Freidline, S. E., Neubauer, S., Skinner, M. M., et al. (2017). New fossils from Jebel Irhoud, Morocco and the pan-African origin of *Homo sapiens*. *Nature*, *546*, 289–292.

Hublin, J.-J., Spoor, F., Braun, M., Zonneveld, F., & Condemi, S. (1996). A Late Neanderthal Associated with Upper Palaeolithic Artefacts. *Nature*, *381*, 224–226.

Hublin, J.-J., Talamo, S., Julien, M., David, F., Connet, N., Bodu, P., et al. (2012). Radiocarbon Dates from the Grotte du Renne and Saint-Césaire Support a Neandertal Origin for the Châtelperronian. *Proceedings of the National Academy of Sciences of the United States of America*, *109*, 18743–18748.

Kandler, A., & Shennan, S. J. (2013). A Non-equilibrium Neutral Model for Analysing Cultural Change. *Journal of Theoretical Biology*, *330*, 18–25.

Kandler, A., & Shennan, S. J. (2015). A Generative Inference Framework for Analysing Patterns of Cultural Change in Sparse Population Data with Evidence for Fashion Trends in LBK Culture. *Journal of the Royal Society Interface*, *12*, 20150905.

Kline, M. A., & Boyd, R. (2010). Population Size Predicts Technological Complexity in Oceania. *Proceedings of the Royal Society, Series B: Biological Sciences*, *277*, 2559–2564.

Krause, J., Lalueza-Fox, C., Orlando, L., Enard, W., Green, R. E., Burbano, H. A., et al. (2007). The Derived FOXP2 Variant of Modern Humans Was Shared with Neandertals. *Current Biology*, *17*, 1908–1912.

Kuhlwilm, M., Gronau, I., Hubisz, M. J., de Filippo, C., Prado-Martinez, J., Kircher, M., et al. (2016). Ancient Gene Flow from Early Modern Humans into Eastern Neanderthals. *Nature*, *530*, 429–433.

Langbroek, M. (2014). Ice Age Mentalists: Debating Neurological and Behavioural Perspectives on the Neandertal and Modern Mind. *Journal of Anthropological Sciences*, *92*, 285–289.

Lieberman, P. (1993). On the Kebara KMH 2 Hyoid and Neanderthal Speech. *Current Anthropology*, *34*, 172–175.

Lipo, C. P., Madsen, M. E., & Dunnell, R. C. (2015). A Theoretically-Sufficient and Computationally-Practical Technique for Deterministic Frequency Seriation. *PLoS One*, *10*(4), e0124942.

Lozano, M., Bermúdez de Castro, J. M., Carbonell, E., & Arsuaga, J. L. (2008). Nonmasticatory Uses of Anterior Teeth of Sima de los Huesos Individuals (Sierra de Atapuerca, Spain). *Journal of Human Evolution*, *55*, 713–728.

Marean, C. W., & Kim, S. Y. (1998). Mousterian Large-Mammal Remains from Kobeh Cave. *Current Anthropology*, *39*, S79–S113.

Mesoudi, A., Chang, L., Murray, K., & Lu, H. J. (2015). Higher Frequency of Social Learning in China than in the West Shows Cultural Variation in the Dynamics of Cultural Evolution. *Proceedings Biological Sciences*, *282*, 20142209.

Muthukrishna, M., & Henrich, J. (2016). Innovation in the Collective Brain. *Philosophical Transactions of the Royal Society of London. Series B, Biological Sciences*, *371*. doi: 10.1098/rstb.2015.0192.

O'Brien, M. J., & Bentley, R. A. (2011). Stimulated Variation and Cascades: Two Processes in the Evolution of Complex Technological Systems. *Journal of Archaeological Method and Theory*, *18*, 309–335.

O'Brien, M. J., Boulanger, M. T., Buchanan, B., Bentley, R. A., Lyman, R. L., Lipo, C. P., et al. (2016). Design Space and Cultural Transmission: Case Studies from Paleoindian Eastern North America. *Journal of Archaeological Method and Theory*, *23*, 692–740.

Pike, A. W. G., Hoffmann, D. L., García-Diez, M., Pettitt, P. B., Alcolea, J., De Balbín, R., et al. (2012). U-Series Dating of Paleolithic Art in 11 Caves in Spain. *Science*, *336*, 1409–1413.

Powell, A., Shennan, S. J., & Thomas, M. G. (2009). Late Pleistocene Demography and the Appearance of Modern Human Behavior. *Science, 324*, 1298–1301.

Premo, L. S. (2014). Cultural Transmission and Diversity in Time-Averaged Assemblages. *Current Anthropology, 55*, 105–114.

Richter, D., & Krbetschek, M. (2015). The Age of the Lower Paleolithic Occupation at Schöningen. *Journal of Human Evolution, 89*, 46–56.

Rots, V., & Plisson, H. (2014). Projectiles and the Abuse of the Use-Wear Method in a Search for Impact. *Journal of Archaeological Science, 48*, 154–165.

Rots, V., Hardy, B. L., Serangeli, J., & Conard, N. J. (2015). Residue and Microwear Analyses of the Stone Artifacts from Schöningen. *Journal of Human Evolution, 89*, 298–308.

Roussel, M., Soressi, M., & Hublin, J.-J. (2016). The Châtelperronian Conundrum: Blade and Bladelet Lithic Technologies from Quinçay, France. *Journal of Human Evolution, 95*, 13–32.

Schoch, W. H., Bigga, G., Böhner, U., Richter, P., & Terberger, T. (2015). New Insights on the Wooden Weapons from the Paleolithic Site of Schöningen. *Journal of Human Evolution, 89*, 214–225.

Serangeli, J., & Conard, N. J. (2015). The Behavioral and Cultural Stratigraphic Contexts of the Lithic Assemblages from Schöningen. *Journal of Human Evolution, 89*, 287–297.

Shennan, S. J. (2011). Descent with Modification and the Archaeological Record. *Philosophical Transactions of the Royal Society of London. Series B, Biological Sciences, 366*, 1070–1079.

Stringer, C. (2014). Why We Are Not All Multiregionalists Now. *Trends in Ecology & Evolution, 29*, 248–251.

Stringer, C. (2016). The Origin and Evolution of *Homo sapiens. Philosophical Transactions of the Royal Society of London. Series B, Biological Sciences, 371*, 20150237.

Stringer, C., & Galway-Witham, J. (2017). Palaeoanthropology: On the origin of our species. *Nature, 546*, 212–214.

Thieme, H. (1997). Lower Palaeolithic Hunting Spears from Germany. *Nature, 385*, 807–810.

Turk, M., & Dimkaroski, L. (2011). Neanderthal Flute from Divje Babe I: Old and New Findings. In B. Toškan (Ed.), *Fragments of Ice Age Environments: Proceedings in Honour of Ivan Turk's Jubilee* (pp. 251–265). Ljubljana, Slovenia: Opera Instituti Archaeologici Sloveniae.

Vale, G. L., Flynn, E. G., Pender, L., Price, E., Whiten, A., Lambeth, S. P., et al. (2016). Robust Retention and Transfer of Tool Construction Techniques in Chimpanzees (*Pan troglodytes*). *Journal of Comparative Psychology, 130*, 24–35.

Welker, F., Hajdinjak, M., Talamoa, S., Jaouen, K., Dannemann, M., David, F., et al. (2016). Palaeoproteomic Evidence Identifies Archaic Hominins Associated with the Châtelperronian at the Grotte du Renne. *Proceedings of the National Academy of Sciences of the United States of America, 113*, 11162–11167.

Wilkins, J., Schoville, B. J., Brown, K. S., & Chazan, M. (2012). Evidence for Early Hafted Hunting Technology. *Science, 338*, 942–946.

Wright, S. (1932). The Roles of Mutation, Inbreeding, Crossbreeding and Selection in Evolution. In D. F. Jones (Ed.), *Proceedings of The Sixth Congress on Genetics*, vol. 1 (pp. 356–366). New York: Brooklyn Botanic Garden.

Wynn, T., & Coolidge, F. L. (2004). The Expert Neandertal Mind. *Journal of Human Evolution, 46*, 467–487.

Wynn, T., & Coolidge, F. L. (2010). Beyond Symbolism and Language. *Current Anthropology, 51*, S5–S16.

4 Why Convergence Should Be a Potential Hypothesis for the Emergence and Occurrence of Stone-Tool Form and Production Processes: An Illustration Using Replication

Metin I. Eren, Briggs Buchanan, and Michael J. O'Brien

In recent years, the concept of evolutionary convergence as an explanatory factor for the emergence and occurrence of flaked stone-tool forms and production processes evident in the archeological record has become widely acknowledged, robustly examined, and thoroughly demonstrated by lithic analysts. From overshot flakes/flaking and bi-pointed bifaces (Boulanger and Eren 2015; Eren, Patten, O'Brien, and Meltzer 2013; Eren, Patten, O'Brien, and Meltzer 2014; Eren, Boulanger, and O'Brien 2015; O'Brien et al. 2014a, 2014b; Straus, Meltzer, and Goebel 2005) to Levallois cores (Adler et al. 2014), Nubian cores (Will, Mackay, and Phillips 2015), Victoria West cores (Lycett 2009, 2011), Acheulean handaxes (Wang, Lycett, von Cramon-Taubadel, Jin, and Bae 2012), and perhaps even fluting (Charpentier 2003; Charpentier et al. 2002; Crassard and Petraglia 2014), the list goes on. Indeed, Straus (2002: 70) goes so far as to state that *before* any sort of human connection (e.g., dispersal, transmission) is posited to explain cases of morphological similarity, "archaeologists must first eliminate the possibility of independent invention (even in cases of [pene-] contemporaneous developments)."

Still, there are those who remain skeptical. Some archeologists have equated prehistoric people with their technology, which inexorably results in a hyperdiffusionist view of the archeological record. If people "are" their technology, and let's say, for example, that a particular technology is found in different geotemporal contexts, then it follows that that technology must have been moved by that group of people—or been shared across the landscape—and could not have been independently invented. In reality, of course, the relationship between prehistoric people and technology is by no means a straightforward, one-to-one relationship.

Perhaps another reason some archeologists ignore evolutionary convergence may have to do with their chosen approach to studying the archeological record. The hitherto popular *chaîne opératoire*, "technological," and "authoritative flintknapper" approaches to flaked stone technology center on subjective, intuitive, and descriptive "readings" of lithic artifacts (Eren et al. 2016). In the same way no two people view the same movie in exactly the same way, practitioners of these nonquantitative approaches tend to emphasize, perhaps unavoidably, differences in artifact form and production process rather than similarity.

Further, since lithic artifact "readings" often describe details and nuances of different stone tools, even if similarities exist—in overall form, for example—these similarities will be ignored. And if similarities are ignored, then inevitably the concept of evolutionary convergence will be ignored as well.

Yet, positively, as the phenomenon of evolutionary convergence becomes increasingly tested for and documented in flaked-stone technology, archeologists can start to ask why convergence occurs. Why, for example, do tool types on opposite sides of an ocean and separated by thousands of years share similar production processes (chapter 1, this volume)? As McGhee (2011; chapter 2, this volume) notes, convergence can occur by means of two processes: functional and developmental constraints. With respect to the former (an *extrinsic* evolutionary constraint [McGhee 2007]), given the same function, natural selection should produce the same tool form or production process to serve that function (McGhee 2011). In the case of flaked-stone technology, function may include tasks such as cutting, shooting, scraping, engraving, striking a large flake, or creating a series of parallel flake scars; attributes such as efficiency, effectiveness, portability, or durability; and social objectives such as costly signaling, establishing group coherence, or aesthetics. Our focus here, however, is to begin to think about the role of developmental constraint (an *intrinsic* evolutionary constraint [McGhee 2007]) in flaked-stone technology. Here, different types of tool forms or production processes humans can develop are limited (McGhee 2011).

As a case study, we examined just how variable stone-tool production flakes were among knapped stone tools of different forms. If the shapes of production flakes generated from different reduction sequences substantially overlapped, then this outcome would potentially signal an important developmental constraint in stone-tool technology. This is because the result would be consistent with the idea that there exists only a limited variability in production-flake shape, and thus the probability for the parallel evolution of novelties is consequently higher than if production-flake shape was unbounded. To investigate this topic, we used modern stone-tool replication. By making replicas of different tools, we could ensure that we had at our disposal entire sets of production flakes. If we had tried to use archeological assemblages, we would have had no way of knowing how biased those assemblages might have been from prehistoric human behavior or other taphonomic processes. Of course, there are potential biases with replication as well (Eren et al. 2016), and we use it informally for illustrative purposes.

Methods and Materials

Over the course of two days one of us (MIE), a highly skilled knapper, produced a Lower Paleolithic–style unidirectional core (figure 4.1), an Acheulean handaxe (figure 4.2), a preferential Levallois core (figure 4.3), an Upper Paleolithic–style blade core (figure 4.4), and a Clovis point (figure 4.5). The knapper also produced a series of flakes by means

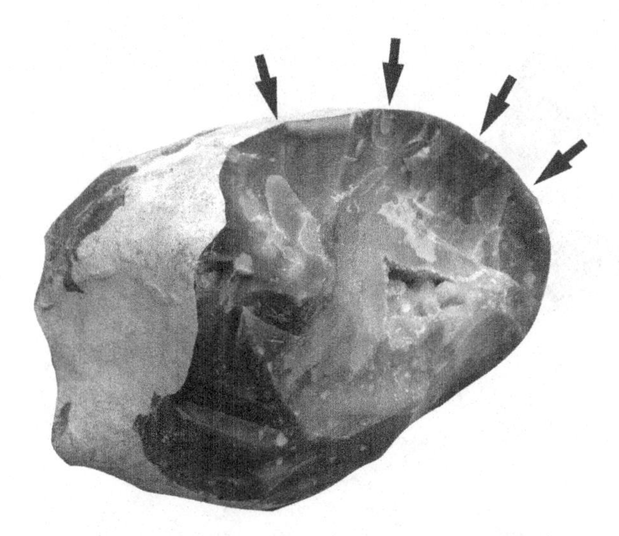

Figure 4.1
"Oldowan-style" unidirectional core reduction.

Figure 4.2
Acheulean handaxe reduction.

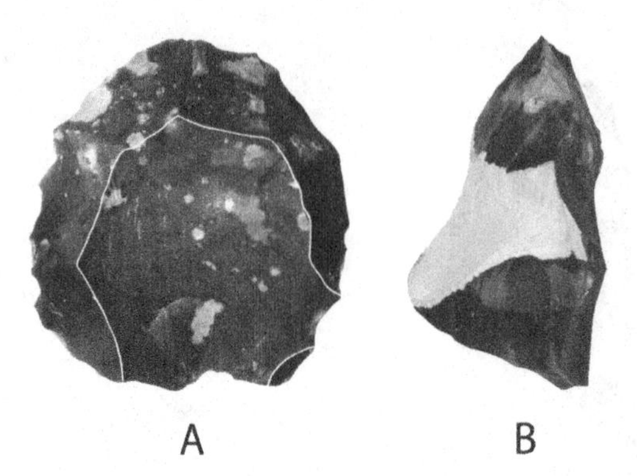

Figure 4.3
Preferential Levallois reduction.

Figure 4.4
Prismatic blade core reduction.

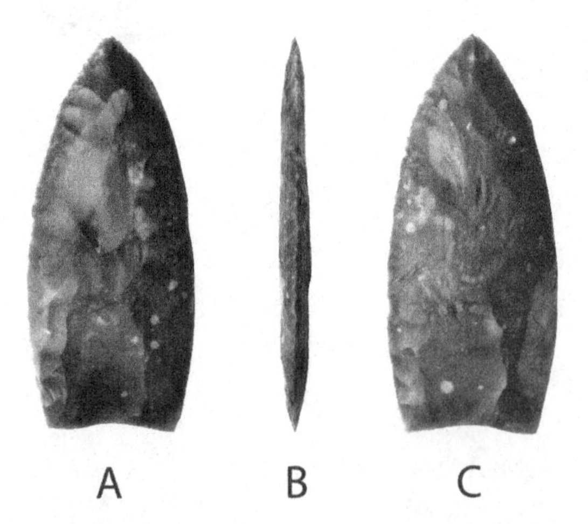

Figure 4.5
Clovis fluted projectile-point reduction.

of bipolar reduction. Each replication was taken "as is"; in other words, we did not, for example, make several Clovis points and pick the "best" one.

Raw material used in the replications was a siliceous flint procured from an inland quarry in the northeast English county of Norfolk. As shown in table 4.1 and figure 4.6, the six flint nodules differed considerably in terms of size and shape. Production took place on a large leather hide so that all flakes could be easily collected and bagged. A different set of production tools was used for each reduction sequence. For the Clovis point and Acheulean handaxe, the knapper used different combinations of antler billets and hammerstones. For the Lower Paleolithic–style unidirectional core, the knapper used a single hammerstone, and for the Levallois core, two hammerstones were used. For the removal of blades from the blade core, the knapper used a French boxwood billet, a sandstone hammerstone/abrader, and a small pebble hammerstone. For the bipolar reduction, a large hammerstone and a granite anvil were used.

The number of flakes produced differed in each reduction sequence (table 4.1). To select a sample of flakes from each reduction sequence for analysis, all flakes greater than 2.5 cm in maximum dimension were placed in line order. Then, using a random number generator, 30 flakes were selected. This procedure resulted in a total sample of 180 flakes (thirty times six reduction sequences).

Seven interlandmark distances were recorded on each flake: maximum dimension, width at 25%, 50%, and 75% of maximum dimension, and thickness at 25%, 50%, and 75% of maximum dimension. Flakes were oriented morphologically such that width at 25% of length was always the larger dimension over width at 75% of maximum dimension. Thicknesses at 25%, 50%, and 75% of maximum dimension were measured orthogonally to their corresponding width measurements. To turn these variables into shape data, we size-adjusted them by dividing each variable in turn by the geometric mean of all variables (Lycett, von Cramon-Taubadel, and Foley 2006). We also recorded flake curvature, which uses Collins's (1999) ratio of maximum dimension divided by maximum distance between a flake's ventral face and a plane connecting the flakes distalmost and proximalmost points.

We carried out several statistical analyses to assess the differences and similarities in the six flake samples. First, we tested the variables measured on the overall sample of flakes to see if they conformed to multivariate normality using Mardia's (1970) multivariate test of skewness and kurtosis and Box's *M* test for the equivalence of the covariance matrices. Both multivariate normality and equal covariance matrices are assumptions for a parametric multivariate analysis of variance (MANOVA) test. Second, we conducted a MANOVA test using reduction type as the grouping variable. Following the MANOVA test, we carried out pairwise comparisons using a Bonferroni correction of the *p*-values. Third, we entered the flake data into an exploratory principal component analysis to reduce variation in the dataset to a smaller number of principal components, which allowed us to visually assess variation.

Table 4.1
Data on the six flint nodules used in the experiment

Reduction sequence	Mass (g)	Maximum dimension (cm)	Width at 25% of maximum dimension (cm)	Width at 50% of maximum dimension (cm)	Width at 75% of maximum dimension (cm)	Thickness at 25% of maximum dimension (cm)	Thickness at 50% of maximum dimension (cm)	Thickness at 75% of maximum dimension (cm)	Number of knapped specimens
Handaxe	2326	23.6	13.5	11.7	10.5	6.3	7.0	5.7	160
Clovis point	1537	17.6	12.7	9.8	8.6	5.6	6.1	7.0	133
Levallois	2076	21.3	11.0	15.2	10.5	6.5	9.7	7.8	72
Bipolar	744	20.8	8.3	10.0	4.9	3.7	3.3	3.4	51
Blade	1266	16.5	13.3	11.7	8.6	6.1	8.5	6.6	111 (47 prismatic blades)
Unidirectional Oldowan style	4110	26.2	12.2	15.9	8.4	11.8	10.8	8.7	108

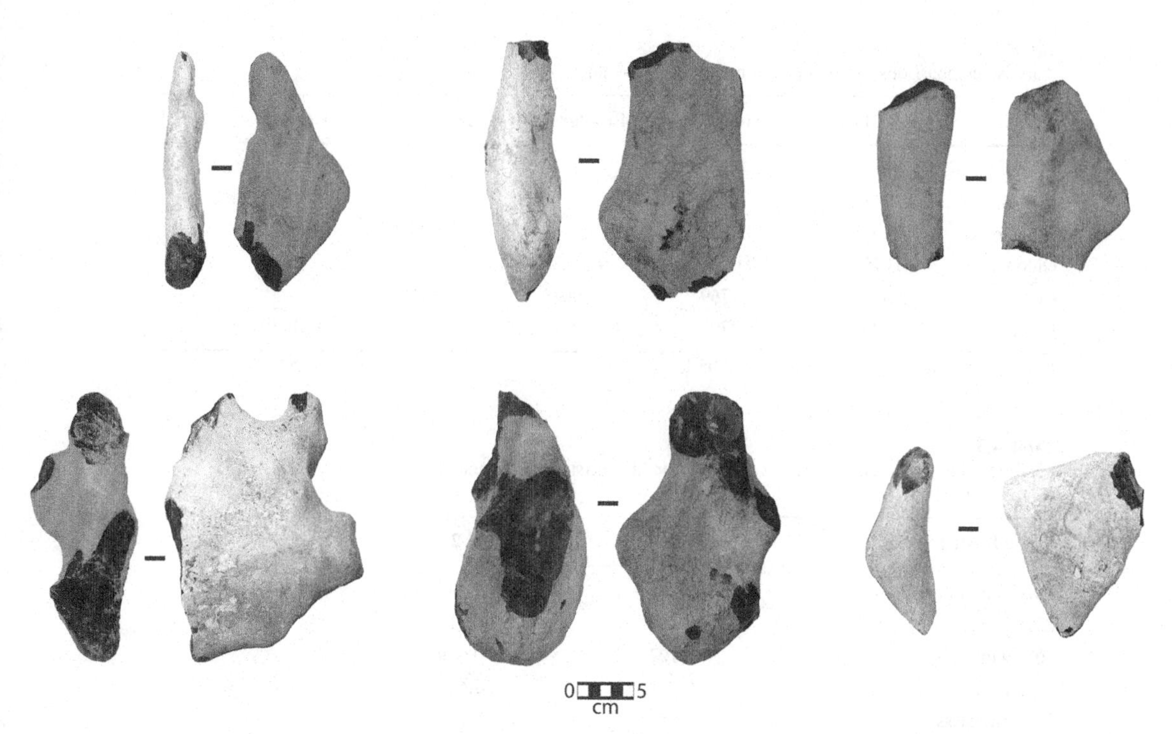

Figure 4.6
Raw flint nodules used in the experiment: bipolar (upper left), Levallois (upper center), Clovis point (upper right), Acheulean handaxe (lower left), Oldowan (lower center), and prismatic blade core (lower right).

Results

The size-adjusted variables from the six different reduction techniques do not conform to underlying multivariate normality ($p < 0.000$ for both Mardia's tests) and unequal variance-covariance matrices. The latter is revealed in the significant results of a Box's *M* test (M = 324.9, F = 2.08, df1 = 140, df2 = 47310, Monte Carlo $p = 0.001$). Given the lack of multivariate normality, we treat the following MANOVA results cautiously. The overall MANOVA results are significant (Wilks' lambda = 0.4488, Pillai trace = 0.6355, $p < 0.000$), but examination of the pairwise comparisons reveals some differences and similarities among the flake samples (table 4.2). Overall, for the 15 pairwise comparisons, there are seven differences and eight similarities. The Levallois sample is different from the bipolar sample, the blade sample is different from the bipolar and Oldowan samples, and the bipolar and Oldowan samples are different from the Clovis and hand-axe samples. However, the Levallois sample is similar to the blade, Oldowan, Clovis, and hand-axe samples. The blade sample is similar to the Levallois, Clovis, and hand-axe samples. Bipolar is similar only to Oldowan. Oldowan is similar to Levallois and Bipolar. Clovis

Table 4.2
Pairwise comparisons using Bonferroni corrected *p*-values

	Levallois	Blade	Bipolar	Oldowan	Clovis	Handaxe
Levallois	-					
Blade	0.3363	-				
Bipolar	0.0197*	< 0.0001*	-			
Oldowan	0.1148	< 0.0001*	9.2963	-		
Clovis	1.0378	3.1769	< 0.0001*	< 0.0001*	-	
Handaxe	5.8429	3.279	0.0001*	0.0009*	0.2493	-

*Differences are significant at the $p = 0.05$ level.

Table 4.3
Loadings for the size-adjusted variables for the first two principal components

Size-adjusted variable	PC 1	PC 2
Maximum dimension	0.968	–0.232
25% width	0.162	0.519
50% width	0.102	0.659
75% width	0.091	0.416
25% thickness	–0.074	–0.133
50% thickness	–0.087	–0.176
75% thickness	–0.072	–0.149

is similar to Levallois, blade, and handaxe, and handaxe is similar to Levallois, blade, and Clovis.

Another way to examine the variability in the flake samples that does not rely on multivariate normality is through principal component analysis. We concentrated on the principal components (PCs) that account for more than 5% of the variation. The results indicate that the first two PCs account for 91.4% of the overall variation in the size-adjusted variables. PC1, with an eigenvalue of 1.31, accounts for 66.8% of the variation, juxtaposing width and maximum dimension with the thickness variables (table 4.3). PC2 has an eigenvalue of 0.48 and accounts for 24.6% of the variation. It is similar to PC1 but combines the maximum dimension with the thickness variables, which are contrasted with the width variables. The bivariate plot of the PC scores for PC1 against PC2 shows considerable overlap (figure 4.7). The Clovis flakes appear to be the most different of all the reduction samples, but there is still substantial overlap with the other flake sets.

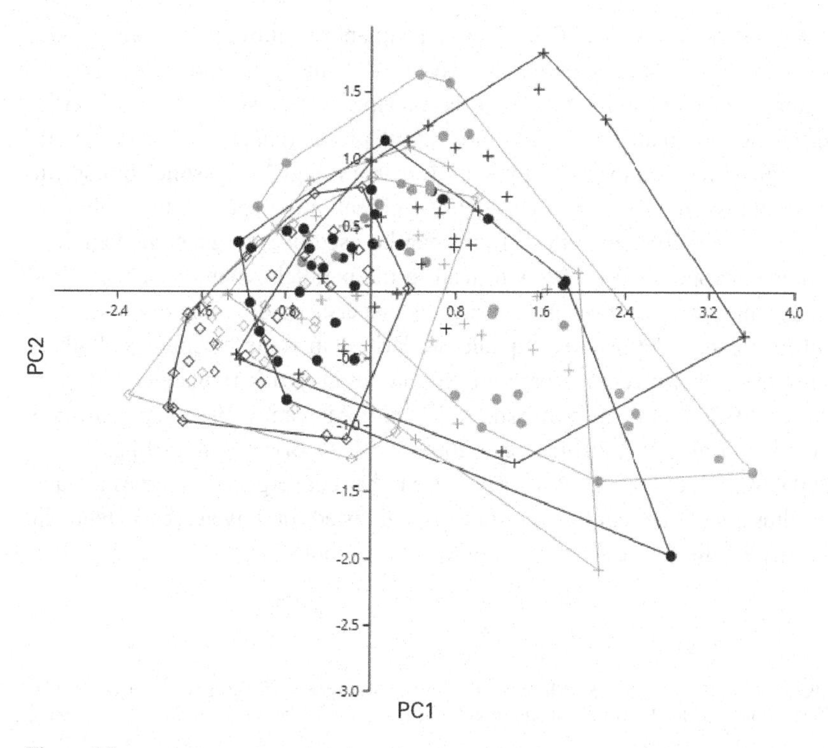

Figure 4.7
Principal component analysis of size-adjusted flake morphometric data (n = 180). PC1 (along x-axis) accounts for 66.8% of the overall variation; PC2 accounts for 24.6% of the overall variation. Key: black circles, Levallois; gray circles, blade; black diamonds, bipolar; gray diamonds, Oldowan; black crosses, Clovis; gray crosses, handaxe.

Conclusion

The notion that evolutionary convergence plays an important, and perhaps at certain times and scales of analysis a predominate, role in the emergence and evolution of stone-tool forms has in recent years steadily gained ground. Here we used stone-tool replication to examine the role of developmental constraint (McGhee 2007), the idea being that the number of different types of forms or production processes humans can develop is limited (McGhee 2011). Our results show that there is substantial overlap among the production-flake morphology of six different stone-tool reduction sequences. While there are some statistical differences among the sets, there is far more similarity in terms of morphological variability. What is striking about this overlap is that not only did the knapper make tools of very different form, but the original stone nodule shapes and sizes were extremely different *and* the knapper used different sets of tools to make the six different forms. Despite all these sources of variability, there is still substantial overlap in production-flake shape.

A knapper cannot strike a spherical flake. Nor can a knapper remove a cylindrical flake from the center of a core. Our results emphasize the fact that conchoidal fracture and flake removal can produce only a limited set of flake shapes. Of course, complex, diverse, and unique forms can be created from a limited set of pieces (chapter 5, this volume), but our results signal that, analogously to the "deep homology" of distinct biological forms sharing generative processes and cell-type specification (chapter 2, this volume), the limited number of possible stone flake forms can be used both to generate complex, diverse, and unique adaptations as well as to lead to the parallel evolution of novelties. We are in no way implying that mechanisms such as dispersal or diffusion did not occur in the past and thus cannot be used to explain similarities in stone technology. Rather, our results suggest, as other recent research does (Adler et al. 2014; Boulanger and Eren 2015; Charpentier 2003; Charpentier, Inisan, and Féblot-Augustins 2002; Crassard and Petraglia 2014; Eren et al. 2013, 2014; Lycett 2009, 2011; O'Brien et al. 2014a, 2014b; Straus et al. 2005; Wang et al. 2012; Will et al. 2015), that convergence deserves a place at the table and thus should be regularly considered and tested for. Given recent trends in archeological analysis, the future for these sorts of studies looks bright.

References

Adler, D. S., Wilkinson, K. N., Blockley, S., Mark, D. F., Pinhasi, R., Schmidt-Magee, B. A., et al. (2014). Early Levallois Technology and the Lower to Middle Paleolithic Transition in the Southern Caucasus. *Science*, *345*, 1609–1613.

Boulanger, M. T., & Eren, M. I. (2015). On the Inferred Age and Origin of Lithic Bi-points from the Eastern Seaboard and Their Relevance to the Pleistocene Peopling of North America. *American Antiquity*, *80*, 134–145.

Charpentier, V. (2003). From the Gulf to Hadramawt: Fluting and Plunging Processes in Arabia. In D. Potts, H. Naboodah, & P. Hellyer (Eds.), *Archaeology of the United Arab Emirates* (pp. 66–71). London: Trident.

Charpentier, V., Inisan, M. L., & Féblot-Augustins, J. (2002). Fluting in the Old World: The Neolithic Projectile Points of Arabia. *Lithic Technology*, *27*, 39–46.

Collins, M. B. (1999). *Clovis Blade Technology: A Comparative Study of the Keven Davis Cache, Texas*. Austin: University of Texas Press.

Crassard, R., & Petraglia, M. (2014). Stone Technology in Arabia. In H. Selin (Ed.), *Encyclopedia of the History of Science, Technology, and Medicine in non-Western Cultures* (pp. 1–5). New York: Springer.

Eren, M. I., Patten, R. J., O'Brien, M. J., & Meltzer, D. J. (2013). Refuting the Technological Cornerstone of the Ice-Age Atlantic Crossing Hypothesis. *Journal of Archaeological Science*, *40*, 2934–2941.

Eren, M. I., Patten, R. J., O'Brien, M. J., & Meltzer, D. J. (2014). More on the Rumor of "Intentional Overshot Flaking" and the Purported Ice-Age Atlantic Crossing. *Lithic Technology*, *39*, 55–63.

Eren, M. I., Boulanger, M. T., & O'Brien, M. J. (2015). The Cinmar Discovery and the Proposed Pre-Late Glacial Maximum Occupation of North America. *Journal of Archaeological Science: Reports*, *2*, 708–713.

Eren, M. I., Lycett, S. J., Patten, R. J., Buchanan, B., Pargeter, J., & O'Brien, M. J. (2016). Test, Model, and Method Validation: The Role of Experimental Stone Artifact Replication in Hypothesis-Driven Archaeology. *Ethnoarchaeology*, *8*, 103–136.

Lycett, S. J. (2009). Are Victoria West Cores "Proto-Levallois"? A Phylogenetic Assessment. *Journal of Human Evolution, 56*, 175–191.

Lycett, S. J. (2011). "Most Beautiful and Most Wonderful": Those Endless Stone Tool Forms. *Journal of Evolutionary Psychology (Budapest), 9*, 143–171.

Lycett, S. J., von Cramon-Taubadel, N., & Foley, R. A. (2006). A Crossbeam Co-ordinate Caliper for the Morphometric Analysis of Lithic Nuclei: A Description, Test and Empirical Examples of Application. *Journal of Archaeological Science, 33*, 847–861.

Mardia, K. V. (1970). Measures of Multivariate Skewness and Kurtosis with Applications. *Biometrika, 36*, 519–530.

McGhee, G. (2007). *The Geometry of Evolution*. Cambridge: Cambridge University Press.

McGhee, G. (2011). *Convergent Evolution*. Cambridge, Mass.: MIT Press.

O'Brien, M. J., Boulanger, M. T., Collard, M., Buchanan, B., Tarle, L., Straus, L. G., et al. (2014a). On Thin Ice: Problems with Stanford and Bradley's Proposed Solutrean Colonisation of North America. *Antiquity, 88*, 606–613.

O'Brien, M. J., Boulanger, M. T., Collard, M., Buchanan, B., Tarle, L., Straus, L. G., et al. (2014b). Solutreanism. *Antiquity, 88*, 622–624.

Straus, L. G. (2002). Selecting Small: Microlithic Musings for the Upper Paleolithic and Mesolithic of Western Europe. In R. G. Elston and S. L. Kuhn (Eds.), *Thinking Small: Global Perspectives on Microlithization* (pp. 69–81). Archeological Papers, no. 12. Washington, D.C.: American Anthropological Association.

Straus, L. G., Meltzer, D. J., & Goebel, T. (2005). Ice Age Atlantis? Exploring the Solutrean-Clovis "Connection." *World Archaeology, 37*, 507–532.

Wang, W., Lycett, S. J., von Cramon-Taubadel, N., Jin, J. J., & Bae, C. J. (2012). Comparison of Handaxes from Bose Basin (China) and the Western Acheulean Indicates Convergence of Form, Not Cognitive Differences. *PLoS One, 7*, e35804.

Will, M., Mackay, A., & Phillips, N. (2015). Implications of Nubian-like Core Reduction Systems in Southern Africa for the Identification of Early Modern Human Dispersals. *PLoS One, 10*, e0131824.

5 Technical Constraints on the Convergent Evolution of Technologies

Mathieu Charbonneau

Observed similarities in two distinct technological traditions can be attributable to different causes. The similarities could have been inherited from a common ancestral tradition, in which case the two similar traditions would be *homologous*. Alternatively, the similarities could be the result of the two traditions independently having invented tools with similar forms and/or functions. This latter situation is one of *convergent evolution*.

With respect to lithic technology, the plausibility of convergent evolution depends on the range of stone-tool forms and functions that can be produced. For instance, if prehistoric knappers could have produced an endless range of forms, the coincidental invention of similar tools would be very surprising. In contrast, should the range be limited to but a few forms, we would expect convergence to occur frequently. My premise here is that the range of possible prehistoric stone tools is located in between these two extremes because of technical constraints—those imposed on the morphological evolution of tools by the specific set of techniques used for their manufacture. Not just any form of stone tool can be produced by any single manufacturing technique, and no manufacturing technique can produce any and all forms of stone tools. Given their technical knowledge and the restricted set of tools in their possession, prehistoric societies could produce only limited stone-tool forms.

The study of technical constraints on technological evolution promises more than marking the boundaries between the space of possible and impossible tool forms. By explicitly addressing the impact of manufacturing techniques on the evolution of tools, I aim to expand the evolutionary models set at the level of tool morphology in order to address important factors in the production and convergent evolution of stone tools. I develop a framework for understanding the evolution of manufacturing techniques and their impact on the evolution of tool technology, stone-based or otherwise.

I take that "the paramount goals of archaeological research are constructing and explaining the evolutionary lineages of cultures as they are represented by artifacts" (Lyman 2001: 70). Consequently, the study of convergent technological evolution asks minimally two types of questions, one related to genealogy and the other to process. Solving the genealogy question means identifying the descent relationships (or lack of) between two similar artifact forms by reconstructing their lineages. To solve it, we rely on methods of phylogenetic ordering, meaning that we identify the most parsimonious evolutionary story tracing the heritable transmission of variant tool forms. The use of cladistics to construct phylogenetic trees has been employed to address this type of question (e.g., Lycett 2009, 2011; Mace et al. 2005; O'Brien and Lyman 2003; O'Brien, Darwent, and Lyman 2001).

Here, I limit discussion to process by addressing the causal, effective story behind the forming of technological phylogenies. To solve the process problem, we need to explain how the evolution of some technological traditions happened to converge and what factors, causal or otherwise, shaped this evolution. Addressing the process question means identifying the different mechanisms involved in shaping technological lineages and balancing the causes that led to the specific evolutionary trajectories, including instances of two independent traditions creating similar technological products.

Morphocentric Models of Technological Evolution

The formal resemblance of tools has long been used by archaeologists as a marker of the passage of time and of cultural affiliation (O'Brien and Lyman 2000). Social learning mechanisms, such as teaching, imitation, and emulation (chapters 3 and 9, this volume) create cultural-inheritance systems that ensure the transmission of information required to reproduce similar tools from one generation to the next. As the transmitted information changes through time, so will corresponding behaviors used to produce the tools and consequently the tools themselves. Overall, we expect a general pattern of descent with modification within the archaeological record. The stability of technological traditions is then explained by the faithful transmission of the information necessary to reproduce similar tools, whereas divergence in technological traditions is explained by the transformation of the transmitted information.

Archaeology's prevalent model of the *process* of technological evolution is one that centers on morphological change in artifacts (e.g., Bettinger and Eerkens 1997, 1999; Buchanan and Hamilton 2009; Eerkens 2000; Eerkens and Bettinger 2001; Eerkens and Lipo 2005, 2008; Hamilton and Buchanan 2009; Kempe, Lycett, and Mesoudi 2012; Lycett 2008; Lyman 2001; Mesoudi and O'Brien 2008a, 2008b; Neiman 1995; Schillinger, Mesoudi, and Lycett 2014; Shennan and Wilkinson 2001; VanPool 2001). These "morphocentric" approaches rely on what is probably the most intuitive and simple model of technological evolutionary change one can imagine, yet it is a very stringent one. It makes three assumptions about the nature and process of technological change.

First, it is generally assumed that the stone-tool manufacturer produces a tool with a certain form because he or she has a mental representation—sometimes referred to as an "ideational unit" (e.g., O'Brien, Lyman, Mesoudi, and VanPool 2010)—that specifies what the final product—the "empirical unit"—should look like:

> [A] manufacturer of, say, projectile points, thinks of his intended creation using ideational units: "I need a 6-inch-long point that is 2 inches wide and has 60-degree notches instead of the usual 40-degree notches." Those units—inches and degrees—cannot be anything else but ideational because we cannot "see" or "feel" them. The manufacturer then uses ideational units to create the object and can also describe the object using ideational units. The actual specimen that he creates—a 6-inch-long projectile point—is an empirical unit in that it can be seen and felt. (O'Brien et al. 2010: 3798)

Implicit here is the idea that should the mental representation of the intended tool be altered, the resulting final form of the manufactured tool would vary accordingly. To build on O'Brien et al.'s example, should the manufacturer intend to produce a projectile point that "has 60-degree notches instead of the usual 40-degree notches," he would, in fact, produce a projectile point that has those characteristics. Consequently, morphocentric models track the transmission of variation in stone-tool form, since what one intends to produce is what one effectively produces. The specific means of production are typically abstracted away, what I have referred to elsewhere as the *congruence assumption*: variation in the mental representation of tools' final form maps isomorphically on the morphological variation of the produced tools (Charbonneau 2015a).

Second, mechanisms introduce variation in stone-tool lineages. In morphocentric models, it is typically assumed that errors in the social learning process alter the form of the tools. The main variation-generating mechanism of technological evolution is that of the accumulation of copying error, or ACE (Eerkens 2000; Eerkens and Bettinger 2001; Eerkens and Lipo; 2005, 2008; Schillinger, Mesoudi, and Lycett 2014). According to the ACE model, novel variation in technological traditions reduces to slight quantitative variations in artifact form that are introduced by the imperfections of human cognition and, as far as techniques are involved, by slight (generally unconscious) mistakes in the manufacturing of tools. For instance, one may misjudge the actual size of a tool one is copying, or one might imperfectly recall the shape of the tool one is copying. Eerkens (2000; see also Eerkens and Bettinger 2001; Eerkens and Lipo 2005, 2008) found that cognition-based copying errors typically alter a morphological character's value by about 5 percent, irrespective of the quantitative character's absolute value. Just as technological traditions are assumed to persist from one generation to the next through the transmission of information about the form of the tools, the variation necessary to fuel the evolutionary process is itself introduced by errors in copying stone-tool forms.

Third, morphocentric models generally assume that tool morphology changes but slightly from one generation to the next. For instance, any morphological differences between a model tool and its copies that are less than or around 5 percent—that is, within the copying-error threshold—can be understood as a gradual change of this kind (Charbonneau 2015a). Consequently, if two different tools are part of the same technological lineage, yet diverge by more than 5 percent in some of their morphological characters, the assumption of gradual morphological evolution implies that we should observe a series of transitory forms linking the two tools, all within the error threshold from one another. The dissimilarity in morphology expected from copying errors is thus expected to give us a measure of the evolutionary distance between any two tool forms.

Novel Forms, Intermediate Forms, and Impossible Forms

Morphocentric models have proven to be invaluable in many case studies. However, these models deal strictly with the evolution of morphological characters of final tool forms, and only those characters that vary on a continuous quantitative dimension. Here, I identify additional aspects of tool morphology that need explanation, such as novel forms, intermediate forms, and impossible forms (chapter 2, this volume). Developing evolutionary models that satisfactorily account for the historical change of these additional forms requires us to attend to the techniques involved in their production. After discussing these additional forms and the importance of technical and material constraints in their making, I develop a general modeling strategy dealing with these constraints, thus complementing the morphocentric approach.

Novel Forms

Studies dealing with processes of tool evolution concern the evolutionary change of tools' morphological characters. Both the copying-error mechanism and the notion of gradual evolution assumed by the morphocentric models are bound to quantitative morphological characters—those that can be quantitatively defined (e.g., size, side angles, and thickness). Consequently, morphocentric models rely on quantitative character spaces to model technological evolution—that is, a set of dimensions, each corresponding to a specific quantitatively describable morphological character of a tool that the archaeologist is interested in. In such morphospaces (McGhee 1999, 2007, 2011; chapter 2, this volume), the range of possible values a given character of this sort can take is usually given by a subset of real numbers or, when more relevant, a subset of natural numbers. A specific tool form, then, is located in a morphospace at the intersection of all its modeled characters' values (O'Brien and Bentley 2011; O'Brien et al. 2016).

Morphocentric models were not built to deal with the generation of novel morphological characters. Indeed, the copying-error mechanism is bound to transform only the value of a tool's morphological character, and for this to happen that character needs to already exist. Copying errors cannot, however, explain why that character was present in the first place, as the mechanism has no stated capacity to generate novel morphological characters. In other words, copying errors is a variation-generating mechanism for *pre-existing, quantitative characters*.

Consider, for instance, the invention of Paleoindian fluted points in North America. A fluted tool is "a bifacial piece from which an elongated flake has been removed along the longitudinal axis, in order to thin one or both faces, without reaching the edges" (Inizan, Reduron-Ballinger, Roche, and Tixier 1999: 136). These modifications of projectile points were not obtained by gradual removal of increasingly elongated flakes through the same techniques used to shape the projectile point in the first place. Rather, following Crabtree (1966), they necessitate their very own special version of pressure flaking. Inizan et al. (1999) identify the production of flutes as requiring a "special technique" of channel flaking (see also Whittaker 1994).

Since morphocentric models were not built to deal with the invention of novel morphological characters, technological convergence at this level calls for an extension of the models. Indeed, in its current form, the morphocentric explanatory strategy is constrained in dealing strictly with cases of homological evolution and cases of parallel evolution for *shared quantitative characters*. Morphocentric models can explain how two homologous characters came to be so similar: the character itself and its specific value in the two novel traditions were inherited from a common tool ancestor that already had that same morphological character, and one with a value similar to that of its descendants. Morphocentric models can also deal with cases of parallel evolution, but only if shared characters converged on the same quantitative value after having first evolved in different directions. For instance, a process of disruptive selection followed by a process of stabilizing selection on copying errors could lead two traditions sharing a common ancestral form to evolve a specific character value in opposite directions and then bring them back around to the same value. However, in both cases, the presence of the similar quantitative character is assumed, not explained.

Intermediate Forms

The archaeological record is populated by partial forms—incomplete, broken, or discarded tools—but morphocentric models are concerned primarily with the evolution of a tool's final form and the specific values that its morphological characters assume. The models are not designed to deal with intermediary forms that a tool takes while it is being manufactured, nor do they account for the manufacturer's mental representations of what forms the tool under production takes at each intermediate step of its production. Moreover, the specific sequence of intermediate forms needs to be taken into account for a complete explanation of final, functional forms, as important evolutionary processes and changes may affect the transitory forms that link a blank core to a functional stone tool. The materials involved in the production of tools tend to follow structured manufacturing patterns of change. This means two things: first, variation in the intermediate forms can have downstream effects on the final form of the functional tool; and second, similar forms can be produced by variant manufacturing routes—that is, the sequence of intermediary forms may vary yet land on the same final, functional morphology.

There are different ways that variation in the manufacturing sequences can be introduced without affecting the general morphology of the resulting tool form. First, two manufacturers may hold different mental representations of the morphological form that some intermediary steps should take, yet they may share similar mental representations of the final form the artifact should take, and thus produce such similar forms. Variation in the mental representations of the desired intermediary steps can be transmitted and serve as the basis for alternative technological traditions. Second, as the manufacture of a stone tool is a fragile process, accidents can happen. The different ways that these manufacturing errors can be solved are, in themselves, specialized techniques. However, different technological traditions—different in terms of the strategies they adopt to solve manufacturing mistakes—may not show up in the final morphological characters that typically are of interest in morphocentric models. Flake scars are studied by archaeologists, but they typically are absent from morphocentric models. The same applies for the marks left by the use of manufacturing strategies to deal with the idiosyncrasies of the raw materials, such as strategies to correct knapping errors or to rejuvenate and reshape old tools.

The production of similar forms might in fact go through divergent manufacturing routes. Variation in the intermediate forms found in the archaeological record may not only serve to show that two similar forms were produced by different manufacturing techniques—thus explicitly showing that the two technological traditions in fact diverge in their manufacturing techniques—but also identify important commonalities in the production of different tool morphologies. In other words, examining the intermediate forms under production can allow us to observe convergence and homologies not only at the level of the final morphological form of tools but additionally at the level of their production techniques. In fact, two final forms may largely differ morphologically (according to the accumulated copying-error metric), yet share important similarities in their intermediate forms and production techniques.

Impossible Forms

Taking into account the manufacturing techniques also allows the study of forms that are not found. The manufacturing techniques available to prehistoric tool makers also delimited the boundaries between possible tool forms, regardless of whether they are actually observed in the record, and impossible forms (chapter 2, this volume). By possible forms, I mean those shapes that could be produced through the use of prehistoric knapping techniques, and by impossible forms those that could not. Consider the trilobate arrowhead, defined by Delrue (2007: 239) as "an arrowhead that has three wings or blades that are usually placed at equal angles (i.e., c. 120°) around the imaginary longitudinal axis extending from the centre of the socket or tang." As far as I could find, there are no stone-based trilobate arrowheads. In contrast, ivory, antler-based, and metal trilobate arrowheads have been found (Delrue 2007). This absence is surprising at first, given the functional advantage of the trilobate arrowheads over two-bladed (bifacial) ones:

> From an archer's point of view trilobate arrowheads are generally more accurate than flat, two-bladed ones. Arrowhead blades act as aerodynamic surfaces, and two-bladed heads are larger and more easily affected by crosswinds than trilobate ones with the same mass. The use of more than two wings (three or four) increases the weight and stabilises the flight of the arrow, two good reasons to use multi-winged arrows. (Delrue 2007: 245)

If Delrue's functional account is correct, then the reason why stone arrowheads were limited to the bilobate form cannot be stated in terms of functional constraints. Trilobate arrowheads have proved to have important functional advantages over bilobate ones, so it is surprising that the trilobate morphology was not exploited to produce stone arrowheads. The morphocentric model can explain this absence only by claiming that it is a historical accident no arrowhead tradition evolved a trilobate shape. However, the most plausible reason for the absence of lithic trilobate arrowheads is simply that one cannot produce such a shape by traditional knapping techniques. This has to do with the constraints imposed by the conchoidal fracturation process exploited by traditional flintknapping techniques. When a knapper produces such fractures on a core through percussion or pressure, the fissures travel roughly parallel to the surface of the core until they reach one of the core's surfaces. Knapping trilobate arrowheads would necessitate that fractures stop somewhere halfway through the core and then come back toward the hammered platform's surface, which contradicts the physical nature of the fracturing process.

Theoretical Technospace

Here I draw on some of my previous work (e.g., Charbonneau 2015a, 2015b, 2016) and that of others (e.g., Mesoudi and O'Brien 2008c) to develop a basic framework for an evolutionary model of technical change—that is, a model of the evolution of hierarchically structured techniques involved in the production of tools. Although I am interested mainly in stone-tool manufacturing techniques and their impact on convergent evolution, the model can readily be extended to techniques involved in the production and evolution of other kinds of tools and cultural material products (e.g., pottery, adhesives, clothing, and buildings) and even to culturally transmitted structured behaviors that do not produce material outcomes (e.g., dance routines, rituals, and, arguably, language). The specific formal framework I adopt is directly inspired by *models of the morphogenesis of form*, or *theoretical developmental morphospaces* (McGhee 2007; chapter 2, this volume). My objective is to offer conceptual and formal tools to examine the different mapping relationships between the evolution of manufacturing techniques and their effects on the evolution of stone tools. The framework offers the additional benefit of allowing straightforward representation of technical constraints and, for the present discussion, of different levels of technological convergence and homology.

Building a Theoretical Technospace

A theoretical technospace is the technique-centered analog of tool morphospace. However, instead of defining the dimensions of the space by the values of quantitative morphological characters that stone tools possess, the dimensions are defined by the different variant states that their manufacturing techniques can take. This implies that we need to first define what a technique consists of and how techniques can vary. From these, I derive a formal representation of technical variation—that is, the technospace proper. The technospace is intended to map all imaginable techniques, possible and impossible. This contrast with *empirical spaces*, which are defined by the range and diversity of observed stone-tool variants. Not only are empirical spaces limited to observed forms, but they need to be continuously redrawn as novel variants are discovered (McGhee 1999).

Techniques as Hierarchically Structured Behaviors

For our current purpose, a technique consists of the specific recipe of decisions and actions as it is enacted in the production of a stone tool (Mesoudi and O'Brien 2008c). Tools are not produced merely by imagining them. A tool manufacturer needs to effectively engage body and mind into actions and deal with materials (and generally other tools) in order to produce a desired final product. A proper understanding of the evolution of techniques must give a central role to these factors (Charbonneau 2015a, 2016).

The hierarchical structure of techniques can be decomposed into assemblies of actions serving intermediary subgoals, each of which must be satisfactorily completed in order for the manufacturer to successfully produce the intended end-product form. These subgoals represent cognitive decisions of how to proceed and whether a specific intermediary manufacturing step has been properly achieved and what to do next. Subgoals can be nested under higher-level subgoals, with the whole structure ultimately resulting in the total hierarchical structure of decision-and-action assemblies that characterize the technique. The main goal of a technique is to produce the intended tool form. All lower-level subgoals are means to satisfy this master goal. The hierarchical structures of techniques are thus functional structures (Charbonneau 2016), which are typically depicted as tree diagrams (figure 5.1).

An important perk of adopting a concept of technique that is hierarchically structured is that it allows us to identify the different kinds of technical variation and thus metrics of evolutionary distances between technical variants, while at the same time allowing us to systematically map technical variation onto artifact morphological variation. We can already identify two main types of differences characterizing technical variation: variation at the level of the actions and variation at the level of the decision nodes, what I refer to as action-level variation and hierarchy-level variation, respectively (Charbonneau 2015b).

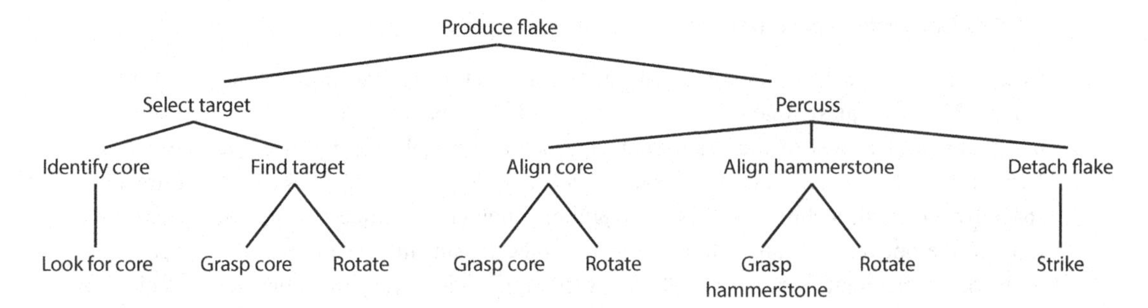

Figure 5.1
The hierarchical structure of the basic flaking unit (after Stout 2011).

Action-Level Variation

Specific actions recruited in the production of a tool serve as the atomic units of technical variation. Actions can vary in two ways. First, they can vary in a discrete manner. A specific action can be replaced by another kind of action (e.g., replacing the grasping of a hammerstone by the pushing away of the hammerstone), or specific actions can be added or subtracted from a technique. Alternatively, actions can vary "internally," through an alteration of the specific value or state that the action takes (e.g., percussing a hammerstone at a 45-degree rather than a 60-degree incidence angle). Between two otherwise identical techniques, both types of action-level variations consist in the difference of at least one of the action units of the two techniques. Different mechanisms can produce these variations. Errors creeping in the imitation of a string of actions or of a specific action assembly can serve as an instance of action-level modification process, as one action could be misinterpreted for another (e.g., hit hammerstone on core at 45 degrees of incidence instead of 50 degrees). Alternatively, a specific action could be intentionally modified.

Hierarchy-Level Variation

Whereas in the case of action-level variation, two techniques differ only in terms of their specific sequence of actions, hierarchy-level differences are set at the level of the decision structure of the techniques. In such cases, not only would the hierarchical structure of the techniques differ, but so would the sequence of the subservient actions. Hierarchy-level modification mechanisms alter subassemblies of a technique's structure. Subassemblies thus can be added, subtracted, concatenated under a new node, or even reshuffled. Two techniques can even be combined together to produce a novel technique (Charbonneau 2016). In all, mechanisms for hierarchy-level modification are those capable of modifying an existing technique by altering it at the level of its decision subassemblies, including its master goal.

Theoretical Technospace Defined

Based on these considerations, we can define a theoretical technospace as the set of all imaginable techniques—that is, all techniques that vary either at the level of their sequence of actions, at the level of their hierarchical structure, or both. Having specified the different kinds of transformation that techniques can undergo, we can construct an abstract space by which all techniques have as neighbors another technique that can be obtained through the operation of any one modification mechanism. In a theoretical technospace, each coordinate point represents a different technique variant, varying either by its precise sequence of actions, by its decision structure, or both. A population undergoing technical evolution can then be represented by a cloud of points exploring the technospace, with each particle of the cloud representing a single individual located at the coordinate of the technical variant he or she possesses. A population moves through technospace by modifying its existing techniques through either action-level or hierarchy-level modification processes.

Technical Constraints

Technical constraints were defined above as constraints imposed on the morphological evolution of tools by the specific set of techniques used for their production. We saw that certain tool forms were possible, in that they could be produced by a given set of techniques (e.g., bilobate arrowheads through prehistoric knapping techniques), whereas others were not (e.g., trilobate arrowheads through the same manufacturing techniques). Theoretical technospaces allow us to represent these constraints as boundaries delimiting the space of possible and impossible tool morphologies, given a specific set of manufacturing techniques. However, as many tool forms can be produced by different techniques, and as different techniques can produce different forms, it is simpler to represent technical constraints on tool forms directly in a technospace. Technical constraints on tool form then become functional constraints on technical variation, where a functional technique is understood as one that can effectively produce its intended, final tool form, as defined by its master goal.

Among all conceivable manufacturing techniques, only some will prove capable of producing the tool form they were intended to produce. By producing a functional form, I mean that a technique with a specific structure is capable of being successfully enacted, such that a manufacturer can go through each step of the technique by satisfying all of its subgoals; that the technique has a clear, recognizable end result; and that the technique, when enacted properly, satisfies its master goal. Dysfunctional techniques, then, are those that fail to satisfy any one of those three conditions. Moreover, we can assume that dysfunctional techniques—those that fail to produce an intended result—will typically fail to be socially transmitted to the next generation, as they offer no functional end result to those enacting its recipe. This is analogous to cases in which a specific developmental regime of an organism leads to an unviable or sterile form (chapter 2, this volume). Consequently, regions of technospace inhabited by dysfunctional techniques of this sort will remain generally empty, as any venture into the dysfunctional regions of technospace will fail to perpetuate itself. For instance, using a stone to produce hard-hammer percussion on the edges of a brittle material (e.g., chert) in order to shape a handaxe is a functional technique. In contrast, trying to shape a similar handaxe but hitting a face instead of an edge will fail to produce conchoidal fractures on the core and will likely result in breakage. In technospace, the former technique will be located inside the boundaries of functional techniques, and the latter will fall outside. By examining which techniques are capable of producing their intended end results and which ones do not, and by identifying the factors involved in shaping the boundaries of these constraints (cognitive, bodily, instrumental, and/or material factors), we can then map the set of techniques that are functionally realizable and those that lead to dysfunctional results.

Mapping Technological Variation onto Morphological Variation

Final Forms

By mapping a specific technical variant onto the material results a technique produces when successfully enacted, we can systematically map technological variation onto possible tool forms. The same kind of reasoning allows us to experimentally examine how varying a technique will effect change on the intended end result. This is what replicative studies of stone tools have been investigating, with a caveat. The caveat is that actualistic studies typically start their investigation with existing tool forms and then reverse-engineer the forms into the sets of techniques that could have produced them (e.g., Crabtree 1966, Pelegrin 2012). The investigation of theoretical technospaces allows us instead to take as a basis a specific technique and examine which forms it can produce, if any. In fact, existing research does just that, not by taking techniques as a basis of study, but rather by examining the effects of varying specific key factors in the realization of manufacturing techniques (e.g., Braun, Plummer, Ferraro, Ditchfield, and Bishop 2009; Dibble and Pelcin 1995; Eren, Lycett, Roos, and Sampson 2011; Eren, Patten, O'Brien, and Meltzer 2014; Magnani, Rezek, Lin, Chan, and Dibble 2014). By combining the use of theoretical technospaces and actualistic studies, we can map variation in technical behaviors with the variation observed in the archaeological record.

Intermediate Forms

The same logic applies to the study of intermediary and retouched forms. The specific forms of the intermediary products can be located in specific regions of morphospace, just as final forms are. However, instead of mapping the final form with the fully enacted technique, we can map a partially enacted technique to the intermediary form it produces. The same goes with retouched forms, but with completely enacted techniques plus rejuvenation steps. The technospace model explicitly represents the process of manufacture at each step of its realization, including the morphology of the untouched core that will be processed. Together with the mapping of technological variation and morphological variation, we can represent the specific morphological trajectory that a tool under production takes. That tool can thus be represented by a trajectory of intermediary forms in morphospace as its manufacture goes on, concluding on the final morphology of the tool. In addition, the regions of morphospace where intermediary forms tend to agglomerate may differ from those regions where final forms find themselves. Technospaces allow us to examine analytically which regions of morphospace are typically represented by intermediary forms.

Convergence in Technospace and Beyond

Here I discuss two types of convergent evolution—accessibility convergence and deep convergence—that are set at the level of the manufacturing techniques rather than at the level of a tool's shape.

Accessibility Convergence

Accessibility convergence refers to the manner in which manufacturing techniques and variation in technospace relate to the morphology of stone tools and its variation. Two similar tool forms can be produced by different manufacturing techniques. We can refer to such situations as a case of *accessibility convergence*—situations where two different techniques share important similarities in the regions of tool morphospace onto which they map. The techniques themselves need not be similar to any degree; they need only to be able to produce (or access) similar tool forms (figure 5.2). Two techniques can thus be understood as accessibility convergent when they share overlapping regions in morphospace.

Consider the following example. Several Maya eccentrics have "holey" shapes: they are flaked, round bifaces with a hole in the middle, similar to a doughnut (e.g., Joyce 1932). Two alternative explanations for the invention of these eccentrics involve taking into account changes in manufacturing techniques. One candidate technique consists of grinding a nonhomogeneous core with an exploitable nonbrittle inclusion (e.g., soft limestone [G. Iannone 1993, personal communication]). Alternatively, one could produce holes in a homogeneous core by using a lapidary drilling technique involving, for instance, a bow drill. The latter technique may have been borrowed from the use of similar techniques used to drill shells and beads and adapted to lithic materials. For this technique, one needs to find the right kind of material to produce the eccentrics. For these alternative techniques, developing the knowledge required to identify promising cores and mastering the perforating skill both depend on the acquisition of novel technical knowledge and expertise.

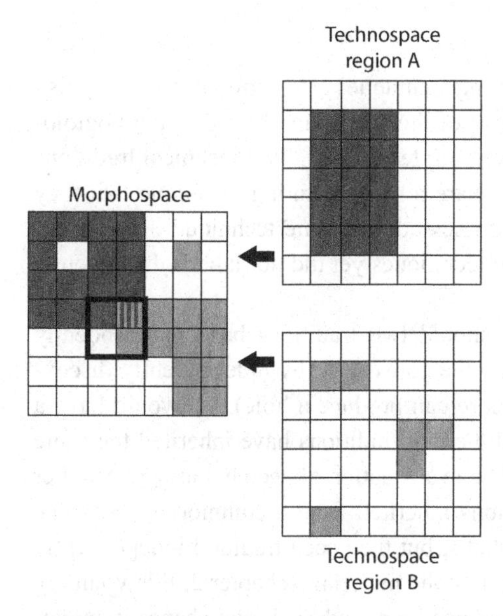

Figure 5.2
Diagram showing two techniques located in different parts of technospace (regions A and B). Techniques found in the two regions may nevertheless map onto tool forms that are close to one another in morphospace, and they can even overlap. The forms producible by the techniques in region A are represented by dark-gray squares, and those by the techniques in region B by light-gray squares. Blank squares in technospace represent technical variants that do not produce any functional results. Blank squares in morphospace are forms that cannot be produced by any of the examined techniques. The black square in morphospace represents a tool form that the techniques in both regions can produce. Assuming an ecological advantage to forms situated inside the box with heavy lines, we should expect convergent evolution in form.

Consider that two different technological traditions could independently invent perforating techniques, but different ones. One tradition may innovate by discovering that the grinding of heterogeneous cores can lead to producing a hole in the core, whereas another one discovers that drilling techniques used to perforate shells can be adapted to lithic materials. Whether or not the two traditions in fact produce similar eccentric forms, they share the potential of stumbling upon similar shapes. Thus, the independent invention of similarly shaped eccentrics could be expected (and would be more probable) simply because the two techniques share the same potential of producible forms. In terms of technospace, the regions of morphospace that both techniques can access overlap to a large degree (Charbonneau 2015a). Using technospaces to identify the technical constraints imposed by different manufacturing techniques also allows us to identify which regions of morphospace both techniques can access, and on which region of morphospace they can converge.

Deep Convergence

Paying attention to the evolution of manufacturing techniques also allows us to distinguish between homology and convergence at the level of the tools' morphologies and homologies and convergences at the level of manufacturing techniques. Two technical traditions can be said to be deeply homologous if they share a manufacturing technique that they inherited from a common ancestor that also possessed that same technique. In contrast, two traditions that have similar manufacturing techniques yet did not inherit them from a common ancestor will be deeply convergent.

In the case of "holey" Mayan eccentrics, should two traditions have independently stumbled upon the same technique to perforate the cores (e.g., by independently discovering that grinding away an inclusion in the core can produce a hole), we would have a case of deep convergence. In contrast, should the two traditions have inherited the same technique from a common ancestor, we would be in a situation of deep homology. Another interesting scenario would be that two traditions inherited from a common ancestor the same bow-drilling techniques for beads and shells, but then each tradition independently transferred that technique to the modification of stone materials (chapter 2, this volume).

Of course, being capable of producing a hole need not lead to similar shapes in manufactured tools. Neither deep convergence nor deep homology—nor deep parallel evolution, for that matter—depends on the morphological similarity of the produced tools. Indeed, this is because of the possible disassociation between technical and morphological traditions. The information concerning a tool form can be transmitted independently of the specific technique used to produce that form. In other words, morphological lineages need not be congruent with technological lineages.

Conclusion

It is not enough to recognize whether morphological similarity is the product of convergence or of homology. One also needs to explain why convergence occurred. I have argued that morphocentric models, the main approach to explaining the process of technological evolution, are not designed to deal with some key questions regarding the archaeological record and the evolution of technological traditions. I showed the importance of addressing how manufacturing techniques constrain both the production and the evolution of tool form. Doing so exposes how the issue of technological convergence is in fact a multilayered process, one in which homologies and analogies are not restricted to similarities in form but also to similarities in the techniques used. I have developed the notion of deep convergence to deal with cases of independently invented yet similar techniques, and that of accessibility convergence to deal with cases of different techniques capable of producing similar forms. Finally, I have offered a basic framework to operationalize the study of technical variation and technical evolution and their impact on the archaeological record.

Further work is required, if only because the framework can prove its usefulness only by being operationalized and effectively used to address specific empirical problems.

Acknowledgments

I thank Briggs Buchanan, Metin I. Eren, and Mike O'Brien for organizing and inviting me to participate in the Altenberg workshop and for their patient and careful work in editing this volume. Thanks to the Konrad Lorenz Institute for both hosting the workshop and also for hosting me for two years in a rich intellectual environment where many of the ideas discussed here germinated. Finally, I thank John Whittaker, Gyles Iannone, and Woody Blackwell for taking the time to guide me and answering my weird questions concerning possible and impossible stone-tool forms. The research leading to these results has received funding from the European Research Council under the European Union's Seventh Framework Programme (FP7/2007-2013) / ERC grant agreement n° 609819 (Somics project).

References

Bettinger, R. L., & Eerkens, J. W. (1997). Evolutionary Implications of Metrical Variation in Great Basin Projectile Points. In C. M. Barton & G. A. Clark (Eds.), *Rediscovering Darwin: Evolutionary Theory in Archeological Explanation* (pp. 177–191). Archeological Papers, no. 7. Washington, D.C.: American Anthropological Association.

Bettinger, R. L., & Eerkens, J. W. (1999). Point Typologies, Cultural Transmission, and the Spread of Bow-and-Arrow Technology in the Prehistoric Great Basin. *American Antiquity*, *64*, 231–242.

Braun, D. R., Plummer, T., Ferraro, J. V., Ditchfield, P., & Bishop, L. (2009). Raw Material Quality and Oldowan Hominin Toolstone Preferences: Evidence from Kanjera South, Kenya. *Journal of Archaeological Science*, *36*, 1605–1614.

Buchanan, B., & Hamilton, M. J. (2009). A Formal Test of the Origin of Variation in North American Early Paleoindian Projectile Points. *American Antiquity*, *74*, 279–298.

Charbonneau, M. (2015a). All Innovations Are Equal, but Some More Than Others: (Re)integrating Modification Processes to the Origins of Cumulative Culture. *Biological Theory*, *10*, 322–335.

Charbonneau, M. (2015b). Mapping Complex Social Transmission: Technical Constraints on the Evolution Cultures. *Biology & Philosophy*, *30*, 527–546.

Charbonneau, M. (2016). Modularity and Recombination in Technological Evolution. *Philosophy & Technology*, *29*, 373–392.

Crabtree, D. E. (1966). A Stoneworker's Approach to Analyzing and Replicating the Lindenmeier Folsom. *Tebiwa*, *9*, 3–39.

Delrue, P. (2007). Trilobate Arrowheads at ed-Dur (U.A.E, Emirate of Umm al-Qaiwain). *Arabian Archaeology and Epigraphy*, *18*, 239–250.

Dibble, H. L., & Pelcin, A. (1995). The Effect of Hammer Mass and Velocity on Flake Mass. *Journal of Archaeological Science*, *22*, 429–439.

Eerkens, J. W. (2000). Practice Makes within 5% of Perfect: The Role of Visual Perception, Motor Skills, and Human Memory in Artifact Variation and Standardization. *Current Anthropology*, *41*, 663–668.

Eerkens, J. W., & Bettinger, R. L. (2001). Techniques for Assessing Standardization in Artifact Assemblages: Can We Scale Material Variability? *American Antiquity*, *66*, 493–504.

Eerkens, J. W., & Lipo, C. P. (2005). Cultural Transmission, Copying Errors, and the Generation of Variation in Material Culture and the Archaeological Record. *Journal of Anthropological Archaeology*, *24*, 316–334.

Eerkens, J. W., & Lipo, C. P. (2008). Cultural Transmission of Copying Errors and the Evolution of Variation in Woodland Pots. In M. T. Stark, B. J. Bowser, & L. Horne (Eds.), *Cultural Transmission and Material Culture: Breaking Down Boundaries* (pp. 63–81). Tucson: University of Arizona Press.

Eren, M. I., Lycett, S. J., Roos, C. I., & Sampson, C. G. (2011). Toolstone Constraints on Knapping Skill: Levallois Reduction with Two Different Raw Materials. *Journal of Archaeological Science*, *38*, 2731–2739.

Eren, M. I., Patten, R. J., O'Brien, M. J., & Meltzer, D. J. (2014). More on the Rumor of "Intentional Overshot Flaking" and the Purported Ice-Age Atlantic Crossing. *Lithic Technology*, *39*, 55–63.

Hamilton, M. J., & Buchanan, B. (2009). The Accumulation of Stochastic Copying Errors Causes Drift in Culturally Transmitted Technologies: Quantifying Clovis Evolutionary Dynamics. *Journal of Anthropological Archaeology*, *28*, 55–69.

Iannone, G. J. (1993). *Ancient Maya Eccentric Lithics: A Contextual Analysis*. Trent, Canada: Trent University.

Inizan, M.-L., Reduron-Ballinger, M., Roche, H., & Tixier, J. (1999). *Technology and Terminology of Knapped Stone*. Nanterre, France: Cercle de Recherches et d'Etudes Préhistoriques.

Joyce, T. A. (1932). Presidential Address. The "Eccentric Flints" of Central America. *Journal of the Royal Anthropological Institute of Great Britain and Ireland*, *62*, xvii–xxvi.

Kempe, M., Lycett, S. J., & Mesoudi, A. (2012). An Experimental Test of the Accumulated Copying Error Model of Cultural Mutation for Acheulean Handaxe Size. *PLoS One*, *7*(11), e48333.

Lycett, S. J. (2008). Acheulean Variation and Selection: Does Handaxe Symmetry Fit Neutral Expectations? *Journal of Archaeological Science*, *35*, 2640–2648.

Lycett, S. J. (2009). Are Victoria West Cores "Proto-Levallois"? A Phylogenetic Assessment. *Journal of Human Evolution*, *56*, 175–191.

Lycett, S. J. (2011). "Most Beautiful and Most Wonderful": Those Endless Stone Tool Forms. *Journal of Evolutionary Psychology (Budapest)*, *9*, 143–171.

Lyman, R. L. (2001). Culture Historical and Biological Approaches to Identifying Homologous Traits. In T. D. Hurt & G. F. M. Rakita (Eds.), *Style and Function: Conceptual Issues in Evolutionary Archaeology* (pp. 69–89). Westport, Conn.: Bergin and Garvey.

Mace, R., Holden, C. J., & Shennan, S. (2005). *The Evolution of Cultural Diversity: A Phylogenetic Approach*. London: University College Press.

Magnani, M., Rezek, Z., Lin, S. C., Chan, A., & Dibble, H. L. (2014). Flake Variation in Relation to the Application of Force. *Journal of Archaeological Science*, *46*, 37–49.

McGhee, G. R. (1999). *Theoretical Morphology: The Concept and Its Applications*. New York: Columbia University Press.

McGhee, G. R. (2007). *The Geometry of Evolution: Adaptive Landscapes and Theoretical Morphospaces*. Cambridge: Cambridge University Press.

McGhee, G. R. (2011). *Convergent Evolution: Limited Forms Most Beautiful*. Cambridge, Mass.: MIT Press.

Mesoudi, A., & O'Brien, M. J. (2008a). The Cultural Transmission of Great Basin Projectile Point Technology I: An Experimental Simulation. *American Antiquity*, *73*, 3–28.

Mesoudi, A., & O'Brien, M. J. (2008b). The Cultural Transmission of Great Basin Projectile Point Technology II: An Agent-Based Computer Simulation. *American Antiquity*, *73*, 627–644.

Mesoudi, A., & O'Brien, M. J. (2008c). The Learning and Transmission of Hierarchical Cultural Recipes. *Biological Theory*, *3*, 63–72.

Neiman, F. (1995). Stylistic Variation in Evolutionary Perspective: Inferences from Decorative Diversity and Interassemblage Distance in Illinois Woodland Ceramic Assemblages. *American Antiquity*, *60*, 7–36.

O'Brien, M. J., & Bentley, R. A. (2011). Stimulated Variation and Cascades: Two Processes in the Evolution of Complex Technological Systems. *Journal of Archaeological Method and Theory*, *18*, 309–335.

O'Brien, M. J., & Lyman, R. L. (2000). *Applying Evolutionary Archaeology: A Systematic Approach*. Plenum, New York: Kluwer.

O'Brien, M. J., & Lyman, R. L. (2003). *Cladistics and Archaeology*. Salt Lake City: University of Utah Press.

O'Brien, M. J., Darwent, J., & Lyman, R. L. (2001). Cladistics Is Useful for Reconstructing Archaeological Phylogenies: Palaeoindian Points from the Southeastern United States. *Journal of Archaeological Science*, *28*, 115–136.

O'Brien, M. J., Lyman, R. L., Mesoudi, A., & VanPool, T. L. (2010). Cultural Traits as Units of Analysis. *Philosophical Transactions of the Royal Society of London. Series B, Biological Sciences*, *365*, 3797–3806.

O'Brien, M. J., Boulanger, M. T., Buchanan, B., Bentley, R. A., Lyman, R. L., Lipo, C. P., et al. (2016). Design Space and Cultural Transmission: Case Studies from Paleoindian Eastern North America. *Journal of Archaeological Method and Theory*, *23*, 692–740.

Pelegrin, J. (2012). New Experimental Observations for the Characterization of Pressure Blade Production Techniques. In P. M. Desrosiers (Ed.), *The Emergence of Pressure Blade Making: From Origin to Modern Experimentation* (pp. 465–500). New York: Springer.

Schillinger, K., Mesoudi, A., & Lycett, S. J. (2014). Copying-Error and the Cultural Evolution of "Additive" vs. "Reductive" Material Traditions: An Experimental Assessment. *American Antiquity*, *79*, 128–143.

Shennan, S., & Wilkinson, J. R. (2001). Ceramic Style Change and Neutral Evolution: A Case Study from Neolithic Europe. *American Antiquity*, *66*, 577–593.

Stout, D. (2011). Stone Toolmaking and the Evolution of Human Culture and Cognition. *Philosophical Transactions of the Royal Society of London. Series B, Biological Sciences*, *366*, 1050–1059.

VanPool, T. L. (2001). Style, Function, and Variation: Identifying the Evolutionary Importance of Traits in the Archaeological Record. In T. D. Hurt & G. F. M. Rakita (Eds.), *Style and Function: Conceptual Issues in Evolutionary Archaeology* (pp. 119–140). Westport, Conn.: Bergin and Garvey.

Whittaker, J. C. (1994). *Flintknapping: Making and Understanding Stone Tools*. Austin: University of Texas Press.

6 Being a Carnivorous Hominin in the Lower Paleolithic: A Biological Perspective on Convergence and Stasis

Daniel E. Lieberman

The arrow of time is challenging to comprehend when experienced from the present, especially for primitive artifacts. When students get their first glimpse of drawers of stone tools from the Lower, Middle, and Upper Paleolithic, they struggle to comprehend just how much more time those collections of Lower Paleolithic (LP) choppers, flakes, and handaxes represent compared to the much more diverse and sophisticated assemblages of Middle Paleolithic and Upper Paleolithic blades, points, and scrapers. If the LP began some 3.3 million years ago (Harmand et al. 2015), then this period of relatively minor technological change endured almost three million years, or almost 90 percent of the Stone Age.

In addition, given the paucity and simplicity of obvious LP tool types, it is difficult to discern how many of the similarities evident among assemblages resulted from convergence, which presumably occurred. The most obvious candidates for convergence are the independent appearances of handaxes on both sides of the Movius line, which at one time appeared to delimit the eastern extent of handaxes in Asia (Lycett and Bae 2010; Wang, Lycett, von Cramon-Taubadel, Jin, and Bae 2012). However, absence of evidence for other examples of convergence over the sparse record of more than two million years in Africa, Europe, and Asia is hard to diagnose as evidence of absence for most LP tools. We are therefore left with the problems of why so little change occurred over the course of the LP and of how many of the similarities evident among assemblages spread across millions of years and miles of time and space have independent origins.

There are two standard hypotheses for the lack of lithic variation over the span of the LP. The more common explanation is that technological change was constrained cognitively. Although chimpanzees do make and use tools on occasion, sometimes by modifying branches and twigs, they appear to be unable to figure out how to modify rocks to create hard, sharp tools (Panger, Brooks, Richmond, and Wood 2002; Toth, Schick, and Semaw 2009). By extrapolation, it is possible that the neurological capacity to make prepared cores had not evolved in the authors of the LP. Another, nonmutually exclusive hypothesis is that social-group sizes were too small and population densities too low to generate the kind of networks that foster innovation and transmission of new ideas (Derex, Beugin,

Godelle, and Raymond 2013; Henrich 2004; Powell, Shennan, and Thomas 2009). Under such conditions, innovations have less chance of arising and spreading.

Although these hypotheses for LP stasis are reasonable, they are difficult to test, and there are three reasons to question them. The first and most obvious rationale for skepticism, illustrated in figure 6.1, is how much brain size increased over the course of the LP. In fact, the average endocranial volume (ECV) of fossil hominin crania (n = 15, none from the genus *Homo*) from the period between 3.3 million years ago and 2.5 million years ago average 450.6 cm^3; ECVs from the period between 2.5 million years ago and 1.7 million years ago (n = 19, mostly from the genus *Homo*) average 611.2 cm^3; and ECVs attributed solely to *Homo* associated with LP assemblages (n = 25) aged less than 1 million years ago average 1,019.4 cm^3 (Lieberman 2011; see also Rightmire 2004). Given the high metabolic cost of increased brain mass, not to mention other challenges such as giving birth to larger-brained infants, the cognitive benefits of these brain-size increases must have outweighed the costs, and it is difficult to believe that hominins with brains of 1,000 cm^3, at the very lower end of the human range, lacked the ability to innovate enough to create new stone-tool forms.

A second reason to be critical of the hypothesis that LP hominins were unable to innovate, at least in terms of lithic technology, is the breadth of habitats they occupied. In addition to adapting to heterogeneous African habitats, the genus *Homo* dispersed during the LP throughout large swathes of Eurasia, occupying an impressive variety of temperate and tropical environments with markedly different fauna and flora, not to mention lithic raw materials (Finlayson 2005). Hominins must have needed to adapt in numerous ways

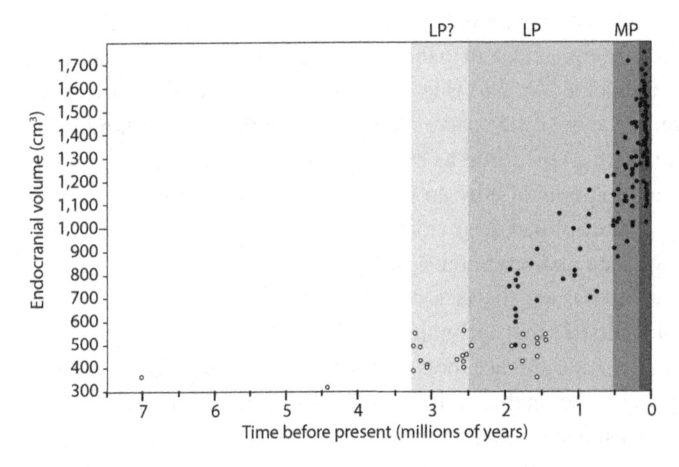

Figure 6.1
Endocranial volume over time in relation to the Lower, Middle, and Upper Paleolithic (data from Lieberman 2011). Species in *Homo* are solid circles, and australopiths and earlier genera are open circles. Given that the existence of stone tools from 3.3 million to 2.6 million years ago is poorly documented, it is unclear whether this period represents the earliest Lower Paleolithic.

both behaviorally and anatomically to these conditions, and it is reasonable to assume that how they made and used stone tools were no exceptions.

Finally, during the LP there were many different species of hominins. At least five well-recognized species of *Homo* are associated with the LP—*H. habilis*, *H. rudolfensis*, *H. erectus*, *H. heidelbergensis*, and *H. floresiensis*—and this is probably an underestimate (Lieberman 2013). Further, if the LP extends back to 3.3 million years ago, then we must add one or more species of australopiths to the list of authors who made LP industries (Harmand et al. 2015). The impressive diversity of these early tool-making species must have encompassed considerable behavioral diversity, some of which was likely manifested in how they fabricated and used stone tools. Although the LP differs technologically from the Middle Paleolithic and Upper Paleolithic, principally from the lack of prepared cores, there is actually considerable variation in the size and shape of flakes, choppers, bifaces, and other tools both within and between assemblages (Stout 2011). It is reasonable to hypothesize—but difficult, if not impossible, to test—that many of these forms probably arose independently many times.

Here I consider another explanation for why we perceive so little difference among lithic assemblages, thus blurring evidence for convergence, across the vast time and space of the LP: *there was little fitness benefit for substantial innovation.* As a general principle, selection (natural or cultural) favors change when it improves reproductive success, but otherwise either constrains changes with negative fitness effects or permits increased variation when changes have neutral effects. Accordingly, whatever biological selection occurred during the LP for increased brain size, improved abilities to walk or run long distances, and other derived features in the genus *Homo*, these adaptations did not lead to any benefit for making stone tools in new forms. Put differently, during the LP there was limited coevolution between biological and cultural change, at least in terms of stone tools.

There are many dimensions of behavior to explore in terms of this hypothesis, but by far the most important is hunting because of its reliance on lithic technology. As the genus *Homo* entered the carnivore guild by becoming scavengers and hunters, what were the major challenges that had to be "solved," and how would these challenges have either favored or constrained changes in LP tool types?

The Challenges of Being an Upright, Bipedal, Carnivorous Ape

Ever since Jane Goodall's first published descriptions of chimpanzees hunting colobus monkeys and other animals (including fellow chimpanzees), it is no longer tenable to consider hunting a uniquely human trait among the apes (Goodall 1986). However, there are several reasons to believe that early hominins were less capable of hunting than chimpanzees and that selection for endurance in hominins initially came at the expense of hunting ability. Most important, chimpanzees hunt using power and speed through sprinting and leaping (Boesch 2002; Gilby et al. 2015), but bipedal hominins can generate only half the locomotor power of quadrupeds such as chimps, making the fastest humans approximately half as fast as chimps, not to mention intrinsically much less stable (Lieberman 2015). It is improbable that any early hominin was capable of the acrobatic feats chimpanzees employ to hunt. In addition, there is clear evidence for selection during human evolution for endurance, which is necessarily traded off with speed and power. This trade-off is most evident in lower-extremity muscle-fiber types, which are predominantly fast-twitch in chimps but slow-twitch in humans (Myatt, Schilling, and Thorpe 2011). Another challenge hominins faced as carnivores was the loss of natural weapons such as fangs, claws, and heavy forelimbs, as well as natural defenses such as thick fur. Yet there is unquestionable evidence for meat eating in the hominin fossil record by 2.6 million years ago, and possibly earlier, and there is solid evidence that, by 2 million years ago, hominins were able to hunt large quadrupeds such as kudu and wildebeest (Diez-Martín et al. 2015; Domínguez-Rodrigo 2002). How did they do it?

Following Darwin (1871), students of human evolution have argued that hominins were able to become carnivores by using technology and cooperation—essentially brains over brawn. According to this line of thinking, early hominin carnivory was made possible by projectiles such as spears and later by more sophisticated tools such as nets, atlatls, and bows and arrows. In addition, hominins are usually inferred (sensibly) to have hunted cooperatively, perhaps through means such as ambushing and driving prey into traps (Stanford 1999). Following from these inferences, almost all debates about the evolution of human hunting have focused on whether early *Homo* hunted or scavenged (Egeland and Barba 2011), but few have considered the question of *how* they hunted or scavenged in the first place.

This question is challenging to answer because the technology available to early hominins was markedly limited compared to that available to ethnohistorically documented hunter-gatherers, which have informed many assumptions about early hominin hunting. For one, the oldest known stone spear points are 500,000 years old (Wilkins, Schoville, Brown, and Chazan 2012; chapter 9, this volume), which means that perhaps the most lethal weapons available to LP hunters were sharpened sticks, clubs, and rocks. Not only is there is little ethnographic evidence of humans successfully employing such weapons to kill large animals, there is also reason to doubt their effectiveness. Weapons such as spears usually kill prey animals through hemorrhaging and other tissue damage caused by lacerations of the weapon's sharp edges within a wound (Shea 2016). In addition, stone points are much sharper than untipped spears and thus more effectively penetrate thick hides and bones that protect the vital organs of animals. Throwing rocks may ward off some animals, but they are obviously ineffective for killing at a distance. Ethnographic evidence indicates that modern human hunters armed with tipped spears and other weapons rarely get closer than 8 meters to large animal prey to avoid being injured or killed by horns, kicks, bites, and other defenses (Churchill 1993). Given the high fitness cost of being injured by an animal's kick, hunting from close quarters was more a cost than a benefit from the perspective of natural selection.

The other major set of problems LP hominins would have faced as carnivores are digestive. Hominins like apes lack the carnassial postcanine teeth of carnivores and instead have low-crested (bunodont) molars that are notoriously ineffective for chewing meat. Researchers report that chimpanzees require as much as 11 hours to chew a 5-kilogram colobus monkey (Wrangham and Conklin-Brittain 2003), and humans chewing raw game routinely complain of the inability to break the food into small enough particles to swallow (see below). It is thus unclear how hominins were able to efficiently and effectively ingest meat prior to the invention of cooking.

Yet early *Homo* evidently succeeded in overcoming these challenges using LP technology. To be sure, cooperation must have been important, but in a series of papers my colleagues and I have argued that three additional adaptations were also critical: endurance running, throwing, and mechanical food processing (Bramble and Lieberman 2004; Roach, Venkadesan, Rainbow, and Lieberman 2013; Zink and Lieberman 2016). And, from the perspective of LP change and convergence, it is possible that these adaptations were so effective that they spurred little selection for modest technological innovation. To explore this hypothesis, I review each of these sets of adaptations for carnivory and consider how they might have constrained or encouraged the development of varied types of LP tools.

Endurance Running, Scavenging, and Persistence Hunting

As noted above, even the fastest humans are comparatively slow sprinters, with maximum running speeds of a little more than 10 meters per second, about half that of equal-sized quadrupedal cursors such as digs or antelopes (Garland 1983). Human sprinters, moreover, cannot maintain maximum speeds for as long as most quadrupedal cursors, making it extremely improbable that hominins were ever able to hunt by outrunning their prey. In contrast, humans are extraordinary endurance runners able to run long distances under aerobic capacity. Most important, reasonably fit but otherwise average human runners can easily run on a regular basis distances of more than 10 kilometers in hot conditions at speeds that require cursors to have to gallop rather than trot (Bramble and Lieberman 2004; Carrier 1984). These performance capabilities are important because most quadrupeds can run long distances only with a trotting gait because the see-saw motions of galloping cause oscillatory movements of the viscera that collide with the diaphragm, preventing them from cooling via panting (Bramble and Jenkins 1993). All quadrupeds cool primarily through panting, and most quadrupeds cannot cool at all through sweating (horses have some capabilities, but these are less effective than those in humans; Schmidt-Nielsen

1990). As a result, galloping quadrupeds rapidly overheat in hot temperatures. Amateur human runners in annual "man versus horse" races in Arizona and Wales thus can match and, when conditions are hot, often outrun thoroughbred horses over long distances.

Human abilities to run long distances relatively rapidly in the heat made possible two different strategies to enter the carnivore guild. The first is to scavenge. All carnivores, including top predators such as lions, scavenge to some extent, and humans are no exception. However, scavenging presents two important challenges: first, the ability to locate and get to carcasses before they are eaten or monopolized by other scavengers and second, the ability to compete with other predators. Neither of these challenges is trivial in the relatively open African habitats in which the genus *Homo* apparently evolved because scavengeable carcasses are evanescent resources that are hotly contested by a wide range of carnivores, ranging from vultures to foxes, hyenas, and lions (Van Valkenburgh 2001). In this context, humans with the ability to run long distances in the heat have a distinct advantage over hyenas and lions by power scavenging, in which humans see vultures circling in the distance or heading toward prey, both indicators of the location of carcasses, and then run to the carcass before hyenas or lions get there (Liebenberg 2006). The ability to outrun these major competitors when it is hot thus enables human scavengers to outcompete hyenas and lions in hot, diurnal conditions. Power scavenging has been documented among Hadza hunters (O'Connell, Hawkes, and Blurton-Jones 1988) and among female Bushmen hunters in the Kalahari (Shostak 1981).

Running may have been selected initially for scavenging, but it is arguably even more important for hunting. Although almost all carnivores need to run, humans probably evolved to hunt by means of a special method known as persistence hunting (PH). In PH, hunters pursue an animal—usually the largest possible because bigger animals overheat more easily than small ones—in the middle of the day in hot conditions until the prey collapses from heat stroke. Since almost all quadrupeds can gallop faster than any human, PH usually works through a combination of chasing and tracking (Liebenberg 2001). During the chasing phase, the runners cause the animal to gallop and thus overheat, at which point it usually tries to hide and cool down. While the prey rests and pants to dump heat, the runners track the animal (often at a walk) and attempt to find and chase their prey again before it has restored a normal core temperature. Over time and distance, usually between 15 and 30 kilometers (Liebenberg 2006), the prey's core temperature continues to rise, eventually causing the animal to collapse from heatstroke. At this point, the hunters can easily dispatch the prey at close quarters without any risk of being kicked or gored and with simple technology such as a thrusting spear or even just a rock.

Ethnohistoric records indicate that PH was practiced not just in Africa but also almost universally among hunter-gatherers in the Americas as well as in Australia and elsewhere (e.g., Bennett and Zingg 1935; Carrier 1984; McCarthy 1957; Nabokov 1987; Schapera 1930). In addition, PHs are not as extreme as many imagine. The PHs documented by Liebenberg (2006) averaged 27.8 kilometers, with hunters running only about half the time at a moderate speed, about a 10-minute mile, typical of the speed used by amateur ultrarunners. Moreover, according to Liebenberg (2006), the success rate for PHs was approximately 75 percent, considerably higher than hunting with a bow and arrow. Persistence hunters report that the most challenging component of PH is not the running but the ability to track the prey.

Of course, it impossible to test when PH first evolved, because it leaves no direct archaeological traces. However, persistence hunters report that they prefer to chase large, prime-aged prey because they overheat more easily (Liebenberg 2001), which is exactly the nature of prey found in many Lower Paleolithic sites such as FLK-ZINJ (Domínguez-Rodrigo 2002), that are incorrectly cited as evidence *against* the PH hypothesis (Pickering and Bunn 2007). However, many of the adaptations that enable humans to run long distances can be detected from the skeleton, and these adaptations strongly indicate that by the time of *H. erectus*, hominins had evolved many if not all the capacities of modern humans to do endurance running.

These features, summarized in figure 6.2, include feet with a spring-like arch; a short tuber calcaneus; short toes; long legs; an elongated Achilles tendon; an expanded cranial portion of the gluteus maximus; a narrow waist that enables independent counterrotations of the thorax and pelvis; enlarged hindlimb and intervertebral joints; low, wide shoulders that are decoupled from the head; a nuchal ligament that helps stabilize the head, which is

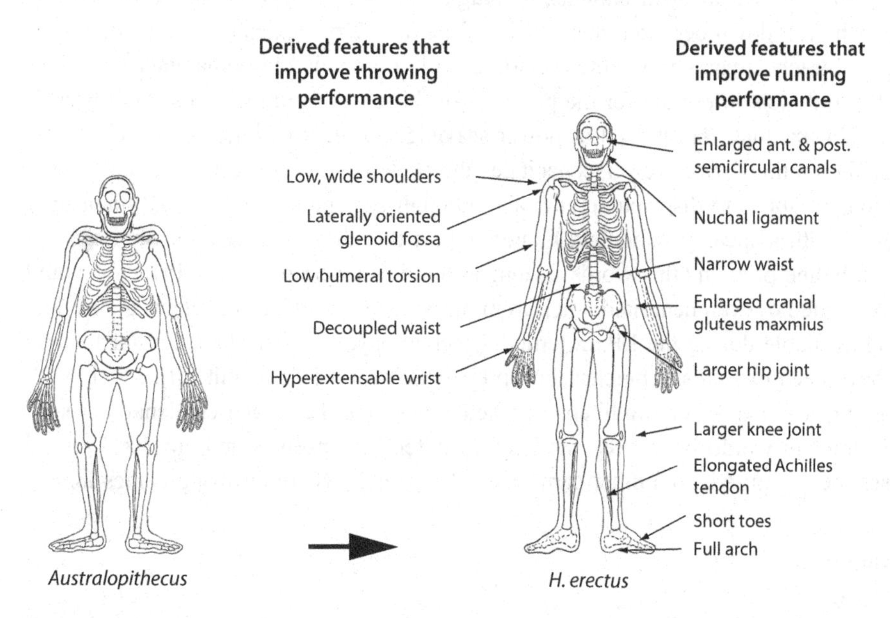

Figure 6.2
Some (not all) of the derived features in *H. erectus* that improve performance in throwing and endurance running (after Bramble and Lieberman 2004; Roach et al. 2013).

more balanced; enlarged anterior and posterior semicircular canals that stabilize the gaze during head-pitching motions caused by running; and elaborated heat-dumping capabilities in the pharynx (Bramble and Lieberman 2004; Lieberman 2011; Lieberman, Raichlen, Pontzer, Bramble, and Cutright-Smith 2006; Raichlen, Armstrong, and Lieberman 2011; Rolian, Lieberman, Hamill, Scott, and Werbel 2009). Importantly, many of these novel features (e.g., short toes, a short tuber calcaneus, an elongated Achilles, an expanded gluteus maximus, enlarged anterior and posterior semicircular canals, and the nuchal ligament) would have had no benefit for walking, indicating selection specifically for running (Bramble and Lieberman 2004). Unfortunately, we do not yet have direct evidence for when hominins gained the ability to sweat and lost fur, but a number of indirect sources of evidence indicate that these changes had occurred by the time of appearance of *H. erectus* (Lieberman 2013). All in all, it is reasonable to infer that *H. erectus* was a capable endurance runner and would have been able to engage in persistence hunting.

Although PH might have been an important method of hunting, it almost surely was not the only method employed by LP hunters. In fact, there are several constraints on PH. The first is that hunters need water, especially before a hunt (blood can be drunk after a hunt), but all hominins are obligate drinkers. Second, PH relies on driving animals to hyperthermia, and thus is most effective in hot conditions, and could have been practiced only during warm summers in temperate climates. Finally, PH requires the ability to track and is most effective in open habitats, although the success of Tarahumara persistence hunters, who run down deer in piñon forests of the northern Mexican Sierra, underscores how skilled trackers can employ the method even in relatively closed habitats.

Finally, and most important for the purposes of this essay (and this volume), it is critical to emphasize that PH (as well as power scavenging) requires little to no technology. Because the method effectively incapacitates the prey, persistence hunters don't require any serious weapons to dispatch an animal. Although Bushmen reportedly kill their prey most often with a spear, Tarahumara hunters report that they often use locally available rocks to kill their prey. In other words, as long as hominins were relying on PH, they would not have needed or even benefited much from more-sophisticated projectile weapons than anything available during the LP. In terms of technology, a bigger challenge for hunters would have been to skin and butcher their prey, which can be done quite effectively with LP tools (Shea 2016). It is more likely, as noted above, that the greater challenge for early hominin hunters would have been the need to defend themselves and any hard-earned carcasses they acquired from other carnivores. And this is where throwing comes in.

Throwing

Humans are the only species that can throw with both accuracy and speed. Other primates, including chimpanzees, can throw objects underhanded with accuracy but slowly

(Westergaard, Liv, Haynie, and Suomi 2000), and, when displaying, sometimes hurl branches and other objects overhand with little accuracy. Humans alone generate fast, on-target throws by sequentially generating torques in a whiplike fashion starting in the legs, then the hip, spine, shoulders, elbow, and finally the wrist (Fleisig, Andrews, Dillman, and Escamilla 1995; Hirashima, Kadota, Sakurai, Kudo, and Ohtsuki 2002). Accuracy is largely a function of the neural control required to move the arm in the plane of the throw and, most especially, to release the projectile at the right moment. Power comes from the torque generated at each joint, which accelerates each sequential segment of the body, accumulating kinetic energy (the product of mass times the square of velocity) that is imparted to the projectile at the moment of release. Although every joint of the kinetic chain helps accumulate this energy, the largest contribution comes from the shoulder (Hirashima, Kudo, Watarai, and Ohtsuki 2007). About half the power generated at the shoulder comes from muscles that flex and internally rotate the humerus, but as shown by Roach and colleagues (2013), the other half comes from unique adaptations in the human shoulder for elastic energy storage. When a throwing human cocks her arm by externally rotating the humerus, elastic energy is stored in the muscles and connective tissues that surround the joint, which then recoil rapidly like a catapult, helping internally rotate the humerus at velocities approaching 9,000 degrees per second, the fastest movement in the human body (Pappas, Zawacki, and Sullivan 1985).

Understanding the biomechanics involved in throwing allows us to make some inferences about the evolutionary history of throwing, given that many of features that make humans capable of fast, powerful throws are preserved in the skeleton and hence in the fossil record. These features (figure 6.2) include a laterally oriented shoulder joint; low, wide shoulders; low humeral torsion; a decoupled waist that can rotate out of phase with the torso; and a hyperextensible wrist. Although not all of these features evolved at the same time, and some may have evolved for functions other than throwing, they were all present in *H. erectus*, suggesting that this species was a capable thrower, unlike apes or australopiths (Roach and Lieberman 2014; Roach et al. 2013).

Evidence for the evolution of throwing capabilities in the genus *Homo* raises two hypotheses relevant to the lack of change in stone tools evident over the LP. One possibility is that throwing capabilities evident by *H. erectus* were selected to help hominins hunt with untipped spears, such as those preserved at Schöningen, Germany, which date to about 300,000 years ago (Thieme 1997), or with even more simple projectiles such as rocks. Although this hypothesis is difficult to test, ethnographic and experimental studies have questioned the efficacy of such weapons for hunting because the force required to kill large animals would have limited accuracy from a distance (as in javelin throwing) and thus probably required hominins to get dangerously close to their prey (Churchill 1993; Wilkins, Schoville, and Brown 2014). If so, there would have been a strong selective benefit to stone points, which would have made projectiles considerably more lethal when thrown from a distance. The alternative hypothesis is that hominins primarily threw for defense to ward off carnivores. Under such a scenario, there would have been much less selection for more sophisticated and lethal projectile technology.

Chewing

A final, less appreciated challenge for hominins to become carnivores is chewing meat. Humans today eat mostly domesticated animals that have been raised to yield tender flesh, and even then we process almost all the meat we eat prior to ingestion. Extra-oral food processing takes two forms: cooking (thermal processing) and a variety of mechanical processing methods such as slicing, pounding, and grinding. Cooking, which yields many benefits, has received considerable attention, and may have started as early as a million years ago, but archaeological evidence for cooking does not become common until less than 500,000 years ago, essentially coincident with the Middle Paleolithic (Gowlett and Wrangham 2013; Shimelmitz et al. 2014).

In contrast, mechanical processing is obviously much older, going back at least to the beginning of the LP, but until recently the importance of LP processing techniques have not been studied. Yet raw, unprocessed game meat is notoriously difficult for humans and other primates to chew because it is both tough and viscous and thus cannot be fractured with low-crested (bunodont) primate molars (Lucas 2004). For example, as I noted earlier, chimpanzees, which have maximum bite forces that are much greater than those exhibited by humans (Eng, Lieberman, Zink, and Peters 2013), have been documented to spend between 5 and 11 hours to chew 4-kilogram animals such as colobus monkeys (Wrangham and Conklin-Brittain, 2003). Another challenge is the large size of unprocessed meat particles swallowed, presenting an increased risk of asphyxiation, a common cause of accidental death. Large particles also limit digestive efficiency, which is a function of surface area to volume ratio (Lucas 2004). Experiments have shown that humans fed raw game are unable to reduce the size of the food particles and end up swallowing large boluses of meat with much lower caloric yields (Boback et al. 2007; Carmody, Weintraub, and Wrangham 2011).

To test the effects of LP food-processing techniques on chewing performance, Zink and Lieberman (2016) compared bite-force production and the size of swallowed food particles for both meat and tubers in a large sample of human subjects. These studies found that even simple Oldowan flakes had substantial effects on the ability to chew meat, primarily by reducing the number of chews to swallow by 25 percent, the force per chew by 33 percent, and particle size by a whopping 40.5 percent (Zink and Lieberman 2016). Overall, slicing meat using the simplest LP technology had major effects on the ability of hominins to chew uncooked game, which helps explain why the oldest evidence for meat-eating and the LP seem to be coincident. Moreover, until the invention of cooking, it is unlikely that innovation in tool forms had much effect on hominin masticatory performance. Put differently, simple flakes are so effective at cutting meat, there would have been little selection for novel tool forms.

Implications

The transition from *Australopithecus* to *Homo* encompassed many anatomical and behavioral changes that, overall, signal the origin of the hunting-and-gathering mode of subsistence. This way of life involves not just hunting but also extractive foraging, tool use, and intense levels of cooperation. Although there is debate about the relative importance of meat eating during the Paleolithic, the archaeological record provides incontrovertible evidence that hominins entered the carnivore guild during the LP. Regardless of how much early *Homo* scavenged versus hunted, getting access to animal carcasses must have presented substantial challenges to these hominins, who lacked the speed, strength, natural weapons, and defenses present in other carnivores. As discussed above, the problem of how to become a part-time carnivore was evidently solved by a combination of selection for new capabilities such as endurance running and throwing, but also required cultural evolution, especially the stone tools necessary to butcher animals and facilitate effective digestion before the invention of cooking.

It is therefore likely that selection for hunting was a major early impetus for coevolution. In particular, selection for the anatomical and physiological adaptations that made hunting possible favored the cultural evolution of lithic technology, which in turn led to additional selection on the human body (Henrich 2015). But if early carnivory fostered coevolution, then why didn't LP toolmakers develop a greater and more sophisticated variety of tools, especially weapons, notably spear points? Why did these simple but transformative innovations take more than two million years to appear? Further, although there is variation in the LP, how many LP tool types such as handaxes, cleavers, and picks were invented independently?

Obviously, no one knows to what extent the LP stasis was a result of cognitive constraints or small population sizes, but the relatively large brains of these hominins, in combination with their temporal, ecological, and taxonomic diversity, calls these hypotheses into question. As I have argued, it is also worth considering whether early hunting methods, including persistence hunting as well as butchering and food preparation, were so effective and so unreliant on technology that they created little benefit for major innovations in terms of tool types. Stated simply, *H. erectus* might have been so good at running down prey, warding off other carnivores by throwing sticks and rocks, and using flakes and other LP tools to butcher and process carcasses, that there was little additional benefit from developing novel technologies.

I emphasize that this is just a hypothesis, and one that is difficult to test. One prediction is that if LP hunting and butchering methods were sufficiently effective to benefit minimally from lithic innovation, then there would be little selection on tool form. When selection is low or absent, variation tends to increase, leading to the expectation of an increase in overall variation among LP tool forms over time. This hypothesis merits testing. However, the phenomenon of biological systems performing so well that they minimize the selective benefit of innovations is probably so common we rarely even think about them. For example, until the relatively recent invention of the oven, roasting technology was largely unchanged since the invention of cooking. In addition, apart from using shoes, people today still walk and run as their ancestors did a million years ago, and the human foot, like those of other animals, is obviously still well adapted for being barefoot in a wide variety of habitats. Although shoes were probably invented in the Upper Paleolithic (Trinkaus 2005), the two basic kinds of footwear—sandals and moccasins—underwent very little change for tens of thousands of years until the Industrial Revolution (Connolly et al. 2011).

That said, innovations do occur, raising the question of why it took until the Middle Paleolithic for breakthroughs such as stone points and cooking. Did these innovations occur because hominins were finally smart enough to invent them? Or did they occur because of increases in population growth, by chance, or other factors? This question remains a central problem of human evolutionary biology and archaeology, but its answer will also require understanding instances of long-term stasis such as the LP when change did not occur.

Acknowledgments

I thank Briggs Buchanan, Metin Eren, and Mike O'Brien for the invitation to think about this problem and to the Konrad Lorenz Institute for sponsoring the thought-provoking workshop that stimulated and influenced this paper. I am very grateful to the many colleagues with whom I have worked on the evolution of running, throwing, and chewing, including Dennis Bramble, Carolyn Eng, Herman Pontzer, Mike Rainbow, David Raichlen, Neil Roach, John Shea, Madhu Venkadesan, and Katie Zink.

References

Bennett, W. C., & Zingg, R. M. (1935). *The Tarahumara: An Indian Tribe of Northern Mexico*. Chicago: University of Chicago Press.

Boback, S. M., Cox, C. L., Ott, B. D., Carmody, R., Wrangham, R. W., & Secor, S. M. (2007). Cooking and Grinding Reduces the Cost of Meat Digestion. *Comparative Biochemistry and Physiology. Part A, Molecular & Integrative Physiology*, *148*, 651–656.

Boesch, C. (2002). Cooperative Hunting Roles among Taï Chimpanzees. *Human Nature*, *13*, 27–46.

Bramble, D. M., & Jenkins, F. A., Jr. (1993). Mammalian Locomotor-respiratory Integration: Implications for Diaphragmatic and Pulmonary Design. *Science*, *262*, 235–240.

Bramble, D. M., & Lieberman, D. E. (2004). Endurance Running and the Evolution of Homo. *Nature*, *432*, 345–352.

Carmody, R. N., Weintraub, G. S., & Wrangham, R. W. (2011). Energetic Consequences of Thermal and Nonthermal Food Processing. *Proceedings of the National Academy of Sciences of the United States of America*, *108*, 19199–19203.

Carrier, D. R. (1984). The Energetic Paradox of Human Running and Hominid Evolution. *Current Anthropology*, *24*, 483–495.

Churchill, S. E. (1993). Weapon Technology, Prey Size Selection and Hunting Methods in Modern Hunter-Gatherers: Implications for Hunting in the Palaeolithic and Mesolithic. In G. L. Peterkin, H. M. Bricker, & P. Mellars (Eds.), *Animal Exploitation in the Later Palaeolithic and Mesolithic of Eurasia* (pp. 11–24). Archeological Papers, vol. 4. Washington, D.C.: American Anthropological Association.

Connolly, T. J., Hlavacek, P., & Moore, K. (2011). *10,000 Years of Shoes*. Eugene: University of Oregon Press.

Darwin, C. R. (1871). *The Descent of Man and Selection in Relation to Sex*. London: Murray.

Derex, M., Beugin, M.-P., Godelle, B., & Raymond, M. (2013). Experimental Evidence for the Influence of Group Size on Cultural Complexity. *Nature*, *503*, 389–391.

Diez-Martín, F., Sánchez Yustos, P., Uribelarrea, D., Baquedano, E., Mark, D. F., Mabulla, A., et al. (2015). The Origin of the Acheulean: The 1.7 Million-Year-Old Site of FLK West, Olduvai Gorge (Tanzania). *Scientific Reports*, *5*, 17839.

Domínguez-Rodrigo, M. (2002). Hunting and Scavenging by Early Humans: The State of the Debate. *Journal of World Prehistory*, *16*, 1–54.

Egeland, C. P. M. D.-R., & Barba, M. (2011). The Hunting-versus-Scavenging Debate. In M. Domínguez-Rodrigo (Ed.), *Deconstructing Olduvai: A Taphonomic Study of the Bed I Sites* (pp. 11–22). Dordrecht, Netherlands: Springer.

Eng, C. M., Lieberman, D. E., Zink, K. D., & Peters, M. A. (2013). Bite Force and Occlusal Stress Production in Hominin Evolution. *American Journal of Physical Anthropology*, *151*, 544–557.

Finlayson, C. (2005). Biogeography and Evolution of the Genus *Homo*. *Trends in Ecology & Evolution*, *20*, 457–463.

Fleisig, G. S., Andrews, J. R., Dillman, C. J., & Escamilla, R. F. (1995). Kinetics of Baseball Pitching with Implications about Injury Mechanisms. *American Journal of Sports Medicine*, *23*, 233–239.

Garland, T. (1983). The Relation between Maximal Running Speed and Body Mass in Terrestrial Mammals. *Journal of Zoology*, *199*, 1557–1570.

Gilby, I. C., Machanda, Z. P., Mjungu, D. C., Rosen, J., Muller, M. N., Pusey, A. E., et al. (2015). "Impact Hunters" Catalyse Cooperative Hunting in Two Wild Chimpanzee Communities. *Philosophical Transactions of the Royal Society of London. Series B, Biological Sciences*, *370*, 20150005.

Goodall, J. (1986). *The Chimpanzees of Gombe: Patterns of Behavior*. Cambridge, Mass.: Harvard University Press.

Gowlett, J., & Wrangham, R. W. (2013). Earliest Fire in Africa: Towards the Convergence of Archaeological Evidence and the Cooking Hypothesis. *Azania*, *48*, 5–30.

Harmand, S., Lewis, J. E., Feibel, C. S., Lepre, C. J., Prat, S., Lenoble, A., et al. (2015). 3.3-Million-Year-Old Stone Tools from Lomekwi 3, West Turkana, Kenya. *Nature*, *521*, 310–315.

Henrich, J. (2004). Demography and Cultural Evolution: How Adaptive Cultural Processes Can Produce Maladaptive Losses: The Tasmanian Case. *American Antiquity*, *69*, 197–214.

Henrich, J. (2015). *The Secret of Our Success: How Culture Is Driving Human Evolution, Domesticating Our Species, and Making Us Smarter*. Princeton, NJ: Princeton University Press.

Hirashima, M., Kadota, H., Sakurai, S., Kudo, K., & Ohtsuki, T. (2002). Sequential Muscle Activity and Its Functional Role in the Upper Extremity and Trunk during Overarm Throwing. *Journal of Sports Sciences, 20*, 301–310.

Hirashima, M., Kudo, K., Watarai, K., & Ohtsuki, T. (2007). Control of 3D Limb Dynamics in Unconstrained Overarm Throws of Different Speeds Performed by Skilled Baseball Players. *Journal of Neurophysiology, 97*, 680–691.

Liebenberg, L. (2001). *The Art of Tracking: The Origin of Science*. Claremont, South Africa: Philip.

Liebenberg, L. (2006). Persistence Hunting by Modern Hunter-Gatherers. *Current Anthropology, 47*, 1017–1025.

Lieberman, D. E. (2011). *The Evolution of the Human Head*. Cambridge, Mass.: Harvard University Press.

Lieberman, D. E. (2013). *The Story of the Human Body: Evolution*. Pantheon, New York: Health and Disease.

Lieberman, D. E. (2015). Human Locomotion and Heat Loss: An Evolutionary Perspective. *Comprehensive Physiology, 5*, 99–117.

Lieberman, D. E., Raichlen, D. A., Pontzer, H., Bramble, D. M., & Cutright-Smith, E. (2006). The Human Gluteus Maximus and Its Role in Running. *Journal of Experimental Biology, 209*, 2143–2155.

Lucas, P. W. (2004). *Dental Functional Morphology: How Teeth Work*. Cambridge: Cambridge University Press.

Lycett, S. J., & Bae, C. J. (2010). The Movius Line Controversy: The State of the Debate. *World Archaeology, 42*, 521–544.

McCarthy, F. D. (1957). *Australian Aborigines: Their Life and Culture*. Melbourne: Colorgravure.

Myatt, J. P., Schilling, N., & Thorpe, S. K. (2011). Distribution Patterns of Fibre Types in the Triceps Surae Muscle Group of Chimpanzees and Orangutans. *Journal of Anatomy, 218*, 402–412.

Nabokov, P. (1987). *Indian Running: Native American History and Tradition*. Santa Fe, N.M.: Ancient City.

O'Connell, J. F., Hawkes, K., & Blurton-Jones, N. G. (1988). Hadza Scavenging: Implications for Plio-Pleistocene Hominid Subsistence. *Current Anthropology, 29*, 356–363.

Panger, M. A., Brooks, A. S., Richmond, B. G., & Wood, B. (2002). Older Than the Oldowan? Rethinking the Emergence of Hominin Tool Use. *Evolutionary Anthropology, 11*, 234–245.

Pappas, A. M., Zawacki, R. M., & Sullivan, T. J. (1985). Biomechanics of Baseball Pitching: A Preliminary Report. *American Journal of Sports Medicine, 13*, 216–222.

Pickering, T. R., & Bunn, H. T. (2007). The Endurance Running Hypothesis and Hunting and Scavenging in Savanna-Woodlands. *Journal of Human Evolution, 53*, 438–442.

Powell, A., Shennan, S., & Thomas, M. G. (2009). Late Pleistocene Demography and the Appearance of Modern Human Behavior. *Science, 324*, 1298–1301.

Raichlen, D. A., Armstrong, H., & Lieberman, D. E. (2011). Calcaneus Length Determines Running Economy: Implications for Endurance Running Performance in Modern Humans and Neandertals. *Journal of Human Evolution, 60*, 299–308.

Rightmire, G. P. (2004). Brain Size and Encephalization in Early to Mid-Pleistocene *Homo*. *American Journal of Physical Anthropology, 124*, 109–123.

Roach, N. T., Venkadesan, M., Rainbow, M. J., & Lieberman, D. E. (2013). Elastic Energy Storage in the Shoulder and the Evolution of High-Speed Throwing in *Homo*. *Nature, 498*, 483–486.

Roach, N. T., & Lieberman, D. E. (2014). Upper Body Contributions to Power Generation during Rapid, Overhand Throwing in Humans. *Journal of Experimental Biology*, *217*, 2139–2149.

Rolian, C., Lieberman, D. E., Hamill, J., Scott, J. W., & Werbel, W. (2009). Walking, Running and the Evolution of Short Toes in Humans. *Journal of Experimental Biology*, *212*, 713–721.

Schapera, I. (1930). *The Khoisan Peoples of South Africa: Bushmen and Hottentots*. London: Routledge and Kegan Paul.

Schmidt-Nielsen, K. (1990). *Animal Physiology: Adaptation and Environment* (4th ed.). Cambridge: Cambridge University Press.

Shea, J. J. (2016). *Tools in Human Evolution: Behavioral Differences among Technological Primates*. Cambridge: Cambridge University Press.

Shimelmitz, R., Kuhn, S. L., Jelinek, A. J., Ronen, A., Clark, A. E., & Weinstein-Evron, M. (2014). "Fire at Will": The Emergence of Habitual Fire Use 350,000 Years Ago. *Journal of Human Evolution*, *77*, 196–203.

Shostak, M. (1981). *Nisa: The Life and Words of a! Kung Woman*. Cambridge, Mass.: Harvard University Press.

Stanford, C. B. (1999). *The Hunting Apes: Meat Eating and the Origins of Human Behavior*. Princeton, N.J.: Princeton University Press.

Stout, D. (2011). Stone Toolmaking and the Evolution of Human Culture and Cognition. *Philosophical Transactions of the Royal Society of London. Series B, Biological Sciences*, *366*, 1050–1059.

Thieme, H. (1997). Lower Palaeolithic Hunting Spears from Germany. *Nature*, *385*, 807–810.

Toth, N., Schick, K., & Semaw, S. (2009). The Oldowan: The Tool Making of Early Hominins and Chimpanzees Compared. *Annual Review of Anthropology*, *38*, 289–305.

Trinkaus, E. (2005). Anatomical Evidence for the Antiquity of Human Footwear Use. *Journal of Archaeological Science*, *32*, 1515–1526.

Van Valkenburgh, B. (2001). The Dog-Eat-Dog World of Carnivores: A Review of Past and Present Carnivore Community Dynamics. In C. B. Stanford & H. T. Bunn (Eds.), *Meat-Eating and Human Evolution* (pp. 101–121). Oxford: Oxford University Press.

Wang, W., Lycett, S. J., von Cramon-Taubadel, N., Jin, J. J. H., & Bae, C. J. (2012). Comparison of Handaxes from Bose Basin (China) and the Western Acheulean Indicates Convergence of Form, not Cognitive Differences. *PLoS One*, *7*(4), e35804.

Westergaard, G. C., Liv, C., Haynie, M. K., & Suomi, S. J. (2000). A Comparative Study of Aimed Throwing by Monkeys and Humans. *Neuropsychologia*, *38*, 1511–1517.

Wilkins, J., Schoville, B. J., Brown, K. S., & Chazan, M. (2012). Evidence for Early Hafted Hunting Technology. *Science*, *338*, 942–946.

Wilkins, J., Schoville, B. J., & Brown, K. S. (2014). An Experimental Investigation of the Functional Hypothesis and Evolutionary Advantage of Stone-Tipped Spears. *PLoS One*, *9*(8), e104514.

Wrangham, R. W., & Conklin-Brittain, N. (2003). Cooking as a Biological Trait. *Comparative Biochemistry and Physiology. Part A, Molecular & Integrative Physiology*, *136*, 35–46.

Zink, K. D., & Lieberman, D. E. (2016). Impact of Meat and Lower Palaeolithic Food Processing Techniques on Chewing in Humans. *Nature*, *531*, 500–503.

III EVIDENCE AND OTHER ISSUES

7 Reduction Constraints and Shape Convergence along Tool Ontogenetic Trajectories: An Example from Late Holocene Projectile Points of Southern Patagonia

Judith Charlin and Marcelo Cardillo

With some exceptions (e.g., Lycett 2007, 2009; Lycett, von Cramon-Taubadel, and Gowlett 2010), the study of convergence in stone tools is an issue scarcely addressed in archaeology, at least as an explicit line of research. However, there is much evidence to suggest that this phenomenon is more common than expected and that it can take place at different scales, from particular artifact traits to overall tool shapes, as well as in manufacture techniques.

With respect to Patagonia, where we work, archaeologists tend to focus on processes of divergence throughout the Holocene (Borrero 2001a; Charlin and Borrero 2012; Perez et al. 2011). These studies have focused on cultural and technological divergence in relation to biogeographical barriers such as the Santa Cruz River—the largest drainage in Southern Patagonia (Borrero 1998; Cardillo and Charlin 2016; Franco 2002)—and the Magellan Strait (Borrero 1989–1990; Cardillo, Charlin, and Borrazzo 2015; Charlin, Borrazzo, and Cardillo 2013; Morello et al. 2012) and the creation of the Isla Grande of Tierra del Fuego ca. 8,000 years ago (McCulloch, Clapperton, Rabassa, and Currant 1997; McCulloch, Bentley, Tipping, and Clapperton 2005). In general, morphological and technological similarities among tools are usually explained by population contact and shared technical knowledge (e.g., Nami 1992), but these statements rest more on assumptions than proof.

Our goals here are threefold. First, we discuss the role of reduction as a constraint that channels shape variation and leads to tool convergence. We explore this with a case study of Late Holocene projectile points from Southern Patagonia. Second, we examine the existence of parallelism, a particular case of convergence (McGhee 2011), between two populations of projectile points, one ethnographic and the other archaeological. Both populations share a common ancestry but were later isolated by the formation of the Strait of Magellan. Third, we highlight the utility of geometric morphometric methods (Bookstein 1991; Rohlf and Bookstein 1990) as a tool to study artifact form in general and artifact convergence in particular.

Shape and Evolution

The shape of projectile points is usually used as a tool to characterize variation in time and space. Shape changes over time are often explained as shifts in energy-capture strategies (e.g., Buchanan, Collard, Hamilton, and O'Brien 2011; Hughes 1998; Nelson 1991) or as the existence of particular technological traditions, without necessarily having functional implications (stylistic variation *sensu* Dunnell 1978). Purely stylistic changes (nonfunctional) would explain, at least in part, differences observed across vast spatial scales within the same time period (Binford 1963; Cavalli-Sforza and Feldman 1981). However, it is expected that projectile-point shape (as well as size) as directly related to subsistence reflects, at least partly, functional restrictions (such as cutting capacity, penetration depth, and aerodynamic characteristics), and the tradeoff between techno-units of the same technical system (Hughes 1998; Ratto 1994). In this scenario, when a correlation between different functional and structural factors is expressed, it is expected that shape trajectories will preferentially follow some particular paths in morphospace (*sensu* McGhee 1999) through evolutionary time. Conversely, in the absence of such restrictions, it is expected that all potential shape variation is realized over time (Bookstein 1991; Gould 1989).

One way to address this phenomenon is through the analysis of morphospaces, as they represent the overall design characteristics of projectile points (chapters 2 and 5, this volume). Shape spaces are by nature continuous and multidimensional, and it is common to generate them by means of multivariate methods, in particular geometric morphometrics (Rohlf and Bookstein 1990). If they are estimated by a number of real observations, these spaces are empirical in nature, and their amplitude will be, at least in part, a function of the morphological variation in the dataset (McGhee 1999, 2015). For this reason, morphospace is expected to be affected by the addition of new data. However, as sample size increases, the pattern will turn more robust and representative of the actual morphospace occupied by a set of shapes.

In our case study, we will generate empirical morphospaces using geometric morphometrics. Because of the tradeoffs among different designs, functional requirements, raw materials, as well as the life history of projectile points, some portions of the total design space may be populated over other areas. Thus, it is possible that some parts of shape spaces are denser than others—that is, some designs will be more common than others. In addition, it is expected that classes occupy distinct sectors of morphospace if they follow different allometric trajectories related to their particular life history. If similar constraints act upon different "initial" shapes, convergence can be expected, and, conversely, different constraints can lead to divergence in initially similar shapes.

Reduction as a Constraint

Tool reduction is essentially a continuous process by which the initial form (size and shape) of artifacts is modified in a directional way from the first use to discard. In this sense, making and using stone tools is clearly a process with limits on phenotypic variation. Like ontogeny, reduction is a patterned allometric process that changes in extent and degree. Even though stone tools do not grow—to the contrary, they become smaller with use—reduction is an archaeological analogue of biological growth that works in the opposite direction (Shott 2005).

Nowadays, morphometric variation of stone tools during their use-lives is an accepted fact and a common topic of analysis in archaeology (Buchanan, Eren, Boulanger, and O'Brien 2015; Charlin and González-José 2012; Clarkson 2013; Dibble 1984; Dibble, Schurmans, Iovită, and McLaughlin 2005; Eren et al. 2005; Kuhn 1990; Shott, Hunzicker, and Patten 2007). Although architectural (blank size and shape, quality of lithic raw material, and the like) and functional constraints (performance of a given task) determine the starting tool form, it is the use and resharpening of tools that account for subsequent changes. Resharpening should be considered not only a confounding factor but an integral part of the tool itself. Following Iovită (2010: 236), "a tool can be defined as the *collection of all possible resharpening stages* it undergoes until it is abandoned" (italics in original). Even though stone tools change during their use-lives, they change in one direction, and the possible changes are limited: stone tools are smaller after reduction, and edges dull as use proceeds.

Following Gould (1989) and McGhee (1999), we propose that tool shape is a by-product of different forces that in turn channel shape trajectories through ontogeny. From a design perspective, such forces act as constraints that lead to a directional pattern of evolution. It is hypothesized that under these forces, different initial shapes converge. It is also possible that similar initial shapes can diverge along their ontogenetic trajectories because of different functional requirements, including recycling (figure 7.1).

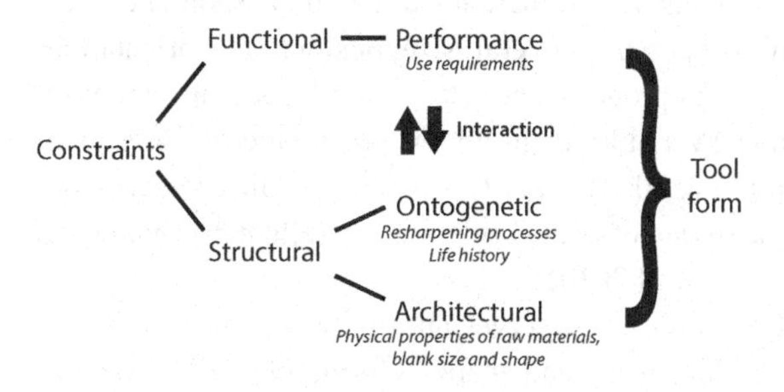

Figure 7.1
Stone-tool constraints model.

As functional and experimental analyses suggest, lithic artifacts are designed to perform specific functions, albeit with varying degrees of specificity (Nelson 1991). Further, as with other technologies, lithic tools are subject to use-wear or gradual reduction in their potential efficiency throughout their life (Shott 2005). Within this context, we believe that resharpening, conducted to extend a tool's use-life, follows a cost-benefit structure and therefore is done in a planned and structured way upon a preexisting form. Thus, performance and ontogenetic trajectories are not independent of each other and connect functional and structural dimensions. Other structural elements, such as raw material and technical system (in particular techno-units, *sensu* Oswalt 1976), also impose constraints on tool design. Selection of particular hafting elements and raw materials can also sort variation in a particular direction, as Ratto (1994, 2003) and Hughes (1998) have shown.

Geometric-Morphometric Methods

Geometric morphometrics (GM) is the statistical analysis of form based on Cartesian landmark coordinates (Bookstein 1991; Mitteroecker and Gunz 2009). This method has been applied increasingly to the study of stone tools over the last several years (Buchanan et al. 2011, 2015; Cardillo 2010; Charlin, Cardillo, and Borrazzo 2014; Iovită 2009, 2010; Lycett et al. 2010; Morales, Soto, Lorenzo, and Vergès 2015; Shott and Trail 2010). The application of GM methods to stone tools allows their physical configuration (their size and shape) to be represented as a mathematical object by means of Cartesian coordinates (Mitteroecker and Hutteger 2009). In GM, the term *shape* denotes the geometric properties of an object invariant to scale, position, and orientation, whereas *form* comprises both its shape and size (Bookstein 1991; Mitteroecker and Gunz 2009).

GM techniques have several advantages over traditional approaches, where shape is usually studied through linear measurements, which in most cases are highly correlated with size (Bookstein 1991). Thus, these measures actually describe form (size plus shape) rather than shape and provide largely redundant information (Iovită 2010; Shott and Trail 2010). For this reason, the separation of shape from overall size, position, and orientation of artifacts through GM methods is a useful property because (1) it allows shape and size to be treated as independent mathematical variables in the absence of allometry (Bookstein 1991; Zelditch, Swiderski, Sheets, and Fink 2004), and (2) it permits detailed description of object size and shape, including an assessment of their relationship (allometric analysis) and controls for size variation (Shott and Trail 2010).

Another important advantage of GM is that it preserves an artifact's geometric information throughout the analysis, and artifact shape and shape deformation can be visualized throughout the analyses (Mitteroecker and Gunz 2009). This is a great advantage over what is produced by traditional approaches, in which the geometric information is lost because shape is studied as linear measures that, in general, are not supported by homologous points (e.g., maximum diameter). In GM, the form of an object is captured by discrete points, referred to as *landmarks* (Bookstein 1991). The most important property of landmarks is their homology, either in light of a biological (ontogenetic or phylogenetic criteria) or geometrical principle (same topological positions relative to other landmarks).

A particular kind of point is a *semilandmark* (Bookstein 1991, 1996–1997). These are points deficient in one coordinate that are used to include information about object curvature, where homologous points are difficult to identify. They are arbitrary points defined in terms of their position on curves and surfaces that are used to capture homologous structures (see different procedures and algorithms for sliding semilandmarks in Bookstein 1996–1997; Gunz and Mitteroecker 2013; Gunz, Mitteroecker, and Bookstein 2005; Perez, Bernal, and González 2006). In the analysis of lithic artifacts, it is common to place landmarks according to topological considerations and to use semilandmarks to capture the curvature of some traits (e.g., the blade in projectile points and the edges in end scrapers and side scrapers [Buchanan et al. 2011, 2015; Cardillo and Charlin 2016; Charlin and González-José 2012; Morales et al. 2015]).

Several different GM methods have been developed to obtain shape coordinates. The method used here is a landmark-based method—the Procrustes method (Rohlf 1990; Rohlf and Slice 1990)—which is the most widespread and best-understood method in terms of mathematical and statistical properties (Mitteroecker and Gunz 2009: 236). In landmark-based methods, shape parameters are estimated by a Procrustes superimposition procedure, which is a least-squares approach that involves the translation, rotation, and scaling of the objects being compared (Mitteroecker and Gunz 2009; Perez et al. 2006). In this way, the shape coordinates obtained are free of variation in size, position, and orientation.

In a GM analysis, the shape of an artifact is represented by a single point in Kendall shape space—a multidimensional and curved space with as many dimensions as number of landmarks (Rohlf 1999). Shapes from Kendall space must be projected to a Euclidean tangent space to do standard statistical analyses. The point of tangency is at the reference shape, which is usually the average shape of the sample, also referred to as the consensus shape (the shape whose sum of squared distance to the other shapes is minimal). Euclidean distances in tangent space closely resemble the distance in Kendall shape space (Mitteroecker and Gunz 2009).

Case Study: Southern Patagonia Projectile Points

The first systematic archaeological research in Southern Patagonia began with Junius Bird's work at two caves—Pali Aike and Fell—in Chilean Patagonia (Bird 1938, 1946,

1988; figure 7.2). Based mainly on the stratigraphic evidence from these caves, Bird proposed a regional settlement sequence from ca. 11,000 years ago to historical times. According to artifact types and faunal remains, five prehistoric periods (I–V) were identified. The size and shape of projectile points and scrapers were key traits used to discriminate among periods (Bird 1938, 1946, 1988).

Periods IV and V were the latter periods of the prehistoric sequence, both dating to the Late Holocene. The *Onas*, known also as *Selk'nam*, were an ethnic group of pedestrian hunter-gatherers with a subsistence economy focused on land resources, mainly guanaco hunting. They occupied northern Tierra del Fuego during historical times (Borrero 2001b; Chapman 1986). Bird (1938, 1946, 1988) pointed out the similarity between historical Ona arrows and those recovered in Late Holocene archaeological sites from the mainland, and he named archaeological projectile points of period V (ca. 700 years ago) "Ona points." Moreover, he assumed type V points were arrow points based on ethnographic analogy and because they were smaller than older archaeological points (Charlin and González-José 2012). Based on observed similarities between the type V points and Ona arrows, Bird maintained that the Ona had inhabited Southern Patagonia in previous times, a claim strongly questioned afterward (Borrero 1989–1990; however, see Goñi 2013). It is worth mentioning that until ca. 8,000 years ago, the island of Tierra del Fuego was intermittently connected to the continent by a land bridge that today is an interoceanic passage, the Strait of Magellan (McCulloch et al. 1997). This situation persisted until the start of the marine incursion around 8,300–7,500 years ago (McCulloch et al. 2005). Therefore, the populations inhabiting Tierra del Fuego and southern continental Patagonia shared a common ancestry (González-José et al. 2004; Lalueza, Pérez-Pérez, Prats, Cornudella, and Turbón

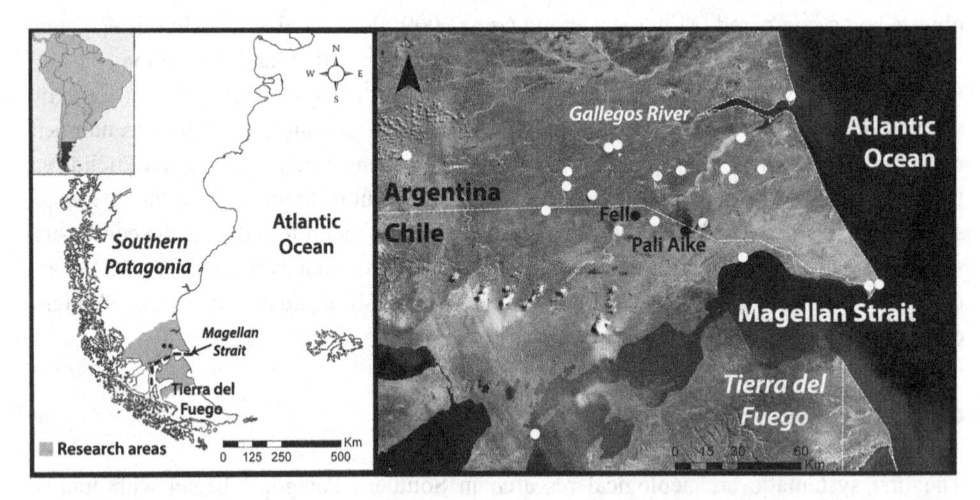

Figure 7.2
Maps showing Southern Patagonia and locations of projectile-point samples included in this study.

1997). Even though the early archaeological evidence on the island is scarce, it shows that the projectile-point technology—“fishtail” points—and the subsistence economy were the same on both sides of the Magellan Strait (Bird 1988; Massone 1987; Nami 1985–1986; Prieto 1991).

Later cultural evolution north and south of the Magellan Strait is still a topic of debate, and different models have been proposed (Borrero 1989–1990; Goñi 2013). From a biogeographical perspective and using a vicariance model, Borrero (1989–1990) noted the development of a cultural divergence of the southern populations after the formation of the Magellan Strait. This model predicts genetic and cultural changes in the populations on both sides of the strait over a long temporal scale. In contrast, a new migration model (Goñi 2013) has been proposed that endorses the similarity in cultural traits between the south of the continent and the north of the island, as proposed earlier by Bird.

Several works citing different evidence support the divergence hypothesis (Béguelin and Barrientos 2006; González-José et al. 2004; Morello et al. 2012). Here it is worth noting that previous studies comparing the composition of stone-tool assemblages (Cardillo et al. 2015) and the size and shape of late projectile points (Charlin et al. 2013) from southern continental Patagonia and northern Tierra del Fuego have found statistical differences. However, direct comparison between type V points from the mainland and the ethnographic Ona arrows from Tierra del Fuego has yet to be performed. For this reason, a sample of 50 Ona arrows has been included here with the aim of testing the existence of parallelism.

Materials and Methods

Our sample comprises 237 archaeological and ethnographic points. The archaeological sample comprises 187 Southern Patagonia stemmed projectile points from the Late Holocene (3,500–200 years ago) assigned to types IV and V (de Azevedo, Charlin, and González-José 2014). The ethnographic sample comprises 50 Ona arrowheads from the late nineteenth and early twentieth centuries (Charlin, Augustat, and Urban 2016). All projectile points were complete or exhibited only slight damage.

Digital Dataset

The digital sample of projectile points comprises standardized photographs and scanned published illustrations and photographs from the literature. All images were kept constant in their orientation and size (100 dpi). Raw images were compiled in tpsUtil (version 1.68; Rohlf 2016a), and the scaling and location of morphometric points was done in tpsDig2 (version 2.25; Rohlf 2016b).

Measurement of Reduction

To control the effects of reduction on the shape of projectile points, the following variables were measured using interlandmark Euclidean distances and trigonometry: tip angle, mean angle of the shoulders (both in plain view), and blade and stem length. The latter two were used to compute the blade length to stem length ratio, a reduction index for stemmed projectile points (its values are inversely proportional to reduction [Iriarte 1995]). All interlandmark Euclidean distances were measured using the PAST program (version 2.17c, Hammer, Harper, and Ryan 2001), and reduction variables were imported as covariates into MorphoJ (version 1.06d; Klingenberg 2011).

Size and Shape Variables

Twenty-one discrete points (seven landmarks and 14 semilandmarks) were digitized on the outline contour of each projectile point (figure 7.3). We reduced the number of semilandmarks on the stem used in this study relative to previous studies in order to include the ethnographic sample. These specimens are hafted, but fortunately in all cases the stem is larger than the attached shaft, and thus a small portion of the points' lateral margins are exposed. To capture the stem form without bias, we reduced the number of semilandmarks on the stem from seven to four.

Semilandmarks were slided using the minimum bending-energy criterion (Bookstein 1996–1997) in tpsRelw (version 1.62; Rohlf 2016c). Landmark configurations were then superimposed using a generalized Procrustes analysis (Rohlf and Slice 1990). Size was calculated as the centroid size, which is the square root of the summed square distances between each landmark coordinate and the centroid (Zelditch et al. 2004).

Multivariate Statistical Analyses

Principal component analysis (PCA) on Procrustes coordinates (covariance matrix) was performed in order to visualize main shape variation among the three groups of points. Canonical variate analysis (CVA) on shape coordinates (pooled within-group covariance matrix) was performed to find the shape features that best distinguished groups. The existence of mean shape differences was tested by permutation (10,000 iterations) using Mahalanobis distance. Both procedures were done in MorphoJ.

CVA was repeated in PAST in order to obtain the percentages of misclassification (not given by MorphoJ) and to identify the predicted group allocation by specimen. Each specimen was assigned to the group with the minimal Mahalanobis distance to the group mean. Group assignment was cross-validated by a leave-one-out (jackknifing) procedure (Hammer 2017). Finally, the percentage of specimens correctly allocated to their original group by the discriminant function and the jackknifing procedure was calculated. We used the percentage of misclassification given by the jackknifing procedure as a proxy of shape convergence at the assemblage level.

Figure 7.3
Location of landmarks and semilandmarks in the contour of projectile points. The large dots are landmarks, and the small dots are semilandmarks. The specimen shown is an Ona point (no. 1974) in the Mallman Collection at the Ethnologisches Museum, Berlin.

To assess the existence of correlation among shape, size, and reduction variables, we computed a multivariate regression of shape on centroid size, tip angle (TA), shoulders mean angle (SMA), and blade/stem length index (IBS) using MorphoJ (analyses not shown). After this first evaluation, and according to the relationship between shape and each independent variable, centroid size was left aside as an explanatory variable, given that the percentage of explained shape variance was low (6 percent).

In order to reduce variation and sum the behavior of the reduction variables in a single one (the reduction component), we performed a PCA on TA, SMA, and IBS. Afterward, a simple regression between the first shape principal component (PC) and the reduction component was carried out to examine the cases of shape convergence in relation with the previous results obtained from CVA.

Finally, reduction trajectories by group were independently estimated by means of ordinary least-squares regression to then compare the reduction trajectories among-groups through the regression slopes. These latter analyses were done using R (R Core Development Team 2005).

Results

The first two PCs explain 88.2 percent of the total shape variation in the point sample (figure 7.4). PC1 accounts for 76.2 percent of the variation and shows shape changes from points characterized by elongated and narrow blades with smaller stems and acute angles in the shoulders and the tip in the positive scores to points with shorter and wider blades,

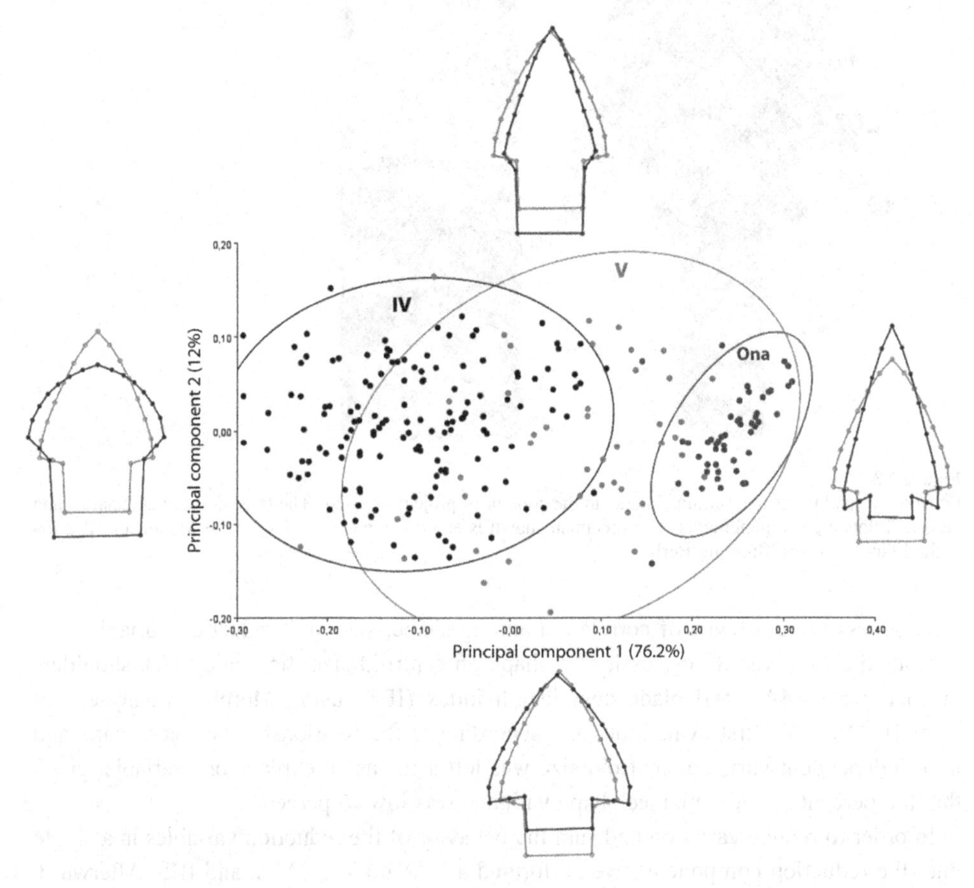

Figure 7.4
Principal component analysis on Procrustes coordinates. Gray point outlines on the margins represent consensus shape, and black point outlines represent target shape according to maximun values in each axis.

rounded tips, shoulders with obtuse angles, and bigger stems in the negative scores. On this axis, the three groups of points, although overlapping, show a spatial arrangement from right to left, with Ona points having the highest positive values and the type IV points having the highest negative values. The second PC explains 12 percent of the shape variation and shows a combination of the previous patterns: points with elongated and narrow blades but with bigger stems in the positive values and points with shorter and wider blades but smaller stems in the negative ones.

Even though there is overlap in the samples, especially between types IV and V, a permutation test (10,000 rounds) showed the existence of mean shape differences among the three groups of points based on Mahalanobis distance (figure 7.5; table 7.1). As a consequence, the first conclusion is that the claimed shape similarity between the ethnographic Ona points from Tierra del Fuego and their counterparts in the archaeological record of southern continental Patagonia, the type V points, is not supported.

To know how well this differentiation in groups works, we repeated this analysis in PAST to obtain a misclassification table from both the linear classifier and a cross-validation procedure (jackknifing). According to the linear classifier, 93.25 percent of cases were correctly allocated to the a priori defined group (table 7.2), whereas the cross-validation procedure correctly classified a lower percentage of 85.23 percent (table 7.3).

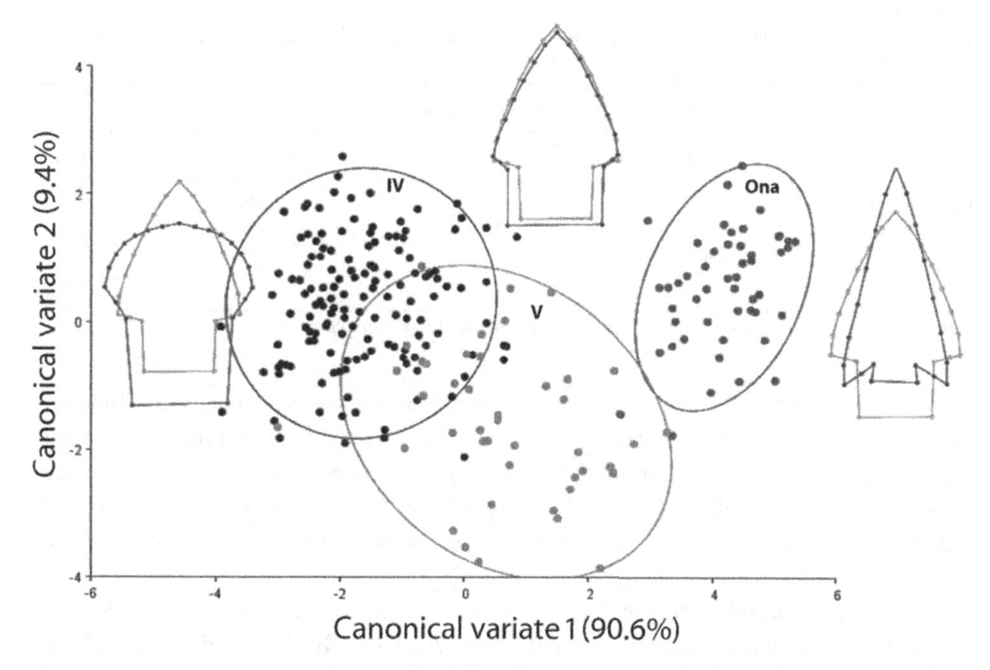

Figure 7.5
Canonical variate analysis on Procrustes coordinates. Gray point outlines on the margins represent consensus shape, and black point outlines represent target shape.

Table 7.1
Mahalanobis distance among groups and *p*-values from permutation tests (10,000 rounds) in parentheses

Groups	IV	Ona
Ona	5,84 (< .0001)	
V	2,98 (< .0001)	4,10 (< .0001)

Table 7.2
Frequencies and percentages (between parentheses) of misclassification from the discriminant function

	Predicted groups from linear classifier			
Given groups	**Ona**	**IV**	**V**	**Total**
Ona	49	0	1	50
IV	0	134 (93%)	10 (7%)	144
V	0	5 (12%)	38 (88%)	43
Total	49	139	49	237

Confusion matrix: rows = given groups; columns = predicted groups from classifier.

Table 7.3
Frequencies and percentages (between parentheses) of misclassification from the cross-validation procedure (jackknifing)

	Predicted groups from cross-validation			
Given groups	**Ona**	**IV**	**V**	**Total**
Ona	48	0	2 (4%)	50
IV	0	129 (90%)	15 (10%)	144
V	4 (9%)	14 (33%)	25 (58%)	43
Total	52	143	42	237

Confusion matrix: rows = given groups; columns = predicted groups from classifier.

In both cases, the type V points are those that show the highest percentages of misclassification (12 percent in the first case and 42 percent in the second). From the discriminant function analysis, they were confused only with type IV points, but from the jackknifing procedure, 33 percent of the cases were confused with the type IV points and 9 percent with the Ona points. Based on the last procedure, we can estimate 15 percent of shape convergence in the sample.

The reduction variables show their maximum correlation with the first PC (the reduction PC; table 7.4), which explains 83 percent of the variation. IBS is negatively correlated with the first PC (its values are inversely proportional to reduction), whereas both TA and

Table 7.4
Correlation among reduction variables and principal components

Loadings	PC 1	PC 2	PC 3
IBS	-0.92	0.18	0.34
TA	0.91	-0.25	0.31
SMA	0.89	0.44	0.034

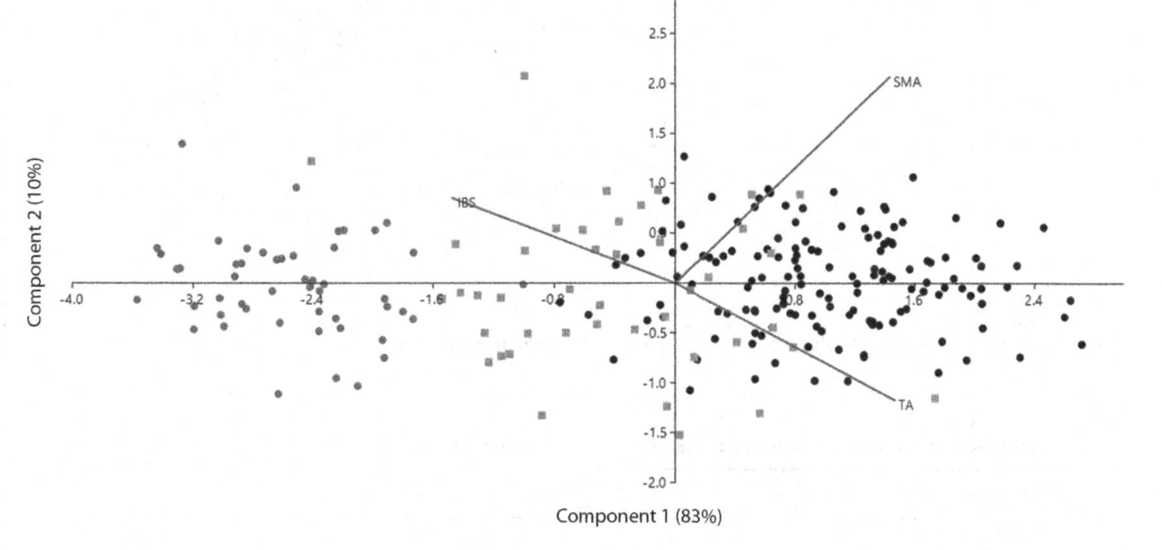

Figure 7.6
Principal component analysis on reduction variables. Black dots represent type IV points, gray squares represent type V points, and gray dots represent Ona points.

SMA are positively correlated. Thus, positive scores for the first PC are occupied by more reduced points, whereas fewer reduced points occupy the negative values (figure 7.6).

To know if shape convergence is related to reduction, we performed a least-squares regression of the shape PC on the reduction PC scores (figure 7.7). The results show that 95.66 percent of shape changes are predicted by reduction (table 7.5). The permutation test of the null hypothesis of independence between shape and reduction shows a significant value ($p < 0.0001$; number of randomization rounds: 10,000). The Ona points are the least-reduced assemblage, which is expected, considering they come from museum collections comprising items probably made for trade or bartering with colonialists and thus were unused. On the other hand, the type IV points are the most-reduced assemblage, and the type V points are in the middle of both groups. The overlap of the different types

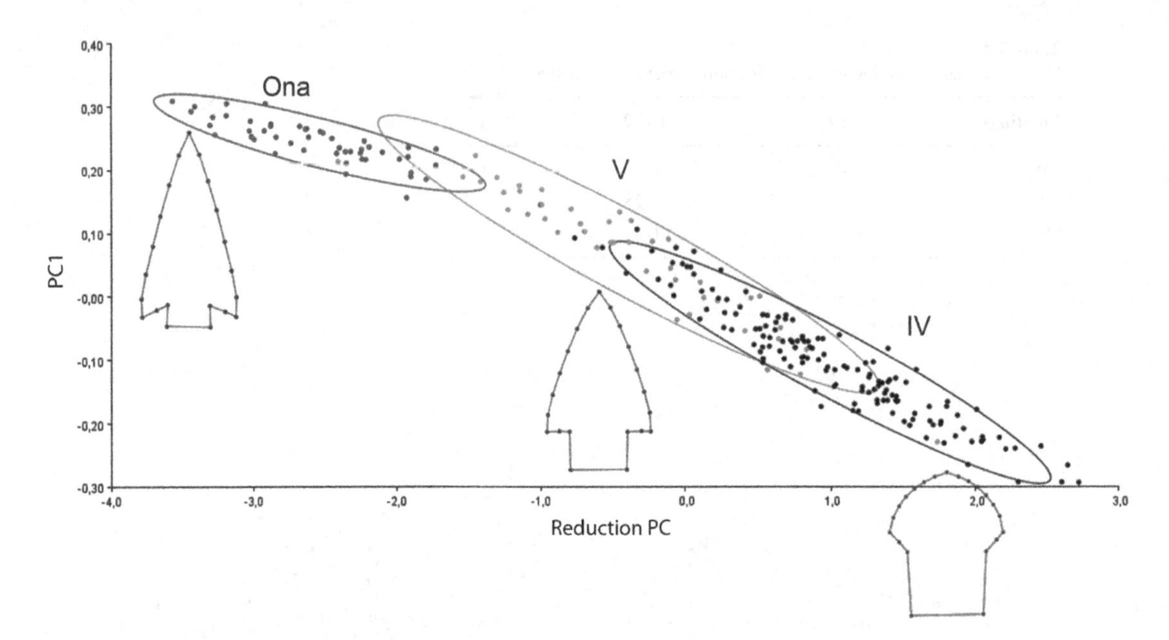

Figure 7.7
Simple regression of first shape principal component on reduction component.

Table 7.5
Sum of squares (SS) in the regression analysis of shape on reduction PC

Sum of squares	
Total SS	6.36
Predicted SS	6.08
Residual SS	0.27

of points shows that the less-reduced type V points converge with Ona points. The graph depicts a lower percentage of overlap between Ona and type V points and a larger overlap between the two archaeological types, as shown by the misclassification percentage of the jackknifing procedure.

Finally, linear-reduction trajectories were estimated by means of ordinary least-squares regression for each point class separately. Then, differences among slope coefficients and the pooled standard error of the slope were compared pairwise, and the null hypothesis of similarity in trajectories was tested. In figure 7.8, the slopes are plotted in the same Cartesian coordinate reference space.

Results suggest that type IV and V points follow almost the same allometric trajectories ($F = 0.08$, $p = 0.79$), and both trajectories are quite different from Ona points ($F = 53$,

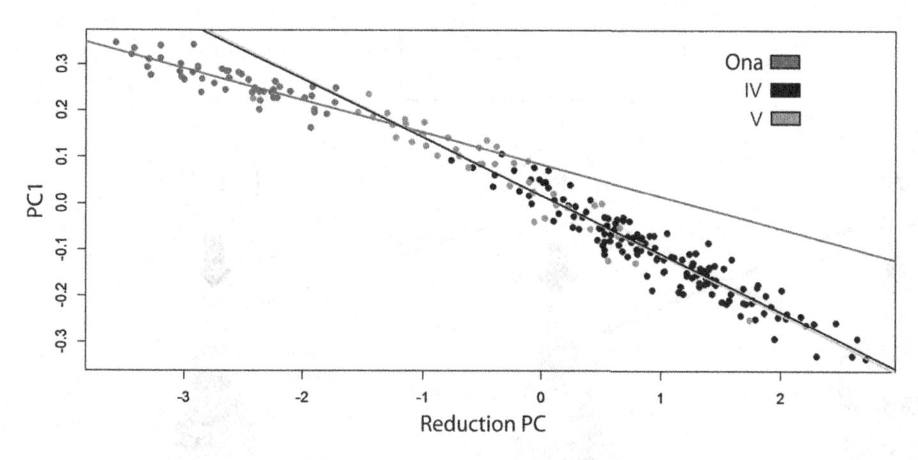

Figure 7.8
Comparison of reduction trajectories among groups of points through least-square regression.

$p < 0.001$). Thus, shape convergence between both archaeological types is attributable to the same allometric trajectories.

Discussion

Geometric morphometric analyses applied on a sample of 237 archaeological and ethnographic projectile points from Late Holocene Southern Patagonia showed that tool reduction accounts for most shape variation. Applying McGhee's model (2011, 2015) to stone tools, we suggest reduction is the interplay between developmental and functional constraints on projectile points. Therefore, reduction constraints impose some limitations in phenotypic variability, leading to shape convergence among different types of projectile points, especially between types IV and V, although according to typological (Bird 1988) and morpho-functional analyses (González-José and Charlin 2012; Ratto 1994), they were used in different weapon systems. In this way, convergence can be seen as a by-product of the ontogenetic trajectory of stone tools during their use-lives.

Based on shape comparisons, we can reject the assumed shape similarity between ethnographic Ona arrowheads from Tierra del Fuego and archaeological type V projectile points from the mainland. Ona arrowheads show a distinct reduction trajectory that could be explained in two alternative ways. On the one hand, we can assume the reduction trajectory of Ona points is not completely reflected in the analyzed sample, given that it comes from museum collections dating to the late nineteenth and early twentieth centuries, when the production of tools for trade or bartering with colonialists was common (see Borrero and Franco 2001; Harrison 2006; Torrence 2000). If this is the case, the constraints on ethnographic points were very different from those that were involved in the performance

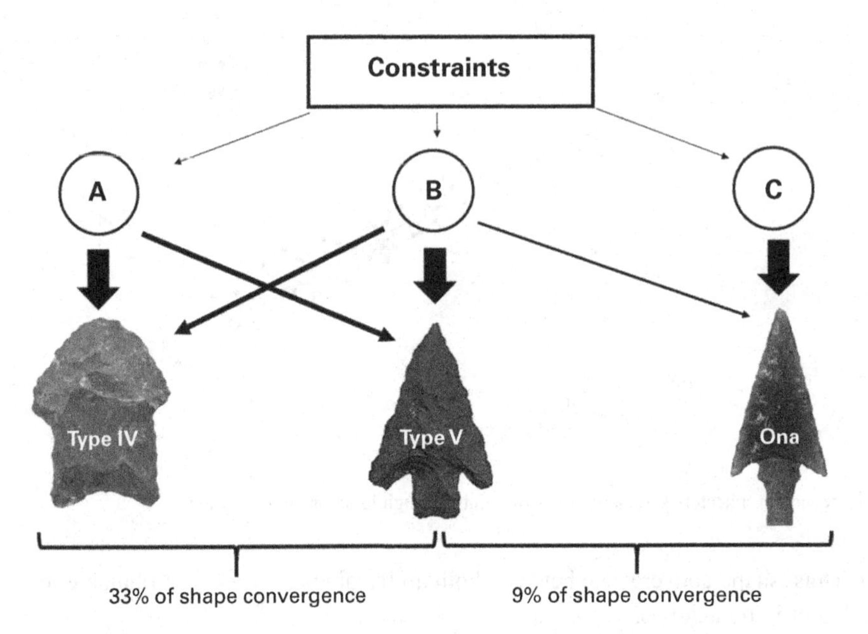

Figure 7.9
Constraints model and the case of Late Holocene projectile points of Southern Patagonia. The relative width of arrows indicates degree of constraint intensity.

of hunting weapons. On the other hand, and following the proposal of Iovită (2009, 2010), the few changes in shape related to reduction in the Ona sample can be understood as an isometric trajectory of change—that is, a shape-conserving trajectory. In contrast, types IV and V show the typical example of allometric trajectories.

Coming back to the constraints model (figure 7.1), we can conclude that different constraints operated on archaeological and ethnographic points, leading to different ontogenetic trajectories. The level or degree of convergence could be related with the quantity or intensity of constraints (figure 7.9). In this way, we can identify three theoretical situations: one in which points share the same constraints but have different intensity (the case of point types IV and V, which have 33 percent shape convergence); a second in which they share a lower number of constraints (the case of point types V and Ona, with 9 percent shape convergence); and a third in which the points do not share any constraints and diverge absolutely (exemplified by point types IV and Ona).

Conclusion

Our analysis of Late Holocene projectile points estimated a 15 percent rate of shape convergence through reduction. Further analyses that focus on a larger time scale will be useful for controlling the occurrence of shape convergence by reduction along longer sequences. In this way, we will be able to estimate its effects and variation in a comparative framework. Taking advantage of the output of GM methods for shape analysis, together with experimental studies, archaeologists should be able to start to assess how much phenotypic variability is the result of selection and how much is channeled by reduction and other constraints, similar to what many biological studies have already done (Brakefield 2006; Sommer 1999).

In a phylogenetic context, ontogenetic constraints that generate allometric trajectories within a class may have a hereditary component and act as heterochronies (McKinney and McNamara 1991). These "microevolutionary processes" would generate differences in the timing of life-history events in different classes, in relative independence from other constraints, and thus generate macroevolutionary patterns. Conversely, if purely adaptive or functional constraint is the main force, convergence and parallelism is expected. To explore these processes, reduction measured from different indices could be taken as a comparative variable within a phylogenetic context, by means of an independent-contrasts approach or phylogenetic regression models (Harvey and Pagel 1991; Revell 2010). We expect to find null allometric slopes for a set of historically related classes in such cases. Beyond the potential role of these processes in particular cases, as we discuss here, understanding the patterns of convergence through allometric models can provide us with valuable information to understand the process of evolution in a macroevolutionary scale.

References

Béguelin, M., & Barrientos, G. (2006). Variación Morfométrica Postcraneal en Muestras Tardías de Restos Humanos de Patagonia: Una Aproximación Biogeográfica. *Intersecciones en Antropología*, *7*, 49–62.

Binford, L. R. (1963). "Red Ocher" Caches from the Michigan Area: A Possible Case of Cultural Drift. *Southwestern Journal of Anthropology*, *19*, 89–109.

Bird, J. (1938). Antiquity and Migrations of the Early Inhabitants of Patagonia. *Geographical Review*, *28*, 250–275.

Bird, J. (1946). The Archaeology of Patagonia. In J. H. Steward (Ed.), *Handbook of South American Indians: The Marginal Tribes* (pp. 17–24). Washington, D.C.: Smithsonian Institution, Bureau of American Ethnology.

Bird, J. (1988). *Travels and Archaeology in South Chile*. Iowa City: University of Iowa Press.

Bookstein, F. L. (1991). *Morphometric Tools for Landmark Data. Geometry and Biology*. New York: Cambridge University Press.

Bookstein, F. L. (1996–1997). Landmarks Methods for Form without Landmarks: Morphometrics of Group Differences in Outline Shape. *Medical Image Analysis*, *3*, 225–243.

Borrero, L. A. (1989–1990). Evolución Cultural Divergente en la Patagonia Austral. *Anales del Instituto de la Patagonia*, *19*, 133–140.

Borrero, L. A. (1998). *Arqueología de la Patagonia Meridional*. Buenos Aires: Búsqueda de Ayllu.

Borrero, L. A. (2001a). Cambios, Continuidades, Discontinuidades: Discusiones sobre Arqueología Fuego-Patagónica. In E. Berberián & A. Nielsen (Eds.), *Historia Argentina Prehispánica* (pp. 815–838). Córdoba, Argentina: Editorial Brujas.

Borrero, L. A. (2001b). *Los Selk'nam (Onas). Evolución Cultural en Tierra del Fuego*. Buenos Aires: Galerna-Búsqueda de Ayllu.

Borrero, L. A., & Franco, N. (2001). Las Colecciones Líticas del Museo Británico. *Anales del Instituto de la Patagonia, 29*, 207–210.

Brakefield, P. (2006). Evo-Devo and Constraints on Selection. *Trends in Ecology & Evolution, 21*, 362–368.

Buchanan, B., M. Collard, M. J. Hamilton, & M. J. O'Brien. (2011). Points and Prey: A Quantitative Test of the Hypothesis That Prey Sizes Influences Early Paleoindian Projectile Point Form. *Journal of Archaeological Science, 38*, 852–864.

Buchanan, B., Eren, M. I., Boulanger, M. T., & O'Brien, M. J. (2015). Size, Shape, Scars, and Spatial Patterning: A Quantitative Assessment of Late Pleistocene (Clovis) Point Resharpening. *Journal of Archaeological Science: Reports, 3*, 11–21.

Cardillo, M. (2010). Some Applications of Geometric Morphometrics to Archaeology. In A. Elewa (Ed.), *Morphometrics for Nonmorphometricians* (pp. 325–355). Heidelberg, Germany: Springer.

Cardillo, M., & Charlin, J. (2016). Morphological Diversification of Stemmed Projectile Points of Patagonia (Southernmost South America): Assessing Spatial Patterns by Means of Phylogenies and Comparative Methods. In E. Delson & E. Sargis (Eds.), *Multidisciplinary Approaches to the Study of Stone Age Weaponry* (pp. 261–272). New York: Springer.

Cardillo, M., Charlin, J., & Borrazzo, K. (2015). Artifactual and Environmental Related Variations in Fuego-Patagonia. In M. J. Shott (Ed.), *Works in Stone: Contemporary Perspectives on Lithic Analysis* (pp. 162–177). Salt Lake City: University of Utah Press.

Cavalli-Sforza, L. L., & Feldman, M. W. (1981). *Cultural Transmission and Evolution: A Quantitative Approach*. Princeton, N.J.: Princeton University Press.

Chapman, A. (1986). *Los Selk'nam. La Vida de los Onas*. Buenos Aires: Emecé Editores. (First published 1982.)

Charlin, J., & Borrero, L. A. (2012). Rock Art, Inherited Landscapes and Human Populations in Southern Patagonia. In J. McDonald & P. Veth (Eds.), *A Companion to Rock Art* (pp. 381–398). Chichester, England: Wiley-Blackwell.

Charlin, J., & González-José, R. (2012). Size and Shape Variation in Late Holocene Projectile Points of Southern Patagonia: A Geometric Morphometric Study. *American Antiquity, 77*, 221–242.

Charlin, J., Borrazzo, K., & Cardillo, M. (2013). Exploring Size and Shape Variations in Late Holocene Projectile Points from Northern and Southern Coasts of Magellan Strait (South America). In F. Djinjian & S. Robert (Eds.), *Understanding Landscapes, from Land Discovery to Their Spatial Organization* (pp. 39–50). BAR International Series, no. 2541. Oxford: Archaeopress.

Charlin, J., Cardillo, M., & Borrazzo, K. (2014). Spatial Patterns in Late Holocene Lithic Projectile Point Technology of Tierra del Fuego (Southern South America): Assessing Size and Shape Changes. *World Archaeology, 46*, 78–100.

Charlin, J., Augustat, C., & Urban, C. (2016). Metrical Variability in Ethnographic Arrows from Southernmost Patagonia: Comparing Collections from Tierra del Fuego at European Museums. *Journal of Anthropological Archaeology, 41*, 313–326.

Clarkson, C. (2013). Measuring Core Reduction Using 3D Flake Scar Density: A Test Case of Changing Core Reduction at Klasies River Mouth, South Africa. *Journal of Archaeological Science, 40*, 4348–4357.

De Azevedo, S., Charlin, J., & González-José, R. (2014). Identifying Design and Reduction Effects on Lithic Projectile Point Shapes. *Journal of Archaeological Science*, *41*, 297–307.

Dibble, H. L. (1984). Interpreting Typological Variation of Middle Paleolithic Scrapers: Function, Style, or Sequence of Reduction? *Journal of Field Archaeology*, *11*, 431–436.

Dibble, H., Schurmans, U., Iovită, R., & McLaughlin, M. (2005). The Measurement and Interpretation of Cortex in Lithic Assemblages. *American Antiquity*, *70*, 545–560.

Dunnell, R. C. (1978). Style and Function: A Fundamental Dichotomy. *American Antiquity*, *43*, 192–202.

Eren, M. I., Domínguez-Rodrigo, M., Kuhn, S. L., Adler, D. S., Le, I., & Bar-Yosef, O. (2005). Defining and Measuring Reduction in Unifacial Stone Tools. *Journal of Archaeological Science*, *32*, 1190–1201.

Franco, N. 2002. *Estrategias de Utilización de Recursos Líticos en la Cuenca Superior del Río Santa Cruz*. Ph.D. dissertation, University of Buenos Aires. Buenos Aires.

Goñi, R. 2013. Reacomodamientos Poblacionales de Momentos Históricos en el Noroeste de Santa Cruz. Proyecciones Arqueológicas. In A. Zangrando, R. Barberena, A. Gil, G. Neme., M. Giardina, L. Luna, C. Otaola, et al. (Eds.), *Tendencias Teórico-Metodológicas y Casos de Estudio en la Arqueología de la Patagonia* (pp. 389–396). San Rafael, Argentina: Museo de Historia Natural de San Rafael, San Rafael, Argentina.

González-José, R., & Charlin, J. (2012). Relative Importance of Modularity and Other Morphological Attributes on Different Types of Lithic Point Weapons: Assessing Functional Variations. *PLoS One*, *7*(10), e48009.

González-José, R., Martínez-Abadías, N., van der Molen, S., García-Moro, C., Dahinten, S., & Hernández, M. (2004). Hipótesis acerca del Poblamiento de Tierra del Fuego–Patagonia a partir del Análisis Genético-Poblacional de la Variación Craneofacial. *Magallania*, *32*, 79–98.

Gould, S. J. (1989). A Developmental Constraint in Cerion, with Comments on the Definition and Interpretation of Constraint in Evolution. *Paleobiology*, *43*, 516–539.

Gunz, P., & Mitteroecker, P. (2013). Semilandmarks: A Method for Quantifying Curves and Surfaces. *Hystrix*, *24*, 103–106.

Gunz, P., Mitteroecker, P., & Bookstein, F. L. (2005). Semilandmarks in Three Dimensions. In D. Slice (Ed.), *Modern Morphometrics in Physical Anthropology* (pp. 73–98). Chicago: University of Chicago Press.

Hammer, Ø. (2017). *Paleontological Statistics version 3.15. Reference Manual.* http://folk.uio.no/ohammer/past/past3manual.pdf.

Hammer, Ø., Harper, D. A. T., & Ryan, P. D. (2001). PAST. Paleontological Statistics Software Package for Education and Data Analysis. *Palaeontologia Electronica*, *4*, 9.

Harrison, R. (2006). An Artefact of Colonial Desire? Kimberley Points and the Technologies of Enchantment. *Current Anthropology*, *47*, 63–88.

Harvey, P. H., & Pagel, M. D. (1991). *The Comparative Method in Evolutionary Biology*. Oxford: Oxford University Press.

Hughes, S. S. (1998). Getting to the Point: Evolutionary Change in Prehistoric Weaponry. *Journal of Archaeological Method and Theory*, *5*, 345–408.

Iovită, R. (2009). Ontogenetic Scaling and Lithic Systematics: Method and Application. *Journal of Archaeological Science*, *36*, 1447–1457.

Iovită, R. (2010). Comparing Stone Tool Resharpening Trajectories with the Aid of Elliptical Fourier Analysis. In S. J. Lycett & P. Chauhan (Eds.), *New Perspectives on Old Stones: Analytical Approaches to Palaeolithic Technologies* (pp. 235–253). New York: Springer.

Iriarte, J. (1995). Afinando la Puntería: Tamaño, Forma y Rejuvenecimiento en las Puntas de Proyectil Pedunculadas del Uruguay. In M. Consens, J. M. López Mazz, & M. Del Carmen Curbelo (Eds.), *Arqueología en el Uruguay: 120 Años despues* (pp. 142–151). Montevideo: Editorial SURCOS.

Klingenberg, C. (2011). MorphoJ: An Integrated Software Package for Geometric Morphometrics. *Molecular Ecology Resources*, *11*, 353–357.

Kuhn, S. L. (1990). A Geometric Index of Reduction for Unifacial Stone Tools. *Journal of Archaeological Science*, *17*, 583–593.

Lalueza, C., Pérez-Pérez, A., Prats, E., Cornudella, L., & Turbón, D. (1997). Lack of Founding Amerindian Mitochondrial DNA Lineages in Extinct Aborigines from Tierra del Fuego–Patagonia. *Human Molecular Genetics*, *6*, 41–46.

Lycett, S. J. (2007). Why Is There a Lack of Mode 3 Levallois Technologies in East Asia? A Phylogenetic Test of the Movius-Schick Hypothesis. *Journal of Anthropological Archaeology*, *26*, 541–575.

Lycett, S. J. (2009). Are Victoria West Cores "Proto-Levallois"? A Phylogenetic Assessment. *Journal of Human Evolution*, *59*, 175–191.

Lycett, S. J., von Cramon-Taubadel, N., & Gowlett, J. (2010). A Comparative 3D Geometric Morphometric Analysis of Victoria West Cores: Implications for the Origins of Levallois Technology. *Journal of Archaeological Science*, *37*, 1110–1117.

Massone, M. (1987). Los Cazadores Paleoindios de Tres Arroyos (Tierra del Fuego). *Anales del Instituto de la Patagonia*, *17*, 47–60.

McCulloch, R., Clapperton, C., Rabassa, J., & Currant, A. (1997). The Natural Setting: The Glacial and Post-Glacial Environmental History of Fuego-Patagonia. In C. McEwan, L. A. Borrero, & A. Prieto (Eds.), *Patagonia* (pp. 12–31). London: British Museum Press.

McCulloch, R., Bentley, M. J., Tipping, R. M., & Clapperton, C. (2005). Evidence for Late-Glacial Ice Dammed Lakes in the Central Strait of Magellan and Bahía Inútil, Southernmost South America. *Geografiska Annaler*, *87A*, 335–362.

McGhee, G. R. (1999). *Theoretical Morphology: The Concept and Its Applications*. New York: Columbia University Press.

McGhee, G. R. (2011). *Convergent Evolution: Limited Forms Most Beautiful*. Cambridge, Mass.: MIT Press.

McGhee, G. R. (2015). Limits in the Evolution of Biological Form: A Theoretical Morphologic Perspective. *Interface Focus*, *5*, 20155534.

McKinney, M. L., & McNamara, K. J. (1991). *Heterochrony: The Evolution of Ontogeny*. New York: Plenum.

Mitteroecker, P., & Gunz, P. (2009). Advances in Geometric Morphometrics. *Evolutionary Biology*, *36*, 235–247.

Mitteroecker, P., & Hutteger, S. (2009). The Concept of Morphospaces in Evolutionary and Developmental Biology: Mathematics and Metaphors. *Biological Theory*, *4*, 54–67.

Morales, J. I., Soto, M., Lorenzo, C., & Vergès, J. M. (2015). The Evolution and Stability of Stone Tools: The Effects of Different Mobility Scenarios in Tool Reduction and Shape Features. *Journal of Archaeological Science: Reports*, *3*, 295–305.

Morello, F., Borrero, L. A., Massone, M., Stern, C., Garcia-Herbst, A., McCulloch, R., et al. (2012). Hunter-Gatherers, Biogeographic Barriers and the Development of Human Settlement in Tierra del Fuego. *Antiquity*, *86*, 71–87.

Nami, H. (1985–1986). Excavaciones Arqueológicas y Hallazgo de una Punta de Proyectil "Fell I" en la "Cueva del Medio" Seno de Ultima Esperanza, Chile. *Anales del Instituto de la Patagonia*, *16*, 103–110.

Nami, H. (1992). Noticia Sobre la Existencia de Técnica "Levallois" en Península Mitre, Extremo Sudoriental de Tierra del Fuego. *Anales del Instituto de la Patagonia, 21*, 73–80.

Nelson, M. (1991). The Study of Technological Organization. *Journal of Archaeological Method and Theory, 3*, 57–100.

Oswalt, W. H. (1976). *An Anthropological Analysis of Food-Getting Technology*. New York: Wiley.

Perez, S. I., Bernal, V., & González, P. (2006). Differences between Sliding Semi-landmark Methods in Geometric Morphometrics, with an Application to Human Craniofacial and Dental Variation. *Journal of Anatomy, 208*, 769–784.

Perez, S. I., Lema, V., Diniz-Filho, J. A. F., Bernal, V., González, P. N., Gobbo, D., et al. (2011). The Role of Diet and Temperature in Shaping Cranial Diversification of South American Human Populations: An Approach Based on Spatial Regression and Divergence Rate Tests. *Journal of Biogeography, 38*, 148–163.

Prieto, A. (1991). Cazadores Tempranos y Tardíos en la Cueva Lago Sofía 1. *Anales del Instituto de la Patagonia, 20*, 75–100.

R Development Core Team. (2005). *R: A Language and Environment for Statistical Computing, Reference Index Version 2.2.1*. R Foundation for Statistical Computing, Vienna, Austria. http://www.R-project.org.

Ratto, N. (1994). Funcionalidad vs. Adscripción Cultural: Cabezales Líticos de la Margen Norte del Estrecho de Magallanes. In J. L. Lanata & L. A. Borrero (Eds.), *Arqueología de Cazadores—Recolectores: Límites, Casos y Aperturas* (pp. 105–120). Buenos Aires: Programa de Estudios Prehistóricos.

Ratto, N. 2003. *Estrategias de Caza y Propiedades de Registro Arqueológico en la Puna de Chaschuil (Dpto. de Tinogasta, Catamarca, Argentina)*. Ph.D. dissertation, University of Buenos Aires. Buenos Aires.

Revell, L. J. (2010). Phylogenetic Signal and Linear Regression on Species Data. *Methods in Ecology and Evolution, 1*, 319–329.

Rohlf, F. J. (1990). Rotational Fit (Procrustes) Methods. In F. J. Rohlf & F. L. Bookstein (Eds.), *Proceedings of the Michigan Morphometics Workshop* (pp. 227–236). Special Publication, no. 2. Ann Arbor, Mich.: University of Michigan Museum of Zoology.

Rohlf, F. J. (1999). Shape Statistics: Procrustes Superimpositions and Tangent Spaces. *Journal of Classification, 16*, 197–223.

Rohlf, F. J. (2016a). *Tps Utility Program Version 1.68*. Department of Ecology and Evolution, Stony Brook University, Stony Brook, New York.

Rohlf, F. J. (2016b). *TpsDig2 Version 2.25*. Department of Ecology and Evolution, Stony Brook University, Stony Brook, New York.

Rohlf, F. J. (2016c). *Relative Warps Version 1.62*. Stony Brook, New York: Department of Ecology and Evolution, Stony Brook University.

Rohlf, F. J., & Bookstein, F. L. (Eds.). (1990). *Proceedings of the Michigan Morphometrics Workshop*. Special Publication, no. 2. Ann Arbor, Mich.: University of Michigan Museum of Zoology.

Rohlf, F. J., & Slice, D. (1990). Extensions of the Procrustes Method for the Optimal Superimposition of Landmarks. *Systematic Zoology, 39*, 40–59.

Shott, M. J. (2005). The Reduction Thesis and Its Discontents: Overview of the Volume. In C. Clarkson & L. Lamb (Eds.), *Lithics "Down Under": Recent Australian Approaches to Lithic Reduction, Use and Classification* (pp. 109–125). BAR International Series, no. 1408. Oxford: BAR Publishing.

Shott, M. J., & Trail, B. W. (2010). Exploring New Approaches to Lithic Analysis: Laser Scanning and Geometric Morphometrics. *Lithic Technology, 35*, 195–220.

Shott, M. J., Hunzicker, D. A., & Patten, B. (2007). Pattern and Allometric Measurement of Reduction in Experimental Folsom Bifaces. *Lithic Technology*, *32*, 203–217.

Sommer, R. (1999). Convergence and the Interplay of Evolution and Development. *Evolution & Development*, *1*, 8–10.

Torrence, R. (2000). Just Another Trader? An Archaeological Perspective on European Barter with Admiralty Islanders, Papua New Guinea. In R. Torrence & A. Clarke (Eds.), *The Archaeology of Difference: Negotiating Cross-Cultural Engagements in Oceania* (pp. 104–141). London: Routledge.

Zelditch, M. L., Swiderski, D. L., Sheets, D. H., & Fink, W. L. (2004). *Geometric Morphometrics for Biologists: A Primer*. New York: Elsevier.

8 Convergence and Continuity in the Initial Upper Paleolithic of Eurasia

Steven L. Kuhn and Nicolas Zwyns

Most archaeologists recognize the potential for convergence and parallelism in lithic technology. Particular methods, techniques, and forms reappear too often in assemblages around the world to think otherwise. Forms of lithic technology such as pressure blade manufacture, bifacial shaping, and even Levallois occur time and time again, often in places widely separated in time and space.

If convergence is so likely in lithic technology, then why even take the trouble to discuss it? As one of the participants at the Konrad Lorenz Institute's 2016 Altenberg Workshop in Theoretical Biology commented, biologists treat convergence "like gravity." Sometimes its effects are obvious, sometimes they are not, but it is always there. The question for biologists is not whether or how often convergence happens, but whether it is the best explanation for a particular set of facts. The same has not been true for archaeologists. Instead, the history of the discipline is marked by pendular swings between broad acceptance of *either* convergence or phylogeny as the preferred explanations for similarities in artifact form and concomitant minimization of the other (chapter 1, this volume).

Here we begin with a brief and admittedly idiosyncratic review of the concept of convergence in lithic technology, why we should expect it to occur, and why archaeology sometimes has had trouble coming to terms with the phenomenon. We argue that *a priori* views on the likelihood of convergence have varied according to the broader ways archaeologists think about culture change. The notion of investigating convergence and phylogenetic connections as equally likely alternative explanations for cultural similarities is comparatively novel, and methods for addressing the problem are not well known or widely accepted. Using a project explicitly designed to test models of phylogeny and convergence as an example, the second part of this chapter underlines a few conceptual issues to be addressed in identifying convergence. With a broad but discontinuous distribution across Eurasia, early Upper Paleolithic assemblages known as Initial Upper Paleolithic are a good test case for applying alternative models contrasting diffusion, migration, and independent developments. The data collection is still underway, so this chapter focuses on strategies of assembling a database appropriate for phylogenetic analysis.

Why Convergence? (And Why Not?)

It is our impression, and the impression of many contemporary researchers, that the global record of stone artifacts is full of examples of convergence in both form and procedure. Not only is convergence apparent, it is also likely on theoretical grounds. A range of factors can promote convergence in lithic technology. The possible design space (*sensu* McGhee 2011; chapter 2, this volume; O'Brien et al. 2016) for lithic artifacts may be large, but it is neither infinite nor continuous (chapter 4, this volume). The range of practical methods for making usable blanks is limited, dictated by the fracture mechanics of isotropic, vitreous materials (e.g., Dibble and Pelcin 1995; Eren and Lycett 2012; Eren, Roos, Story, von Cramon-Taubadel, and Lycett 2014; Lin, Rezek, Braun, and Dibble 2013; Pelcin 1997a, 1997b). And although ranges of variation in particular methods may overlap, there are some parts of the design space where no one ventures, where it is impossible to produce useful artifacts (chapters 2 and 5, this volume). Fracture properties of stone also lead to a clear pattern of contingency in lithic technology. A decision early in the process will constrain options farther down the line. For instance, the initial choice to maximize sizes of flakes by removing them from the broadest part of a core inevitably requires further adjustments in how surfaces are shaped and how striking platform angles are maintained. Likewise, the range of shapes that are suitable for conducting specific tasks is also limited, although use-wear (e.g., Andrefsky 1997; Barton, Olszewski, and Coinman 1996; Dinnis, Pawlik and Gaillard 2009) and ethnographic evidence (White 1967; White and Thomas 1972) should remind us that these limitations are less stringent than we sometimes imagine.

Not all of the constraints on the shapes of artifacts and nature of manufacture processes are mechanical. Some methods may just be more easily learned than others. The degree to which learnability contributes to redundancy remains uncertain in other domains of human cultural transmission (Rafferty, Griffiths, and Ettlinger 2011, 2013; Smith 2011), although it is certainly the case that novices pick up some ways of working stone more readily than other ways. Some procedures or methods may also be more "evolvable," in that they are more easily derived from preexisting practices or from no practices at all. The persistence of generalized, simple Mode 1 core-and-flake technologies across Africa and Eurasia for several million years does not necessarily imply direct lineages of cultural transmission. Instead, the same forms and techniques could have emerged repeatedly and independently as hominins applied a very basic understanding of fracture mechanics to the manufacture of sharp-edged flakes (Tennie, Call, and Tomasello 2009). Likewise, the broad geographic association between Acheulean bifacial shaping and Levallois technology in the Middle Pleistocene, not to mention the morphological similarities between a Levallois core's flaking surface and one face of a large bifacially worked piece, has led some researchers to conclude that Levallois may have its roots in bifacial shaping (White and Ashton 2003), with the potential to emerge more than once from an existing base of biface production (chapter 13, this volume).

Although it makes sense to consider both phylogeny and convergence in explaining why artifacts made at different times and places resemble one another, for much of the field's history the "or" question seldom came up. Throughout the late nineteenth century and the first several decades of the twentieth century, cultural phylogeny was the default explanation for similarities in artifacts. There were exceptions (Steward 1929), but the majority of practitioners believed that resemblances between artifacts from different localities occurred mainly as the consequence of culture contact if not the movement of peoples. So-called hyperdiffusionists were prepared to draw connections between extraordinarily distant locales based on what they perceived as "improbable" similarities in language or material culture (Gladwin 1947; Smith 1929).

This situation changed radically in the 1960s with the appearance of the "New Archaeology," with its materialist, adaptationist focus and its roots in the neo-evolutionism of Marshall Sahlins, Elman Service, and others (e.g., Sahlins and Service 1960). Starting in the late 1960s, most North American archaeologists approached technology as an economic or tactical response to adaptive challenges. From this perspective, just about everything was an example of convergence or parallelism, even if not identified as such. Tool makers in the past kept returning to a limited range of solutions because they were optimally effective. New World Paleoindians used bifacial reduction because it fit so well with their high-mobility lifestyle (Kelly 1988). The bipolar technique re-appeared whenever people were forced to deal with small packages of raw material (e.g., Andrefsky 2005). Meanwhile, explanations based on migration or diffusion were out of favor, seen as the tools of retrograde, culture-historical approaches.

As the topical focus of this chapter suggests, the situation today is different. Both convergence and direct cultural connections are considered viable alternatives to account for the distributions of similar material-culture traits. Several developments have made convergence and phylogeny into alternative hypotheses rather than foundational assumptions. In the Americas, the gradual spread of Darwinian or Neo-Darwinian evolutionary theory, which emphasizes phylogeny in equal measure with (or more than) adaptation in explaining variation in artifacts, has played a major role in adjusting preconceptions. In Europe, the near-universal acceptance of the *chaîne opératoire* approach has, for very different reasons, focused the attention of archaeologists on learning frameworks and cultural inheritance, as well as on function and economy, to explain similarities in how people made things and what they made. Meanwhile, the ascendance of genetic evidence, whether from ancient or contemporary populations, has provided direct information about movement, mixing, and extinction of populations, bringing demography back to the center of thinking about dynamics in human prehistory. Consequently, researchers are actively debating whether globally distributed phenomena such as Levallois (Kuhn 2013; Villa 2009; White and Ashton 2003) or bifacial shaping (Moncel et al. 2016) have single origins or developed independently in multiple places, something that would not have occurred 25 or 50 years ago.

Although these days it is more important to be able to distinguish convergence from phylogenetic effects, archaeologists still have limited options when it comes to methods for distinguishing between the two. In biology, demonstrating that a trait is a homology or homoplasy sometimes requires sophisticated, fine-grained analyses of anatomical or molecular evidence, but it is not always so difficult. It can be enough to know whether the most recent common ancestor (MRCA) possessed the traits shared by the species in question. For example, we know moles and mole rats evolved their peculiar morphologies independently because insectivores and rodents diverged a long time ago, and the MRCA did not possess the specialized fossorial adaptations that make these species superficially similar. Thinking about evolution as the production and elimination of diversity, as biologists have thought about it for a century and half, naturally leads to models of evolutionary change that are full of branching nodes, and hence MRCAs, for just about any pair of taxa. In contrast, for most of the history of archaeology, models of change have been progressive and essentially unilinear, with one less-developed form leading to another, more-developed one. With linear schemes, either all ancestors are common or none are. There is no way to work back to the branching point to see whether the MRCA possessed the traits in question because MRCAs are not really part of the equation.

The result for archaeology is that approaches to distinguishing convergence or parallelism from similarity due to shared ancestry are typically informal. In the absence of a ready means for defining phylogeny, judgments depend on the prior plausibility of arguments for historical connections or on the estimated chance of shared features arising independently. Of course, it is then impossible to specify how likely it is that particular traits will arise independently or how far away in space and time is *too far* to hypothesize a plausible historical connection. For example, it has been asserted that the presence of similar forms of geometric backed artifacts in the Howieson's Poort of South Africa, the Uluzzian in southern Europe, and various late Pleistocene assemblages in south Asia tracks a discrete hominin dispersal event around 50,000–45,000 years ago (Mellars 2006; Mellars, Goric, Carre, Soaresg, and Richards 2013). Not surprisingly, other researchers strongly dispute the model (Korisettar 2007; Petraglia et al. 2009). Whether one accepts that argument depends on whether one believes that the sets of procedures identified are too unique and too derived to have re-emerged on their own, or whether the distances and time spans involved are just too vast to posit any direct historical connection between the respective groups. Advocates on both sides of the argument can usually mount strong rhetorical defenses of their views, and because it comes down to belief and believability, nothing is resolved.

The solution to this dilemma is for us to model phylogeny first. If we have a model for how various archaeological cultures, assemblages, or artifacts are related, we can make a systematic attempt to assess the probabilities that particular traits are homologies or homoplasies. Although evolutionary biologists have developed a number of tools for this purpose, and many of the authors in this volume have pioneered their application in archaeology, some archaeologists are skeptical of borrowing biological methods without a few adjustments to the underlying models. Important challenges to this theoretical framework include appropriate definitions of homology and homoplasy in archaeology and procedures for selecting traits for analysis. More fundamentally, we have to ask ourselves about the mechanisms leading to the transmission or re-invention of a trait and how different these mechanisms are from the processes of biological evolution. We will return to these questions at the end of the chapter, but the next order of business is to introduce the central research question.

What, When, and Where Is the Initial Upper Paleolithic?

The questions we seek to answer concern the vast global distribution of an early Upper Paleolithic (UP) technological phenomenon known as the Initial Upper Paleolithic (IUP). Understanding why this set of technological behaviors is so widespread, and whether it tracks migration or independent invention, has implications not only for dispersal of modern humans but also for understanding other widespread technological patterns or archaeological cultures. As a term, IUP was first used by Marks (1983) to refer to a single assemblage, the material from layer 4 at Boker Tachtit in Israel. Marks viewed layer 4 as representing the end of the gradual evolution of bidirectional, Levallois point production to a more typical UP type of volumetric, unidirectional blade production. Almost 20 years later, Kuhn, Stiner, and Güleç (1999) applied the term somewhat more broadly, using it to describe assemblages from the lower levels (F—I) at Üçağızlı cave in Turkey as well as assemblages from Ksar Akil and other sites in the Levant with roughly similar sorts of blank production technology and retouched tool forms. The rationale for expanding the use of the name was two-fold. First, the IUP seemed a better name than Emiran, the older designation (see below), because many cave assemblages lacked diagnostic Emireh points. Some researchers referred to these assemblages as "transitional," but this presumed an undocumented phylogenetic link between the late Middle Paleolithic and the later UP.

Subsequently, the use of IUP has been expanded even more. It has been applied globally to describe any early UP assemblages with apparent elements of Levallois-like blade production. Other scholars have branded these "lepto-Levalloisian" industries (Svoboda 2015), but the term has never been widely used. The IUP now encompasses other "named" industries, including the Emiran (Mediterranean Levant), Bacho Kiran (Bulgaria), and the Bohunican (Czech Moravia) as well as a range of assemblages from northeast Asia.

Expanding the geographic scale over which the designation IUP is used raises new questions about the nature and significance of the phenomenon. When the term was applied to the Levantine material alone, we knew that it represented a limited span of time and geography, and it was safe to assume a level of cultural continuity among the makers of various assemblages. However, when we think about the IUP as a global phenomenon, extending from the central Levant through central Asia and into Mongolia and western China, we cannot assume the same explanation for similarities among these assemblages. Most importantly, the term IUP when applied globally should not be taken to represent a culture or any other cultural unit as we might define it today. It simply recognizes global commonalities in lithic assemblages. These commonalities might well reflect a broad sharing of practices across this vast area, but the shared features of the assemblages could represent the outcomes of other processes as well.

We note that Sinitsyn (2003) and others (e.g., Goebel and Aksenov 1995; Hublin 2012) have used the term Initial Upper Paleolithic to refer to the first UP assemblages in any region, whatever form they may take. This is a different definition from the one used here. The assemblages we call IUP are united by a series of shared technical features not typically present in other semicontemporaneous assemblages. They fit Sinitsyn's definition, in that they are always the earliest form of UP wherever they occur, but the definition hinges on material-culture traits rather than on chronostratigraphic position. Many other assemblages that fit Sinitsyn's definition would not meet the technological criteria specified here.

What, then, defines the IUP? It its broadest sense, the term refers to early UP assemblages with features of Levallois technology in blade production but having essentially UP retouched-tool inventories. Levallois elements include the use of hard-hammer percussion, frequent platform faceting, and cores with comparatively flat faces of detachment. Many products would be classified as Levallois points and blades even though the mode of production itself is not typical Levallois. More conventional prismatic cores, complete with opening and lateral crests, are often present, but these cores were also exploited using hard-hammer percussion. There is a great deal of variation on this basic theme of core reduction. Unidirectional (single platform) reduction dominates some assemblages, whereas bidirectional reduction is common in others. In the Levant there may be a temporal element, with bidirectional reduction occurring earlier, but this is not generalizable. In some places, knappers habitually exploited one narrow core margin as well as the broad face of a core to make blades (Zwyns 2012; but see Roussel 2013). Zwyns, Rybin, Hublin, and Derevianko (2012) suggested that this kind of blade technology, combined with a specific mode of producing small blades from so-called "burin cores" (on sections of large blades of flakes), could be considered as a regional signature for the IUP in northeast Asia.

Associated assemblages of retouched tools sometimes contain distinctive, apparently derived artifact forms. In the Levant, the IUP is identified most closely with Emireh points—Levallois points with bifacially thinned bases. In fact, what we are calling IUP was originally called Emiran in the Levant. However, while widespread, Emireh points have seldom been recovered from well-controlled contexts and are associated almost exclusively with open-air sites. In the northern Levant, earlier IUP assemblages contain numerous *chanfreins*, a distinctive form of lateral burin, but again, they are confined roughly to the area between Haifa and Antakya. Beyond this, distinctive "type fossils" are scarce, though it is entirely possible that more will be identified in the future. For now, though, the great majority of shaped tools in IUP assemblages are fairly generic, including typical UP forms such as endscrapers, burins, and truncations as well as MP types such as sidescrapers and denticulates.

Some, but not all, IUP assemblages are associated with ornaments, bone tools, and other nonlithic features of stereotypical UP cultures. Ksar Akil and Üçağızlı caves in the northern Levant yielded large assemblages of shell beads as well as some shaped bone awls or points (Kuhn et al. 2009; Stiner, Kuhn, and Güleç 2013). Ornaments or bone tools, or both, have been recovered from IUP layers in sites in the Levant, Bulgaria, Siberia, the Trans-Baikal, and Mongolia (Buvit, Terry, Izuho, and Konstantinov 2015; Kuhn and Zwyns 2014; Rybin 2014, 2015). However, many IUP sites have not yielded such objects. In part, this is a function of preservation: artifacts of perishable materials are seldom recovered at the open-air sites from which many IUP assemblages are derived. Variation in the abundances of such artifacts may also be influenced by site function, a topic about which we know comparatively little.

Paleolithic assemblages showing the basic set of technological features described above show a remarkably broad distribution across temperate Eurasia. Assemblages fitting our definition of the IUP have been reported from more than a dozen sites in the central and northern Levant (Israel, Lebanon, Jordan, Syria, and southern Turkey (summarized by Leder 2014). The important open-air site complexes of Bohunica and Stránska Skála in the Middle Danube of the Czech Republic are another concentration of early UP localities with Levallois-like blade production. Probably the largest cluster of IUP sites is situated in and around the Selenga River drainage system and around the Southern shore of the Lake Baikal (Derevianko et al. 2007; Gladyshev, Olsen, Tabarev, and Jull 2012; Gladyshev et al. 2013; Lbova 2008; Rybin 2014; Zwyns et al. 2014). Finally, at least three distinct localities within the Shidonggou site complex (western China) contain typical IUP materials (Brantingham, Krivoshapkin, Jinzeng, and Tserendagva 2001; Boëda et al. 2013; Li, Kuhn, Gao, and Chen 2013).

Although the IUP is widespread, its distribution is not continuous. Many of the gaps could be a result of uneven research coverage as well as of discontinuous preservation and exposure of sediments of the appropriate age, but these possibilities cannot account for the absence of IUP assemblages from territories west of the Rhine River, for example.

The IUP also occupies a consistent stratigraphic position relative to other kinds of Paleolithic assemblages. Where it occurs, it is always the earliest form of Upper Paleolithic from a stratigraphic perspective. Typically, it is stratified above the MP and below later UP complexes such as Aurignacian, Gravettian, or Ahmarian. The situation in western China and southern Mongolia is different. In these areas, assemblages identified as IUP are preceded *and* succeeded by flake-based assemblages, which are typical for the early and late Paleolithic in these regions.

The range of radiometric ages obtained for IUP assemblages contrasts with their consistent stratigraphic position. Globally, IUP assemblages date to between rough 45,000 and 30,000 calibrated radiocarbon years ago. Wherever multiple dates are present, they show a wide distribution. For example, the large sets of radiocarbon dates from Mugharet el Hamam in Lebanon (Stutz et al. 2015), Üçağızlı Cave in Turkey (Kuhn et al. 2009), and the Bohunice and Stránská Skála localities in Czech Moravia (Richter, Tostevin, and Škrdla 2008) each spans 10,000 years or more. There is no obvious monotonic geographic trend in radiocarbon dates either. As of this writing, the earliest dates are from Boker Tachtit, Israel (pending results from new investigations), but some Bohunician localities in Czech Moravia (Richter et al. 2008), as well as Kara Bom in Siberia (Goebel 1999; Goebel and Aksenov 1995; Kuzmin 2004) and Tolbor 16 in northern Mongolia (Zwyns et al. 2014), provide similarly early age estimates.

The range of absolute dates for the IUP can be explained by several factors, one of which is that nearly all dates are based on radiocarbon, and most are close to the temporal limits of the method. Also, the record we have is also strongly structured by changes in global climate. The overlap in ages, for example, between open-air sites in south-central Siberia and North Mongolia may be aligned with climatic events. These regions may have been colonized so quickly that the differences in ages between neighboring areas fall within the minimum error ranges of radiocarbon dates. At the same time, we need to recognize that the range of dates, even from single localities, may reflect a significant duration of the phenomenon in question. The development of assemblages with similar characteristics over such a wide area, regardless of how it occurred, cannot have come about instantaneously. It must have been the result of time-transgressive processes. Rather than representing a true transition—a brief period of reorganization between two quasi-stable states—the IUP may also be part of a stable and long-lasting set of technological behaviors.

Several alternative scenarios have been proposed to explain the temporal and spatial distribution of the IUP. Some scholars believe that at least part of the widely distributed IUP assemblages track early routes of dispersal of *Homo sapiens* across Eurasia (Goebel 2015; Hoffecker 2011; Zwyns 2012; see also references in Hublin 2012). In this case, technological similarities between far-flung assemblages represent homologous traits, the consequence of cultural descent brought on by territorial expansion of ancestral populations. It should be noted that we are currently uncertain about which population(s) of Upper Pleistocene hominins actually made the IUP in different areas. The few fragmentary fossils currently known—all from the eastern Mediterranean—generally appear like anatomically modern *H. sapiens,* but a few specimens possess traits linking them to Neanderthals as well (Kuhn et al. 2009). A related hypothesis is that the distribution reflects the diffusion of cultural ideas across a network of preexisting populations (Boëda, Hou, Forestier, Sarel, and Wang 2013). Here again, the shared traits would represent homologies, but in this case, their distribution would result from processes analogous to gene flow rather than from actual population dispersal.

A third option is that there was little cultural continuity across this vast area and that IUP assemblages in widely separated areas are the outcomes of repeated independent convergence on similar technological options. Perhaps the particular form of IUP blade production is a good solution to a specific set of problems in making and using stone tools often faced by hominins during the late Pleistocene. The basic commonalities in IUP blank production could also be an "easy" evolutionary transition between Levallois and prismatic-blade manufacture. In these scenarios, the similarities among assemblages would be homoplasies, consequences of convergence or parallelism (chapter 13, this volume).

The distribution of the IUP in time and space could also reflect a combination of any or all of these processes. For example, there might be multiple centers of independent invention with limited dispersal or diffusion out from them (e.g., Li, Kuhn, Chen, and Gao 2016). The challenge is sorting out which factors are behind the similarities and differences in the cases in question. It is not sufficient to simply argue for the plausibility of direct connections or independent invention. At least some plurality of researchers will find any one of the alternatives to be most likely. We must establish a hypothetical phylogeny linking the many IUP assemblages across Eurasia to see how well it fits any of these possible scenarios.

How Do We Solve the "IUP Problem"?

Fortunately, a range of conceptual and analytical models exist for distinguishing among the effects of population movement, diffusion of cultural knowledge, and independent invention, although it is not our intention to review them here. Our concern here is more basic, namely the selection of appropriate models and variables on which to base the analysis.

Transmission, Parsimony, and Other Models

As some participants in the Konrad Lorenz Institute workshop noted, phylogenetic analyses based on phenotypic traits can be unreliable compared to the gold standard of studies based on genetics (e.g., Collard and Wood 2000). For one thing, phenotypic traits are not what is actually inherited, and many phenotypic characters map imperfectly onto genetic units. To cite a much-discussed example, the phenotypes of chimpanzees are rather similar, but the species' genome is relatively diverse. The opposite can be observed among humans, who show high morphological diversity but low genetic diversity. These differences suggest that selection and especially plasticity play large roles in structuring the phenotypic diversity observed among current and past populations.

Maximum parsimony, one commonly used method in phylogenetic analysis, implies that homologies are more common than convergences. It builds trees with the least complex branching, and the most parsimonious trees minimize the amount of convergence. Although it has proven to be misleading, parsimony may work in situations where rates of change are known. And that could be a major concern when dealing with material culture. Phylogenetic analyses based on morphological traits of extant species are often combined with genetic analysis, where assumptions about coalescence times are based on mutation rates. In such cases, the phenotype is used as proxy for detecting selection, whereas the model of transmission is based on genetic transmission. With the fossil record, traits are supposed to be—at least for the most part—transmitted along with the coded part of the genome, and traits observed on bones are used as a proxy to infer a phylogeny in the absence of genetic data. When dealing with material culture, we have no analogue for the genetic models. It is unlikely that we will ever know exactly what people learned in the past and from whom they learned it. So if we are to ask questions about phylogenetic connections and convergence among archaeological assemblages, we have the choice of either (a) designing an overly simplistic model specific to cultural transmission or (b) using parsimony and assuming that traits were typically transmitted gradually and vertically. Here, we use artifacts as analogues for phenotypic information.

It sounds like a difficult dilemma, but just because the tool is imperfect does not mean we should not use it, especially when it is the only device available. The more important point is that phylogenetic analyses can be used to build hypotheses and compare those with what we know about the dates and locations of assemblages. While we do not have genetics to fall back on, we do have some independent control over relationships among the makers of the various assemblages, namely the dimensions of time and space. This strategy has already been applied convincingly to model the expansion of Paleoindian populations across North America (e.g., Buchanan and Collard 2007; Buchanan and Hamilton 2009; Hamilton and Buchanan 2009; O'Brien, Darwent, and Lyman 2001; O'Brien et al. 2014, 2016). Stratigraphy and direct dates provide us with some idea of the time it takes for traits to evolve (Creanza, Fogarty, and Feldman 2012). Depending on the case, the two variables are unlikely to be constant, which should serve as another warning sign about relying too heavily on maximum parsimony. Using such independent information, we should be able to eliminate some scenarios and promote or improve others. If we find a tree rooted in what are clearly the most recent assemblages, or if stratigraphically older assemblages are consistently identified as derived from stratigraphically younger ones, we know the outcome is wrong—or that we are missing huge chunks of the record. In the case of the IUP, we also have at least one working hypothesis with very explicit predictions for the geographic structure of relationships among assemblages. If the single-origin model is correct—that is, if the IUP tracks a distinct dispersal event that began in the Near East, passed through central and eastern Europe, and ended up in northeast Asia—there are clear expectations about the relationships among assemblages in terms of which are ancestral and which are derived.

What Traits Should We Use?

Another important but sometimes overlooked issue in establishing phylogenetic models involves choice of attributes, or units. Obviously traits should be heritable (learnable), discrete, and/or quantifiable. Just as one can design a study using individual traits, fossils, or entire taxa as units, one can also use individual artifacts or assemblages as units of analysis. Both approaches have advantages and pitfalls. However, in the case of the IUP we have only one real option. Unlike the case of Paleoindian projectile points from North America (Buchanan and Collard 2007; Buchanan and Hamilton 2009; Hamilton and Buchanan 2009; O'Brien et al. 2001, 2014, 2016), there are not enough attribute-rich, distinctive and widespread artifact forms to work with in IUP assemblages. Thus, assemblage-level analysis is the most viable approach.

One potential weakness of such an approach is that one does not know how, or whether, various features relate to one another. To include traits that are co-dependent is to run the risk that these traits are irrelevant for phylogeny (Lieberman 1999). In analyzing individual specimens, one can assume that all observed attributes occurred together at least once. Such an assumption is less secure when using assemblages as the unit of analysis. Assemblages of Paleolithic artifacts are typically defined on geological grounds, and the boundaries between them are determined based on changes in sedimentation or, secondarily, on fluctuating densities of archaeological finds. Unlike individual organisms or (arguably) single artifacts, assemblages were not created as entire units and deposited as single events. Rather, they accumulated over periods of variable duration, ranging from days to centuries. It is possible, even likely, that a single geologically defined assemblage contains artifacts that were never in use at the same time or debris from procedures that were carried out by different groups of people. Of course, large assemblages should also contain a wider variety of features and attributes. The only way to cope with these potential problems is to pay close attention to assemblage "grain," or temporal resolution, and to use various techniques to examine rarefaction effects in the occurrence of specific traits. If the presence of a character is strongly linked to assemblage size it should probably be excluded from analysis.

The second issue with using assemblages as units of analysis is that various character states or complexes of features may not have the same phylogenetic history. In principle, a group could have acquired different technological practices from different sources—for example, adopting projectile-point design from successful neighbors but retaining ancestral practices for producing flake blanks. Some behaviors may even be more likely to flow horizontally, among contemporaries, whereas others tend to be transmitted across generations vertically. Fortunately, this complication can also be an advantage. By collecting information on multiple domains of lithic technology, we can subdivide the attribute list in different ways. If we get different results for core-reduction practices than for techniques of tool shaping, then we know that the pattern of influence and relationships was more complicated. Tostevin (2012) developed such a strategy to differentiate between mechanisms of trait diffusion and adoption. His work compares measures of proximity based on traits easily acquired by simple emulation with measures using features that require greater social intimacy to communicate. Our goal is nothing so fine grained. For the purposes of the study at hand, outcomes from different subsets of the data should produce a more diversified and hence nuanced perspective on potential phylogenetic relationships among the many IUP assemblages.

According to Lieberman (1999), archaic and derived characters on fossils should be distinguished when using maximum parsimony, and they should vary more between than within taxonomic units to be phylogenetically meaningful. We think that the distinction could be especially useful in dealing with material culture. For example, the Levallois method long predates the IUP; it was extraordinarily widespread during the Middle Pleistocene and Upper Pleistocene in Africa and Eurasia. Consequently, it is safe to assume that the Levallois characters in IUP assemblages are archaic, inherited along with other features such as basic understanding of conchoidal fracture. The two sets of inherited features (unlikely to be reinvented) are technological elements that refer to common ancestral assemblages that developed in different times and places. Clearly, they should not be used alone to create new branching in the phylogeny or to support or infer local continuity in a given region. To do so would be equivalent to using pentadactyly to characterize *H. sapiens*. Conversely, specific forms of blade production or other late-appearing elements of the UP toolkit can be assumed to be derived features in some regions. These are precisely the traits that could be relevant to model dispersals or convergences of the IUP phenomenon—granted that, with a model other than parsimony, such distinction becomes irrelevant as long as it takes into account both categories of traits.

Toward the Construction of a Database

Assembling a database useful for the purposes of analyzing a broadly distributed phenomenon such as the IUP presents several challenges. Systematic comparison and analysis requires standardized information and systematic observations of the same set of features and attributes across all cases. Although standardization of nomenclature and descriptive conventions is typical for fields such as comparative biology and taxonomy, it is not typical for Paleolithic studies. The published literature on the UP of Eurasia is the product of dozens of researchers coming from vastly different academic backgrounds working with a wide range of methods and descriptive conventions. It is simply impossible to extract a reliably consistent set of observations from it. Some systems of classification, such as Bordes's (1961) Middle and Lower Paleolithic typology, are widely employed, but even they are not ubiquitous. Moreover, the units of analysis defined in these systems are not necessarily appropriate for the task at hand. Consequently, the first priority of our project is to develop a dataset suitable for phylogenetic analysis of IUP assemblages across the globe.

The database must have several features. First, it needs to be simple. Obtaining data on assemblages scattered from the Near East to North China requires the cooperation of many scholars willing to donate their time so the data collection should not be burdensome. The categories of data must also be widely applicable. Although not all assemblages will contain all of the traits included, it is important that at least a minority do. Consequently, arriving at the list of variables should be a bottom-up rather than a top-down endeavor.

In 2014, we convened a group of experts in the Eurasian early Upper Paleolithic at the Max Planck Institute in Leipzig, Germany. One of the reasons for bringing researchers face to face was to learn about locally and regionally distinctive characteristics and unusual artifact types or technological attributes. After discussing both commonalities and variation in what might be called IUP, one outcome of the discussions was a list of features that we agreed should characterize all, some, and just a few early UP assemblages with Levallois-like blade production (see appendix 8.1). The list includes unusual features that are not consistently reported in published findings and that researchers in different areas may have described differently. Using these features means that we have to assume they are evolutionary relevant. No distinction was made between archaic and derived traits at the outset.

In constructing the database, we decided to use presence or absence rather than quantitative measures. There were several reasons for this choice. First, collecting the data often meant that researchers would need to revisit collections they had already studied to look for things they had not been aware of before. It would be an undue burden on our colleagues to ask them to restudy entire assemblages to get a few counts; presence/absence could be scored with a simple inspection. Reducing all variables to binary also helps to ensure comparability. Finally, we also thought it was important to include a range of contextual information. This includes observations about raw-material quality and abundance. We also asked participants to assess the temporal scale of assemblages, which can influence diversity of contents, as well as the reliability of the stratigraphic determinations and excavation quality. These data will be crucial to evaluating the comparability of samples, to pruning or trimming the master sample for specific analyses, and ultimately for testing alternative hypotheses.

Once the database is complete, the next stage will be to determine which models would be the most efficient in evaluating the role of transmission and convergence in what we call the IUP. Maximum parsimony may offer preliminary clues as to what are derived and what are retained features. The latter might not be useful for identifying new branching events, but they can tell us a lot about ancestral connections between assemblages. Ideally, we should then move away from maximum parsimony and use a model that takes into account the entire set of traits. Another strategy is to test whether different dimensions of the "technological phenotype" produce different results. It is reasonable to posit, following Tostevin (2012), that the forms of retouched tools are easily emulated and/or subject to strong pressures toward convergence as a result of functional constraints. In contrast, fine details of manufacturing technology or arbitrary production steps cannot easily be learned except under conditions of exceptional social proximity. As a consequence, we expect that these two dimensions of material culture could reflect different evolutionary processes and perhaps different rates of change.

Alternative scenarios for the origins of the IUP would have different implications for the relationships between tool forms and technology of production. Under the strongest single-origin model, they should both show similar phylogenetic patterns. Because convergence, through horizontal transmission or functional constraints, is more likely to occur in the forms of shaped tools, we expect departures from a single-origin model to show up first in that dimension. However, discrepancies from a single-origin model in "hidden" aspects of technology are more certain indicators of independent origins. Although not all methodological challenges can be overcome in this way, such an approach should help us to assess the sensitivity of models and variables that are too often used implicitly in our field of research. It should help us to better define key concepts such as the IUP as well as provide a more nuanced and constructive perspective on other very widespread "archaeological cultures" during the Paleolithic and beyond.

Acknowledgments

We thank Mike O'Brien, Briggs Buchanan, and Metin Eren for inviting one of us (SLK) to what was an extraordinarily interesting and stimulating workshop. Although we had been thinking along these lines for a while, the time spent at the Konrad Lorenz Institute clarified many key issues and opened up new avenues of enquiry. We are also grateful to workshop participants for the insights they shared in Vienna and for their comments on our presentation. Prof. Jean-Jacques Hublin, director of the Max Planck Institute for Human Evolutionary Studies in Leipzig, Germany, generously provided funding for the 2014 IUP workshop. Finally, we thank all the participants in that workshop for the data they will soon be sharing.

Appendix 8.1 List of Assemblage-Level Characters

		Features of IUP/Emiran industries		
Site				
Layer				
Coordinates				
Topographic setting				
Stratigraphic reliability		Very good/good/bad/out of context		
Chronological information				
Raw material source		Local/20 km/over 20 km/mixed/unknown		
Workshop		Yes/No		
			Scoring	**Comment**
Lithic reduction	*Blade core*			
		Flat-faced (A3-A4)[1]		
		Narrow-faced (A1-A2)		
		Flat + 1 narrow (A5-A10)[2]		
		Flat + 2 narrow[3]		
		Posterior crest		
		Prismatic		
		Pyramidal		
		Other		
	Flake core			
		Levallois centripetal		
		Levallois preferential		
		Levallois recurrent		
		Discoid		
		Convergent		
		Other		

Blade	
	Unidirectional point/blade
	Bidirectional point/blade
	With orthogonal removals
	Lateral crest/neocrest
	Median crest
	Debordant
	Other
Flake	
	Levallois
	Levallois point
	Debordant
	Bifacial thinning
	core tablet
	Other
Core for small blank	
	prismatic
	carinated core/scraper[4]
	carinated burin (B4)[5]
	narrow-faced
	truncated-faceted
	burin-core (B1-B3; B5-B7)[6]
	atypical burin-core[7]
	prior to blade/point removal[8]
	Other
Platform	
	Plain
	Faceted

		Overhang faceted
		Overhang abraded
		Overhang hammered
		Lip
		Small platform (<4mm thick)
		Other
Tools	*Retouch tools*	
		Endscraper
		Convergent scraper
		Transverse scraper
		Dejete scraper
		Emireh point
		Point/blade with inverse proximal thinning
		Point/blade with inverse proximal truncation
		Burin on oblique truncation
		Perforator (retouched)
		Blank with lateral notch/stem
		Chanfrein
		Denticulate Mousterian point
		Tanged pieces
		Leaf points
	Retouch	
		Abrupt/backing
		Direct proximal thinning
		Inverse proximal thinning
		Bifacial proximal thinning
		Bifacial shaping (preform)
		Carinate (laminar scars)

	Formal bone tools	
		Awls
		Points
		Needle
		Retouchoir
		Others
Others	*Beads*	
		Perforated shell
		Perforated animal tooth
		Perforated atone
		Ostrich eggshell
		Bead blank with natural perforation
		Other

References

Andrefsky, W. J. (1997). Thoughts on Stone Tool Shape and Inferred Function. *Journal of Middle Atlantic Archaeology, 13*, 125–144.

Andrefsky, W. (2005). *Lithics: Macroscopic Approaches to Analysis* (2nd ed.). Cambridge: Cambridge University Press.

Barton, C. M., Olszewski, D. I., & Coinman, N. R. (1996). Beyond the Graver: Reconsidering Burin Function. *Journal of Field Archaeology, 23*, 111–125.

Boëda, E., Hou, Y. M., Forestier, H., Sarel, J., & Wang, H. M. (2013). Levallois and Non-Levallois Blade Production at Shuidonggou in Ningxia, North China. *Quaternary International, 295*, 191–203.

Bordes, F. (1961). *Typologie du Paléolithique Ancien et Moyen*. Publications, no. 1. Bordeaux, France: l'Institut de Préhistoire de l'Université de Bordeaux.

Brantingham, P. J., Krivoshapkin, A. I., Jinzeng, L., & Tserendagva, Y. (2001). The Initial Upper Paleolithic in Northeast Asia. *Current Anthropology, 42*, 735–746.

Buchanan, B., & Collard, M. (2007). Investigating the Peopling of North America through Cladistic Analyses of Early Paleoindian Projectile Points. *Journal of Anthropological Archaeology, 26*, 366–393.

Buchanan, B., & Hamilton, M. J. (2009). A Formal Test of the Origin of Variation in North American Early Paleoindian Projectile Points. *American Antiquity, 74*, 279–298.

Buvit, I., Terry, K., Izuho, M., & Konstantinov, M. V. (2015). The Emergence of Modern Behavior in the Trans-Baikal, Russia: Timing and Technology. In Y. Kaifu, M. Izuho, T. Goebel, H. Sato, & A. Ono (Eds.),

Emergence and Diversity of Modern Human Behavior in Paleolithic Asia (pp. 490–505). College Station: Texas A&M University Press.

Collard, M., & Wood, B. (2000). How Reliable Are Human Phylogenetic Hypotheses? *Proceedings of the National Academy of Sciences of the United States of America, 97*, 5003–5006.

Creanza, N., Fogarty, L., & Feldman, M. W. (2012). Models of Cultural Niche Construction with Selection and Assortative Mating. *PLoS One*, *7*(8), e42744.

Derevianko, A. P., Zenin, A. N., Rybin, E. P., Gladyshev, S. A., Tsybankov, A. A., Olsen, J. W., et al. (2007). The Technology of Early Upper Paleolithic Lithic Reduction in Northern Mongolia: The Tolbor-4 Site. *Archaeology, Ethnology & Anthropology of Eurasia, 29*, 16–38.

Dibble, H. L., & Pelcin, A. (1995). The Effect of Hammer Mass and Velocity on Flake Mass. *Journal of Archaeological Science*, *22*, 429–439.

Dinnis, R., Pawlik, A., & Gaillard, C. (2009). Bladelet Cores as Weapon Tips? Hafting Residue Identification and Micro-wear Analysis of Three Carinated Burins from the Late Aurignacian of Les Vachons, France. *Journal of Archaeological Science*, *36*, 1922–1934.

Eren, M. I., & Lycett, S. J. (2012). Why Levallois? A Morphometric Comparison of Experimental "Preferential" Levallois Flakes versus Debitage Flakes. *PLoS One*, *7*(1), e29273.

Eren, M. I., Roos, C. I., Story, B. A., von Cramon-Taubadel, N., & Lycett, S. J. (2014). The Role of Raw Material Differences in Stone Tool Shape Variation: An Experimental Assessment. *Journal of Archaeological Science*, *49*, 472–487.

Gladyshev, S. A., Olsen, J. W., Tabarev, A. V., & Jull, A. J. T. (2012). The Upper Paleolithic of Mongolia: Recent Finds and New Perspectives. *Quaternary International*, *281*, 36–46.

Gladyshev, G., Jull, A. J. T., Dogandzic, T., Zwyns, N., Olsen, J. W., Richards, M. P., et al. (2013). Radiocarbon Dating of Paleolithic Sites in the Ikh-Tulberiin-Gol River Valley, Northern Mongolia. *Vestnik*, *12*, 44–48.

Gladwin, H. S. (1947). *Men Out of Asia*. New York: Whittlesey House.

Goebel, T. (1999). Pleistocene Human Colonization of Siberia and Peopling of the Americas: An Ecological Approach. *Evolutionary Anthropology*, *8*, 208–227.

Goebel, T. (2015). The Overland Dispersal of Modern Humans to Eastern Asia: An Alternative, Northern Route from Africa. In Y. Kaifu, M. Izuho, T. Goebel, H. Sato, & A. Ono (Eds.), *Emergence and Diversity of Modern Human Behavior in Paleolithic Asia* (pp. 437–452). College Station: Texas A&M University Press.

Goebel, T., & Aksenov, M. (1995). Accelerator Radiocarbon Dating of the Initial Upper Palaeolithic in Southeast Siberia. *Antiquity*, *69*, 349–357.

Hamilton, M. J., & Buchanan, B. (2009). The Accumulation of Stochastic Copying Errors Causes Drift in Culturally Transmitted Technologies: Quantifying Clovis Evolutionary Dynamics. *Journal of Anthropological Archaeology*, *28*, 55–69.

Hoffecker, J. F. (2011). The Early Upper Paleolithic of Eastern Europe Reconsidered. *Evolutionary Anthropology*, *20*, 24–39.

Hublin, J.-J. (2012). Commentary: The Earliest Modern Human Colonization of Europe. *Proceedings of the National Academy of Sciences of the United States of America*, *109*, 13471–13472.

Kelly, R. L. (1988). The Three Sides of a Biface. *American Antiquity*, *53*, 717–734.

Korisettar, R. (2007). Towards Developing a Basin Model for Paleolithic Settlement of the Indian Subcontinent: Geodynamics, Monsoon Dynamics, Habitat Diversity and Dispersal Routes. In M. D. Petraglia & B. Allchin (Eds.), *The Evolution and History of Human Populations in South Asia* (pp. 69–96). Dordrecht, Netherlands: Springer.

Kuhn, S. L. (2013). Roots of the Middle Paleolithic in Eurasia. *Current Anthropology, 8*, S255–S268.

Kuhn, S. L., & Zwyns, N. (2014). Rethinking the Initial Upper Paleolithic. *Quaternary International, 111*, 8404–8409.

Kuhn, S. L., Stiner, M. C., & Güleç, E. (1999). Initial Upper Paleolithic in South-central Turkey and Its Regional Context: A Preliminary Report. *Antiquity, 73*, 505–517.

Kuhn, S. L., Stiner, M. C., Güleç, E., Özer, I., Yılmaz, H., Baykara, I., et al. (2009). The Early Upper Paleolithic Occupations at Üçağızlı Cave (Hatay, Turkey). *Journal of Human Evolution, 56*, 87–113.

Kuzmin, Y. (2004). Origin of the Upper Paleolithic in Siberia. In P. J. Brantingham, S. L. Kuhn, & K. W. Kerry (Eds.), *The Early Upper Paleolithic beyond Western Europe* (pp. 196–206). Berkeley: University of California Press.

Lbova, L. (2008). Chronology and Paleoecology of the Early Upper Paleolithic in the Transbaikal Region (Siberia). *Eurasian Prehistory, 5*, 109–114.

Leder, D. (2014). *Technological and Typological Change at the Middle to Upper Palaeolithic Boundary in Lebanon*. Bonn: Habelt Verlag.

Li, F., Kuhn, S. L., Gao, X., & Chen, F. (2013). Re-examination of the Dates of Large Blade Technology in China: A Comparison of Shuidonggou Locality 1 and Locality 2. *Journal of Human Evolution, 64*, 161–168.

Li, F., Kuhn, S. L., Chen, F., & Gao, X. (2016). Raw Material Economies and Mobility Patterns in the Late Paleolithic at Shuidonggou Locality 2, North China. *Journal of Anthropological Archaeology, 43*, 83–93.

Lieberman, D. E. (1999). Homology and Hominid Phylogeny: Problems and Potential Solutions. *Evolutionary Anthropology, 7*, 142–151.

Lin, S. C., Rezek, Z., Braun, D., & Dibble, H. L. (2013). On the Utility and Economization of Unretouched Flakes: The Effects of Exterior Platform Angle and Platform Depth. *American Antiquity, 78*, 724–745.

Marks, A. E. (1983). The Sites of Boker and Boker Tachtit: A Brief Introduction. In A. E. Marks (Ed.), *Prehistory and Paleoenvironments in the Central Negev, Israel*, (pp. 15–37). Vol. III, The Avdat/Aqev Area, Part 3. Dallas: Southern Methodist University.

McGhee, G. (2011). *Convergent Evolution: Limited Forms Most Beautiful*. Cambridge, Mass.: MIT Press.

Mellars, P. (2006). Going East: New Genetic and Archaeological Perspectives on the Modern Human Colonization of Eurasia. *Science, 313*, 796–800.

Mellars, P., Goric, K. C., Carre, M., Soaresg, P. A., & Richards, M. B. (2013). Genetic and Archaeological Perspectives on the Initial Modern Human Colonization of Southern Asia. *Proceedings of the National Academy of Sciences of the United States of America, 110*, 10699–10704.

Moncel, M.-H., Arzarello, M., Boëda, É., Bonilauri, S., Chevrier, B., Gaillard, C., et al. (2016). Assemblages with Bifacial Tools in Eurasia (third part): Considerations on the Bifacial Phenomenon throughout Eurasia. *Comptes Rendus. Palévol*. doi: 10.1016/j.crpv.2015.11.007.

O'Brien, M. J., Darwent, J., & Lyman, R. L. (2001). Cladistics Is Useful for Reconstructing Archaeological Phylogenies: Palaeoindian Points from the Southeastern United States. *Journal of Archaeological Science, 28*, 1115–1137.

O'Brien, M. J., Boulanger, M. T., Buchanan, B., Collard, M., Lyman, R. L., & Darwent, J. (2014). Innovation and Cultural Transmission in the American Paleolithic: Phylogenetic Analysis of Eastern Paleoindian Projectile-Point Classes. *Journal of Anthropological Archaeology, 34*, 100–119.

O'Brien, M. J., Boulanger, M. T., Buchanan, B., Bentley, R. A., Lyman, R. L., Lipo, C. P., et al. (2016). Design Space and Cultural Transmission: Case Studies from Paleoindian Eastern North America. *Journal of Archaeological Method and Theory, 23*, 692–740.

Pelcin, A. W. (1997a). The Effect of Core Surface Morphology on Flake Attributes: Evidence from a Controlled Experiment. *Journal of Archaeological Science*, *24*, 749–756.

Pelcin, A. W. (1997b). The Formation of Flakes: The Role of Platform Thickness and Exterior Platform Angle in the Production of Flake Initiations and Terminations. *Journal of Archaeological Science*, *24*, 1107–1113.

Petraglia, M. D., Clarkson, C. J., Boivin, N., Haslam, M., Korisettar, R., Chaubey, G., et al. (2009). Population Increase and Environmental Deterioration Correspond with Microlithic Innovations in South Asia ca. 35,000 Years Ago. *Proceedings of the National Academy of Sciences of the United States of America*, *106*, 12261–12266.

Rafferty, A. N., Griffiths, T. L., & Ettlinger, M. (2011). Exploring the Relationship between Learnability and Linguistic Universals. In F. Keller & D. Reitter, *Proceedings of the 2nd Workshop on Cognitive Modeling and Computational Linguistics* (pp. 49–57). Madison, Wis.: Omnipress.

Rafferty, A. N., Griffiths, T. L., & Ettlinger, M. (2013). Greater Learnability Is Not Sufficient to Produce Cultural Universals. *Cognition*, *129*, 70–87.

Richter, D., Tostevin, G., & Škrdla, P. (2008). Bohunician Technology and Thermoluminescence Dating of the Type Locality of Brno-Bohunice (Czech Republic). *Journal of Human Evolution*, *55*, 871–885.

Roussel, M. (2013). Methodes et Rythmes du Debitage Laminaire au Chatelperronien: Comparaison avec le Protoaurignacien. *Comptes Rendus. Palévol*, *12*, 233–241.

Rybin, E. P. (2014). Tools, Beads, and Migrations: Specific Cultural Traits in the Initial Upper Paleolithic of Southern Siberia and Central Asia. *Quaternary International*, *347*, 39–52.

Rybin, E. P. (2015). Middle and Upper Paleolithic Interactions and the Emergences of "Modern Behavior" in Southern Siberia and Mongolia. In Y. Kaifu, M. Izuho, T. Goebel, H. Sato, & A. Ono (Eds.), *Emergence and Diversity of Modern Human Behavior in Paleolithic Asia* (pp. 470–489). College Station: Texas A&M University Press.

Sahlins, M. D., & Service, E. R. (Eds.). (1960). *Evolution and Culture*. Ann Arbor: University of Michigan Press.

Sinitsyn, A. A. (2003). The Most Ancient Sites of Kostenki in the Context of the Initial Upper Paleolithic of Northern Eurasia. In J. Zilhão & F. D'Errico (Eds.), *The Chronology of the Aurignacian and of the Transitional Technocomplexes. Dating, Stratigraphies, Cultural Implications* (pp. 89–107). Lisbon: Instituto Português de Arqueologia.

Smith, G. E. (1929). *The Migrations of Early Culture*. Manchester, England: Manchester University Press.

Smith, K. (2011). Learning Bias, Cultural Evolution of Language, and the Biological Evolution of the Language Faculty. *Human Biology*, *83*, 261–278.

Steward, J. (1929). Diffusion and Independent Invention: A Critique of Logic. *American Anthropologist*, *31*, 491–495.

Stiner, M. C., Kuhn, S., & Güleç, E. (2013). Early Upper Paleolithic Shell Beads at Üçagizli Cave I (Turkey): Technology and the Socioeconomic Context of Ornament Life-histories. *Journal of Human Evolution*, *64*, 380–398.

Stutz, A. J., Shea, J. J., Rech, J. A., Pigati, J. S., Wilson, J., Belmaker, M., et al. (2015). Early Upper Paleolithic Chronology in the Levant: New ABOx-SC Accelerator Mass Spectrometry Results from the Mughr el-Hamamah Site, Jordan. *Journal of Human Evolution*, *85*, 157–173.

Svoboda, J. (2015). Early Modern Human Dispersal in Central and Eastern Europe. In Y. Kaifu, M. Izuho, T. Goebel, H. Sato, & A. Ono (Eds.), *Emergence and Diversity of Modern Human Behavior in Paleolithic Asia* (pp. 23–33). College Station: Texas A&M University Press.

Tennie, C., Call, J., & Tomasello, M. (2009). Ratcheting up the Ratchet: On the Evolution of Cumulative Culture. *Philosophical Transactions of the Royal Society of London. Series B, Biological Sciences*, *364*, 2405–2415.

Tostevin, G. (2012). *Seeing Lithics: A Middle-Range Theory for Testing for Cultural Transmission in the Pleistocene*. Oxford: Oxbow.

Villa, P. (2009). The Lower to Middle Paleolithic Transition. In M. Camps & P. Chauhan (Eds.), *Sourcebook of Paleolithic Transitions* (pp. 265–270). New York: Springer.

White, J. P. (1967). Ethno-archaeology in New Guinea: Two Examples. *Mankind*, *6*, 409–414.

White, J. P., & Thomas, D. H. (1972). What Mean These Stones? Ethnotaxonomic Models and Archaeological Interpretation in the New Guinea Highlands. In D. L. Clarke (Ed.), *Models in Archaeology* (pp. 275–308). London: Methuen.

White, M., & Ashton, N. (2003). Lower Paleolithic Core Technology and the Origins of the Levallois Method in Northwestern Europe. *Current Anthropology*, *44*, 598–608.

Zwyns, N. (2012). *Laminar Technologies and the Onset of the Upper Paleolithic in the Altai, Siberia*. Leiden, Netherlands: Leiden University Press.

Zwyns, N., Rybin, E. P., Hublin, J. J., & Derevianko, A. P. (2012). Burin-Core Technology and Laminar Reduction Sequence in the Initial Upper Paleolithic from Kara-Bom (Gorny-Altai, Siberia). *Quaternary International*, *259*, 33–47.

Zwyns, N., Gladyshev, S. A., Gunchinsuren, B., Bolorbat, T., Flas, D., Dogandžić, T., et al. (2014). The Open-Air Site of Tolbor 16 (Northern Mongolia): Preliminary Results and Perspectives. *Quaternary International*, *347*, 53–65.

9 The Point Is the Point: Emulative Social Learning and Weapon Manufacture in the Middle Stone Age of South Africa

Jayne Wilkins

Some researchers hold that human culture is unique because it accumulates modifications over time, in what is known as the "ratchet effect" (Tennie, Call, and Tomasello 2009; Tomasello 1999; Tomasello, Kruger, and Ratner 1993; Tomasello, Carpenter, Call, Behne, and Moll 2005). One generation does things a certain way, and the next generation does it the same way, except it adds modifications or improvements. The generation after that learns the modified or improved way. There is little loss or backward slippage, and new strategies do not have to be relearned every generation. The process relies on faithful transmission from one generation to the next and inventiveness to generate modifications and improvements.

Faithful transmission results from imitation, through which the learner copies both the end goal and the procedural steps to achieving the goal. Inventiveness results from humans choosing not to or failing to perfectly replicate the teacher or behavioral model. In contrast to imitation, emulation is a learning mechanism in which the learner develops his or her own strategy and technique for accomplishing a copied end goal through trial-and-error learning. Humans rely on both imitation and emulation. Imitation as a learning strategy has been emphasized over emulation in human-evolution research because it is rare in the animal kingdom, it has a later ontological development in human infants, and it often requires active teaching—a strategy unique to humans and some great apes (Tomasello 1999; Tomasello et al. 1993). However, emulation likely was an important source of innovation and selectable variation in hominin behavior. Without copying the strategies of their conspecifics, emulative learners must learn through trial and error which strategies work and which do not. In this situation, individuals have the opportunity to make observations about the immediate environment that will affect the process of accomplishing the end goal and can react accordingly. It represents behavioral plasticity. Emulative learning requires little, perhaps no, direct contact time between the learner and the teacher/model, and learning observations can be at quite a distance, facilitating transmission between more socially distant individuals. At a micro-evolutionary scale, emulation is one mechanism, among many, for generating behavioral variation in technological procedures, which are then subject to selective forces.

At a macro-evolutionary scale, and with specific reference to the archaeological record, emulation can result in patterns of intersite variability that mimic those resulting from convergent evolution. Even with minimal interaction and in the absence of direct cultural inheritance, modern humans have the capacity to independently invent different ways to reach the same shared goal, which means it is difficult to identify the relative extent to which emulation and convergence explain any similarity in form and reduction strategy with respect to stone tools. Thus, investigating the role emulative learning plays in generating lithic-assemblage homogeneity and heterogeneity, and how to identify emulative social learning in the archaeological record, is warranted.

Here I propose three lithic-assemblage characteristics that are consistent with an increased focus on emulative learning over imitative learning: (1) diverse reduction strategies used to produce similar blank types, (2) different blank types used to produce similar retouched tool types, and (3) different end-products used to carry out similar functions. As a case study, I focus on evidence from the Early Middle Pleistocene in Africa as represented at the archaeological site of Kathu Pan 1, Northern Cape, South Africa, where we see intrasite diversity in reduction strategies and diverse processes leading to the same end goal: points used as hafted spear tips. The assemblage characteristics are consistent with the above predictions for emulative learning. I suggest that assemblage diversity in the early Middle Stone Age of Africa may represent an increased emphasis on emulation compared to what occurred in earlier and later time periods, with implications for how we interpret spatial and temporal behavioral variability in Pleistocene Africa and beyond.

Background

Social-Learning Mechanisms

In the animal kingdom, complex patterns of behavior can be instinctual or learned, and learned behaviors can develop through individual trial and error or social mechanisms. Trial-and-error learning—inventiveness—is a characteristic shared with primates, but only humans transmit cultural knowledge across generations (Tomasello et al. 1993). The difference is usually attributed to differences in the degree and nature of social learning in humans, which can be imitative or emulative (e.g., Boyd and Richerson 2005; Tennie et al. 2009; Tomasello 1999; Tomasello et al. 1993, 2005). Imitation is considered an integral component of social learning in humans, where imitation differs from emulation because it focuses on process—the exact manner in which the end goal is accomplished. In contrast, emulation is focused on the end goal, and emulators learn the process of getting there on their own, if indeed they do (Tennie et al. 2009).

Tennie et al. (2009) demonstrate that human learning differs from learning in nonhuman primates. In humans, learning is sometimes more oriented toward process than product, with more emphasis on imitation than emulation. With animals, it often is the reverse. The transmission of nut-cracking behaviors in chimpanzees might be an example of emulation (Tennie et al. 2009). Juvenile chimpanzees learn how to crack nuts in proximity to adults, facilitated by the co-presence of nuts, hammers, and anvils. They try many methods involving these objects until they learn how to extract nuts themselves. In other words, chimpanzees reconstruct the product rather than copy the process leading to it. In a sense, when a chimpanzee learns a new behavior, it actually reinvents the strategy. It observes an effect and then uses its own behavioral strategies to reproduce the effect.

In part because of these differences between chimpanzee and human learning mechanisms, the unique capacity for human imitation has been used to explain cumulative culture change (Boyd and Richerson 2005), but this perspective has been challenged in more recent studies. In some scenarios, emulation can also lead to cumulative culture change in the absence of imitation (Caldwell and Millen 2009; Wasielewski 2014). In experimental "microsociety" studies, groups are asked to complete a task and are given differential access to information. For one task (paper-airplane building), there was performance improvement even in the absence of opportunity for imitation (Caldwell and Millen 2009), but for another task (building a load-bearing structure from reed and clay), imitation was required for performance improvement (Wasielewski 2014). Based on these contrasting results, one might argue that the relative importance of imitation as a mechanism for cumulative cultural evolution is task-dependent. Further, there is systematic variation within species, including humans, in the extent of imitative social learning (Mesoudi, Chang, Dall, and Thornton 2016). For example, some collectivist societies emphasize imitation and social learning in their educational systems, whereas other, more individualistic societies emphasize personal discovery and creativity (see Mesoudi et al. 2016 for a review).

Emulation and Lithic-Technology Studies

Emulation as a learning mechanism is also not often explicitly considered in lithic-technology studies. However, through emulation, individual flintknappers can converge on end-product characteristics, such as point size, shape, and form, and can also converge on the reduction sequences used to produce those end-products, such as the choice to use preferential over recurrent strategies.

Despite this, *chaîne opératoire* and other technological approaches tend to emphasize homogeneity and similarity and are rooted in an imitation-based framework for human learning mechanisms. Often, one single trajectory is presented as the dominant reduction sequence represented by a lithic assemblage. As a result, intersite comparisons often rest on the assumption that the degree of similarity reflects the degree of social intimacy, where increased similarity reflects increased relatedness or interaction (e.g., Tostevin 2007). Similarly, it has been suggested that the choice to use one reduction strategy over another served to communicate information about group and individual identity in the Pleistocene (Gamble 1999). This conceptual linkage of increased social learning and increased homogeneity persists, despite the results of ethnoarchaeological research that questions this social-interaction theory of style; increased social interaction does not always generate increased homogeneity in material culture, and can actually have the opposite effect (e.g., Hodder 1982).

Without making an *a priori* assumption about imitation dominance, lithic-assemblage characteristics can provide insight on the extent to which procedural chains are faithfully copied. In situations where chains are copied because knappers rely on imitative learning mechanisms, one would expect a high degree of technological intra- and inter-assemblage homogeneity. Individual knappers would be employing a dominant *chaîne opératoire* for each end goal, resulting in similarities in the lithic-assemblage characteristics across space and time.

To illustrate these expectations, we can imagine a scenario in which there is a shared tradition of producing bifacially retouched points to use as hafted weapon tips. In a learning environment that emphasizes imitation, each individual knapper has learned and employs the same strategy (or a small set of strategies) to manufacture that style of point. This strategy might look something like the following. First, a blank—a large, pointed blade with parallel dorsal scars—is produced. The blank is removed from a core that has been prepared for that purpose by first generating a crested ridge, which when removed sets up the parallel ridges for blade removal. The platform is not prepared before detachment. Second, the blank is retouched into the final point form. This is accomplished through bifacial retouch, which thins the point and shapes it into the desired form. Earlier stages of the retouch process are carried out with direct soft-hammer percussion, but later stages rely on pressure flaking. Third, the final point is hafted onto the end of a shaft and used as a weapon.

The resulting lithic assemblage contains debris from all stages of this process—core-preparation flakes, crested blades, discarded blade cores, blades, unfinished points that were broken during manufacture, finished points not used or discarded for tool maintenance, and so on. Because all knappers follow the general production strategy, there are few signs of divergence. This procedural homogeneity is at the heart of the techno-typological approach and its goal of generating regional-scale technological sequences.

Here, I suggest there are characteristics of lithic assemblages that are consistent with an increased focus on emulative learning over imitative learning. These characteristics are diverse reduction strategies used to produce similar blank types, different blank types used to produce similar retouched tool types, and different end-products used to carry out similar functions. Assemblages that represent the use of diverse reduction strategies can be recognized through the presence of diverse discarded core types, diverse types of core maintenance flakes and blades, and end-products with diverse scar patterns and platform characteristics. Knappers developing their own strategies will generate assemblages with diverse core and core-maintenance flake types. Assemblages that include retouched pieces on different blank types might also represent an emphasis on emulative over imitative learning. If imitation were emphasized, one would expect particular blank types to be exclusively, or at least preferentially, chosen for the manufacture of particular retouched-tool types. When there is diversity in blank type, it may signal that the knapper was focused on the end goal more than on the procedural process. Assemblages with diverse tool types, manufactured on diverse blank types, that indicate similar functions for the different types is also consistent with emulative learning, because it suggests that the tool action was the primary end goal rather than the exact manner in which that goal is accomplished.

The Significance of the Middle Pleistocene Archaeological Record

The Middle Pleistocene archaeological site of Kathu Pan 1 in the Northern Cape of South Africa is considered here as a case study for investigating emulative learning. The Middle Pleistocene (781,000–126,000 years ago) in Africa is significant for understanding the origins and evolution of *Homo sapiens*. Near the end of the Middle Pleistocene, roughly 195,000–150,000 years ago, *H. sapiens* exhibited an anatomically modern suite of traits (Clark et al. 2003; McDougall, Brown, and Fleagle 2005; White et al. 2003). Prior to the emergence of modern *H. sapiens*, Africa was inhabited by a species of hominin alternatively designated as "archaic" *H. sapiens, H. heidelbergensis*, or *H. rhodesiensis* (Rightmire 2001, 2004; Stringer 2012; chapter 6, this volume). Interestingly, there is no significant technological change represented in the archaeological record in Africa at the time when the first anatomically modern *H. sapiens* appear (see McBrearty and Brooks 2000 for a review).

Based on current evidence, similar technologies—Middle Stone Age (MSA) prepared cores, blades, and points—were used between at least 300,000 years ago and 50,000 years ago, before and after the origins of anatomically modern *H. sapiens.* Many significant technological developments generally associated with modern human behavior, including externally stored symbols and complex projectile technologies, first appeared less than 100,000 years ago (Bouzouggar et al. 2007; Brown et al. 2012; Henshilwood, d'Errico, Vanhaeren, Van Niekerk, and Jacobs 2004; Henshilwood, d'Errico, and Watts 2009; Henshilwood et al. 2011; Mourre, Villa, and Henshilwood 2010). Research on modern human origins in Africa in general represents an attempt to reconcile this temporal disjunction. Much recent work has focused on the MSA postanatomically modern *H. sapiens* and considerably less on the MSA preanatomically modern *H. sapiens*. At least part of the explanation for this is the relative paucity of sites chronometrically dated to the Middle Pleistocene.

Lithic technology plays a prominent role in examinations of hominin behavioral evolution in the Middle Pleistocene, in large part because lithic artifacts are often the only ones preserved at many sites. Some investigations have focused on what technological changes in the Middle Pleistocene prior to modern *H. sapiens* indicate about behavioral and cognitive capacities, with emphasis on hafted-spear technology and new capacities for problem solving and planning (Ambrose 2001, 2010; Wynn and Coolidge 2011), adaptability (Foley and Lahr 1997, 2003; Shea 2011), resource extraction (Foley and Gamble 2009), grammatical language (Ambrose 2001, 2010), and cooperation (Boyd and Richerson 2005).

Some interpretations of Middle Pleistocene hominin behavior address the question of *why* there was technological change at the end of the Acheulean and the beginning of the MSA. A direct cause for technological change is offered by Potts (1998), who argues that increased climatic variability some 700,000 years ago could explain changes in hominin morphology and behavior. The degree of behavioral variability that characterized the shift to flake and blade assemblages in the Middle Pleistocene could have resulted from diverse strategies for adapting to ecological variability (Shea 2011, 2013).

Case Study: Learning Mechanisms at Kathu Pan 1

The first excavations at Kathu Pan 1 (KP1) were carried out between 1979 and 1982 and revealed a long but punctuated Earlier-Middle-Later Stone Age sequence across five strata (Beaumont 1990, 2004). The uppermost deposits, strata 1 and 2, contained few lithic artifacts, consistent with a Later Stone Age designation. Stratum 3 contained artifacts such as points and prepared cores, consistent with a MSA designation. Stratum 4a, which yielded large bifaces, points, and prepared cores, was designated as a Fauresmith-bearing unit. The Fauresmith is described as an industry transitional between the Earlier Stone Age (ESA) and the MSA based on the presence of elements diagnostic of both periods (Herries 2011; Underhill 2011). The underlying stratum, 4b contained abundant handaxes and is consistent with the Acheulean (ESA). Stratum 5 was sterile. The faunal component at KP1 includes remains of Reck's elephant, equids, white rhinoceros, Burchell's zebra, and giant alcelaphines (Klein 1988).

Investigations that began in 2008 provide age estimates of the KP1 deposits (Porat et al. 2010) and more detailed analyses of the lithic assemblages (Porat et al. 2010; Wilkins and Chazan 2012; Wilkins and Schoville 2016; Wilkins, Schoville, Brown, and Chazan 2015). Stratum 4a yielded an optically stimulated luminescence age of 464,000 ± 47,000 years ago and an ESR/U-series age of 542,000 + 140,000/–107,000 years ago (Porat et al. 2010). The overlying stratum 3 gave an optically stimulated luminescence age of 291,000 ± 45,000 years ago, providing a minimum age estimate for the underlying strata. A technological analysis of the stratum 4a assemblage revealed that blades were regularly and purposefully produced using direct hard-hammer percussion from cores prepared by means of centripetal flaking (Wilkins and Chazan 2012). Some of these blades were retouched into points. Multiple lines of evidence suggest that points from stratum 4a were used as hafted spear tips (Schoville, Brown, Harris, Wilkins 2016; Wilkins and Schoville 2016; Wilkins, Schoville, Brown, and Chazan 2012; Wilkins et al. 2015) and that they are among the earliest known evidence for composite hunting technologies. Intentionally modified pieces of ochre, including specularite, were recovered from strata 3 and 4a (Watts, Chazan, and Wilkins 2016), representing some of the earliest known evidence for pigment use.

Prediction 1: Diverse Core-Reduction Strategies

Diverse reduction strategies were used to manufacture blades and flakes in stratum 4a. As previously described (Wilkins 2013; Wilkins and Chazan 2012), some of the reduction is related to Levallois, as defined by Boëda (1995). Cores contain two asymmetrical convex surfaces with roles that were irreversible and a flaking surface that was maintained in a way that created lateral and distal convexities to guide the shock wave of each predetermined blank. Blade-core reduction at KP1 diverges from Boëda's (1995) definition in that the upper exploitation surface of the core carries more volume than the lower platform surface. In other words, the upper surface is more convex than the lower surface, and the lateral convexities are steep and sit lower than the platform. Other discarded blade cores indicate more opportunistic blade production with little or no preparation.

Many flakes in the KP1 assemblage are by-products of the blade-production strategy described above, in which flake removals were used to prepare and maintain core surfaces and platforms. However, there is evidence that flake production was also an intentional and distinct goal of core reduction based on the presence of (1) large Levallois flakes with radial dorsal scars, (2) Levallois cores with preferential and recurrent flake removals, (3) multiple types of non-Levallois cores with flake removals, and (4) retouched pieces made on flake blanks. Flakes and flake fragments make up 43.5 percent of the entire lithic assemblage, representing 73.0 percent of all discarded detached pieces.

Based on the diversity of core types, there were multiple strategies for flake production at KP1 that can be grouped into two general categories (Wilkins 2013): Levallois reduction (*sensu* Boëda 1995) and non-Levallois reduction. Even within the Levallois category, there is diversity. Discarded cores have flake-scar patterns consistent with both preferential and recurrent reduction. Cores that were knapped recurrently were exploited centripetally or unidirectionally. Other cores show minimal amounts of preparation and are consistent with more opportunistic reduction strategies for the removal of a few flakes.

Some cores that were initially used for blade production might have been exploited in a different way for flake production just prior to discard. However, Levallois flake production and blade production seem to have had mainly separate trajectories. Some blade cores could have been transformed into Levallois flake cores later in the reduction sequence, but if this occurred, it was irregular. If blade cores were regularly transformed into Levallois

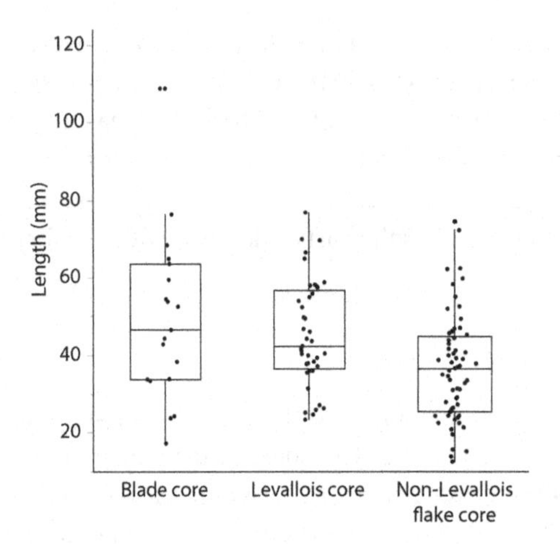

Figure 9.1
Box plot of core size for KP1 stratum 4a assemblage overlain with individual data points. Levallois cores do not have significantly smaller maximum dimensions than blade cores (unequal variance *t*-test, $t = 1.141$, $p = 0.265$).

flake cores, we would expect significant differences in size between the two core types, with Levallois cores exhibiting smaller maximum dimensions than blade cores. Levallois cores do not have significantly smaller maximum dimensions than blade cores (figure 9.1; unequal variance *t*-test, $t = 1.141$, $p = 0.265$).

Prediction 2: Diverse Blank Selection

The most conspicuous category of retouched tools in the stratum 4a assemblage is retouched points (24 percent of retouched pieces, n = 149). The points are 28–122 millimeters long, with a mean length of 70 millimeters. The majority of the retouched points are worked on the dorsal side only, but some show a small amount of additional flaking on the proximal ventral surface. The dorsal retouch is marginal and tends to be concentrated near the tips of the points, and for some complete and nearly complete retouched points it is possible to estimate original blank form as being either flake or blade. Among retouched points for which blank form could be estimated (n = 43), 42 percent (n = 18) were manufactured on flake blanks and 58 percent (n = 25) on blade blanks. Dorsal scars vary, with many showing unidirectional or bidirectional scars, and many showing radial scars. Diverse dorsal scar patterns are consistent with the points being produced using diverse core-reduction strategies. The majority of points were manufactured on banded ironstone (86.2 percent), but there are also points made from black chert (7.6 percent), quartzite (2.3 percent), and other raw-material types (3.8 percent) that are similar in size and shape to the banded ironstone points.

In addition to retouched points, there are many unmodified flakes and blades, often with convergent dorsal scars and pointed tips. These points are similar in size and form to the retouched points and were manufactured on the same range of raw materials. Blanks that were already an appropriate shape were not retouched.

Prediction 3: Convergent Tool Function

A sample of 210 retouched and unretouched points and point fragments in the stratum 4a assemblage was analyzed as a pooled assemblage to test the hypothesis that they were used as hafted-weapon tips. Several lines of evidence supported this hypothesis, including the edge-damage distribution (damage was concentrated at the tips), a high frequency of diagnostic impact fractures (*sensu* Fischer, Vemming Hansen, and Rasmussen 1984), and similarity in size and shape to other points interpreted as weapon tips (Wilkins and Schoville 2016; Wilkins et al. 2012, 2015). These observations are not restricted to any subcategory of point based on the reduction strategies used to produce them. Rather, all subcategories of points—retouched points and unretouched points, points on blade blanks, and points on flake blanks—exhibit traces consistent with use as hafted-weapon tips.

Figure 9.2 presents the edge-damage distribution for different subcategories of points. For all subcategories except for one with a small sample size (nonretouched convergent blades), damage is concentrated at the tip, consistent with use as weapon tips.

Table 9.1 and figure 9.3 present data on diagnostic impact fractures (DIFs) for the different subcategories. DIFs are present in all subcategories. With the exception of irregular points, all subcategories exhibit a frequency within the range observed at residential sites from more-recent archaeological contexts (Fischer et al. 1984; Sano 2012; Villa, Boscato, Ranaldo, and Ronchitelli 2009). The DIF frequency for all subcategories is much higher than and outside of the 95 percent confidence interval of the frequency observed on trampled and unused assemblages (Fischer et al. 1984; Lombard, Parsons, and van der Ryst 2004; Odell and Cowan 1986; Pargeter 2011; Sano 2012; Shea et al. 2001; Wilkins et al. 2012; Yaroshevich, Kaufman, Nuzhnyy, Bar-Yosef, and Weinstein-Evron 2010). Point measurements are also similar among subcategories (table 9.2). Together, these observations are consistent with the prediction that different types of end-products were used as weapon tips.

Discussion

Both emulation and imitation are considered integral components of learning in humans. In Paleolithic and Stone Age research, more emphasis has been placed on imitation as a uniquely human component of learning, and conceptually, imitation is at the root of many *chaîne opératoire* studies and industrial designations. This is because similarities in

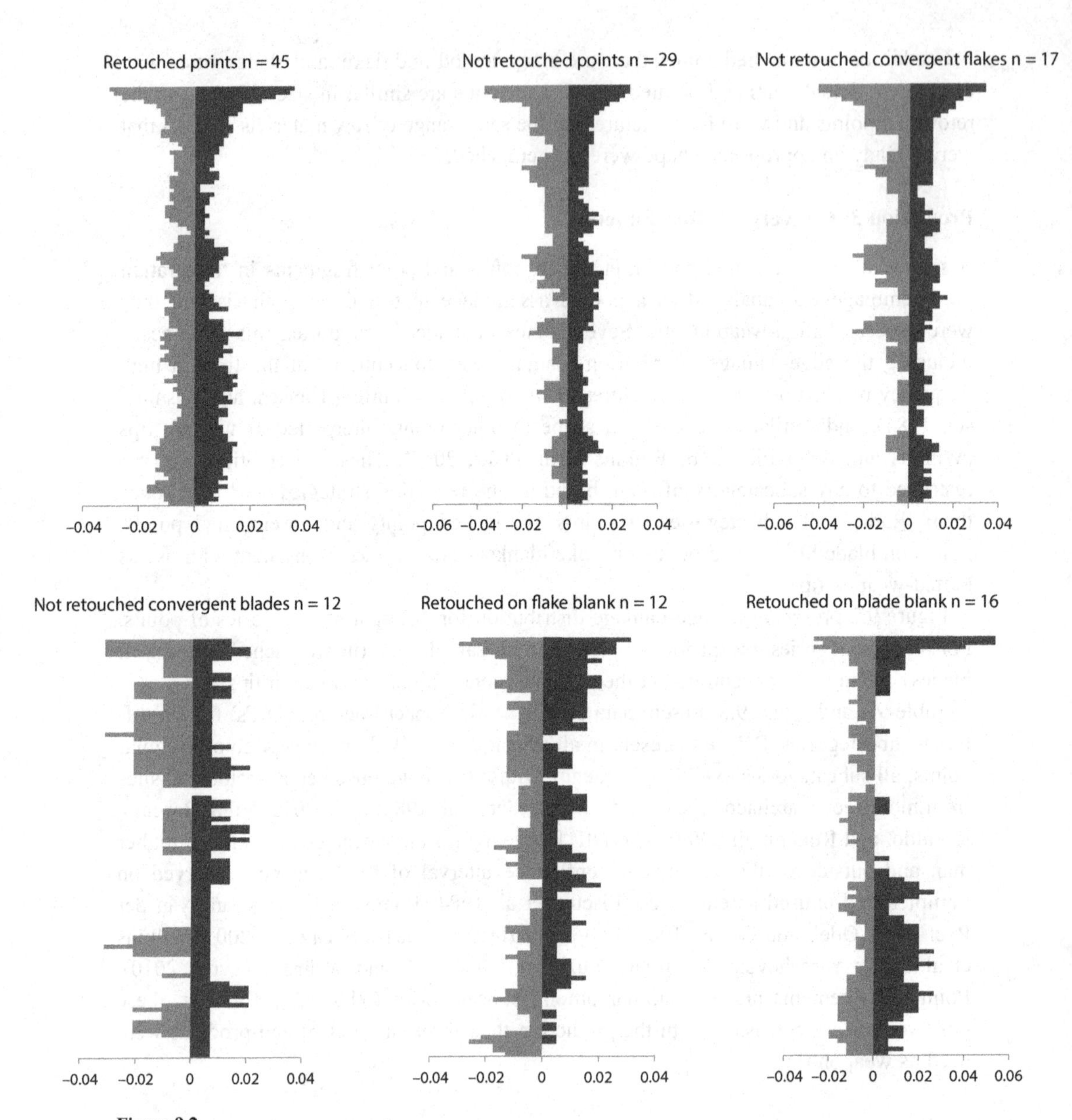

Figure 9.2
Edge-damage distribution for ventral surface of different point subcategories: retouched points and nonretouched points, nonretouched convergent flakes, nonretouched convergent points, retouched points on flake blanks, and retouched points on blade blanks. With the exception of the nonretouched convergent blades, all subcategories exhibit an edge-damage distribution that is concentrated at the tip, consistent with use as weapon tips.

Table 9.1
Counts and frequencies of diagnostic impact fractures on the KP1 points and point fragments, potential armature tips from other archaeological sites, and armature tips used in experimental studies

		DIF count	SSample	Percentage	References
KP1	Retouched	17	120	14.2%	This chapter
	N Not-retouched	9	63	14.3%	
	Retouch flake blank	1	18	5.6%	
	Retouched blade blank	2	25	8.0%	
	Unretouched convergent flake	4	28	14.3%	
	Unretouched convergent blade	5	35	14.3%	
Archaeological	Kill sites	52	121	43.0%	Fischer et al. 1984; Villa et al. 2009
	Residential sites	62	354	17.5%	Fischer et al. 1984; Sano 2012
	MP and MSA	35	776	4.5%	Schoville 2016; Villa et al. 2009; Villa and Lenoir 2006
Experimental	Projectile experiments	220	591	37.2%	Fischer et al. 1984; Lombard et al. 2004; Odell and Cowan 1986; Pargeter 2011; Sano 2012; Shea et al. 2001; Wilkins et al. 2012; Yaroshevich et al. 2010
	Trampling experiments	24	887	2.7%	

DIF, diagnostic impact fracture; MP, Middle Palaeolithic; MSA, Middle Stone Age.

process—reduction strategies—are seen as the result of learning, social proximity, and/or social interaction and are used to define industries and complexes. Although it is likely that both imitation and emulation were important aspects of learning lithic technology during the stratum 4a occupation at KP1, I argue here that emulation may have played a larger role, and I make this argument because there were diverse processes leading to the same end goal, which was use as hafted spear tips. In other terms, multiple *chaînes opératoires* were used to manufacture the same tools.

KP1 is not the only MSA site that documents this kind of pattern. For example, an analysis of the MSA levels at Porc-Epic Cave, Ethiopia, emphasizes the contemporaneity of four main *schema opératoires* (Pleurdeau 2006). Pleurdeau suggests this indicates that MSA toolmakers at the site were flexible to the demands of flake production and capable of employing various and novel behaviors. A variety of exploitation methods was employed during the late ESA and MSA occupations of Kudu Koppie, located in the northern Limpopo Province of South Africa (Wilkins, Pollarolo, and Kuman 2010). During the ESA occupation there, both the lineal and recurrent methods were employed, although the lineal method of prepared-core exploitation clearly dominates the assemblage. A greater variety

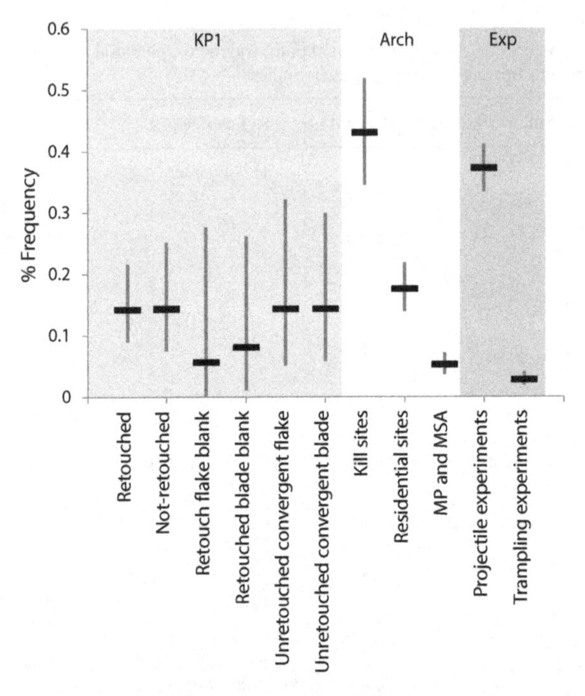

Figure 9.3
Percentage of points with diagnostic impact fractures for the different KP1 point subcategories, for armature tips from other archaeological sites, and for armature tips used in experimental studies (references provided in table 9.1). The central black bar represents the percent frequency, and the horizontal gray bars are the 95 percent confidence interval (CI). Most of the KP1 point subcategories fall outside the 95 percent CI for trampling experiments and overlap with the 95 percent CI for archaeological residential sites.

of exploitation methods was recognized in the later MSA levels of Kudu Koppie. The lineal method was still the most prevalent, but recurrent centripetal, recurrent unidirectional, and recurrent bidirectional cores also formed significant components of the assemblage. There is some evidence at other sites for the diversification of core-reduction strategies through time across the ESA and MSA boundary. At Koimilot, Kenya, the MSA is characterized by the diversification of prepared-core methods compared to what characterizes the ESA (Tryon 2006; Tryon and McBrearty 2006; Tryon, McBrearty, and Texier 2005).

Technological diversity in the MSA appears to exceed the amount of diversity observed between and within Acheulean assemblages. The persistence of the handaxe for at least a million years across much of the Old World is often cited as evidence for a high degree of technological homogeneity resulting from shared traditions (Shipton 2010) or instinct (Corbey, Jagich, Vaesen, and Collard 2016; also see chapters 3 and 6, this volume). There is certainly more variability within and between Acheulean assemblages than text books give credit for, but there is still a fairly robust pattern for increased behavioral variability

Table 9.2
Summary of KP1 point measurements

		Retouched point	Nonretouched convergent flake	Nonretouched convergent blade	Irregular pointed form	All (including fragments)
Mass (g)	n	69	20	18	20	127
	Mean	35.9	27.2	35.4	34.6	34.3
	Min	10.0	8.5	6.0	4.0	4.0
	Max	121.5	72.5	111.0	102.0	121.5
	Std Dev	20.9	16.0	29.0	23.8	22.0
Length (mm)	n	69	20	18	20	127
	Mean	71.0	58.3	78.2	71.1	70.1
	Min	41.3	39.5	49.3	27.8	27.8
	Max	117.8	98.5	117.7	122.8	122.8
	Std Dev	18.5	15.9	19.0	23.7	19.7
Width (mm)	n	69	28	30	26	189
	Mean	40.3	38.6	34.3	38.0	37.7
	Min	24.7	23.4	22.0	20.6	19.6
	Max	62.9	56.1	60.6	59.7	62.9
	Std Dev	8.3	8.3	8.8	9.6	8.8
Length/ Width	n	69	20	18	20	127
	Mean	1.8	1.5	2.3	1.9	1.8
	Min	1.2	1.0	1.7	1.1	1.0
	Max	2.8	1.9	3.0	3.2	3.2
	Std Dev	0.4	0.3	0.3	0.6	0.4
Thickness (mm)	n	69	28	31	27	200
	Mean	12.1	11.2	10.3	11.9	11.1
	Min	5.8	6.4	3.7	5.5	3.7
	Max	20.1	16.9	16.5	27.3	27.3
	Std Dev	3.1	2.6	3.7	4.6	3.3
TCSA (mm^2)	n	69	28	27	26	182
	Mean	249.0	222.9	200.8	242.2	226.1
	Min	86.3	89.9	51.7	63.9	51.7
	Max	591.3	414.6	446.3	814.9	814.9
	Std Dev	101.3	84.9	100.1	146.8	104.2
TCSP (mm)	n	69	28	27	26	182
	Mean	87.5	83.4	77.0	83.5	82.8
	Min	54.1	51.6	45.9	44.6	44.6
	Max	136.2	116.8	127.2	140.6	140.6
	Std Dev	17.3	17.2	18.6	20.8	18.2

StdDev, standard deviation; TCSA, tip cross-sectional area (after Shea 2006); TCSP, tip cross-sectional perimeter (after Sisk and Shea 2009).

represented by the origins of Mode 3 (MSA and Middle Paleolithic) assemblages compared to Acheulean assemblages. Perhaps the cognitive and psychological mechanisms that promoted either imitative or instinctual behaviors were relaxed in the mid-Middle Pleistocene with the origins of MSA technologies.

Mathematical models of culture change demonstrate that diversity increases when innovation rates are high (Kandler and Laland 2009). In these models, there are two types of innovation: those represented by independent invention and those represented by improvement through modification. These two types of innovation show slightly different dynamics. Independent invention generally supports higher levels of diversity than improvement through modification (Kandler and Laland 2009). Diverse core-reduction strategies in the Middle Pleistocene could represent multiple episodes of independent invention of reduction strategies that resulted in points suitable for use as weapon tips and demonstrate the increased importance of emulative learning compared to earlier time periods.

Behavioral strategies of early Middle Pleistocene hominins differed from those in the Late Pleistocene in two main ways that may be related to differences in learning mechanisms. First, there were no long-distance projectiles (Shea 2006), or "complex" tools, defined here as those consisting of more than three parts, or techno-units (Oswalt 1976). Complex tools can be more difficult to replicate and maintain using emulative learning mechanisms alone (Boyd and Richerson 2005; Henrich 2004) and may have required an increased emphasis on imitation compared to earlier time periods. Second, there appears to have been a lesser degree of formal variation in stone tools across time and space. Point styles in the later part of the MSA exhibit regional variability on a continent-wide scale (Clark 1982; McBrearty and Brooks 2000) as well as time-restricted patterning (Wadley 2001). There is general consensus that the spatial and temporal pattern of MSA point styles implies that artifact variability in the MSA is associated with cultural patterns (Clark 1992; Foley and Lahr 2003; Foley and Lahr 1997; McBrearty and Brooks 2000)—an interpretation based on the assumption that point form can be used to communicate messages about ethnic or group affiliation (Wiessner 1983) or that points can serve as important symbolic objects in regional exchange networks (Wilkins 2010). It remains to be shown that the lack of patterning in formal variation in the early MSA during the Middle Pleistocene is not simply a product of the patchy archaeological record, although based on current evidence it seems probable that material culture in the early Middle Pleistocene did not serve to communicate ethnic affiliation or to create and sustain large social networks. Because of that, there might have been less pressure to replicate point style and imitate the manufacturing process exactly.

Humans tend to exhibit high conformity bias, which is the tendency to adopt the behavioral variants of others because to deviate would expose an individual to third-party punishment (Boyd and Richerson 2005; Richerson and Boyd 2005). In mathematical models, the strength of conformity bias has direct consequences for quantitative measures of cultural diversity; high conformity bias leads to less diversity (Kandler and Laland

2009). Point styles in the later MSA are consistent with a fairly high conformity bias if we presume that point form served to communicate group membership and ethnic identity. Individuals conformed to group norms by creating points similar in form to those created by others in their group. However, the earlier MSA, which might lack this kind of temporal and spatial patterning in point form, could reflect a relaxed conformity bias compared to more recent time periods. This would also be consistent with the suggestion above that emulation was more important than imitation with respect to how stone tools were manufactured in the early Middle Pleistocene. Diversity in the *chaînes opératoires* of lithic reduction implies that there was no negative social consequence of emulating, rather than imitating. That is to say, there was no social reason to conform.

If the technological differences highlighted here are related to changing emphases on emulative versus imitative learning, then it is reasonable to ask whether and under what conditions one learning strategy would be more advantageous over another. Population size could play a role. Emulative learning and relaxed conformity bias would be consistent with smaller populations and smaller social networks because conformity is essentially a strategy for establishing, maintaining, and reaffirming bonds among nonkin (Boyd and Richerson 2005). The persistence of complex technologies requires large populations of people to compensate for loss of skills due to imperfect transmission (Henrich 2004). Imitation requires the learner and model to be much more proximate than emulative learning does, and opportunities for this kind of interaction across larger geographic and temporal periods would be reduced when populations are small. Based on genetic studies, Powell, Shennan, and Thomas (2009) estimate that Africa would not have reached the critical effective population for the transmission of certain types of traits until roughly 100,000 years ago.

Environment on both a spatial and temporal scale could also play a role. Toolkit complexity is known to correlate with latitude (Oswalt 1976), and toolkit design is affected by environmental predictability, with “reliable” (highly standardized) tool kits useful in predictable environments and “maintainable” (flexible, expedient) tool kits useful in unpredictable environments (Bleed 1986). Across Africa, environmental variability could have created conditions that differentially selected for emulative over imitative learning. As Potts (1998) pointed out, the last 700,000 years have witnessed increased climatic variability compared to earlier time periods, where fluctuations in global temperature are now more extreme and occur over shorter time periods. A focus on emulative learning might reflect an adaptive response by hominins to this increased climatic variability, encouraging independent invention and innovation so that hominins could occupy shifting environmental conditions.

In his analysis of pastoralist communities in East Africa, Hodder (1982) demonstrated that under conditions of resource stress, communities use material objects to create an “us/them dichotomy.” This response served as a strategy for justifying competition for limited resources, even between communities that lived in close proximity and had a high degree of interaction. Material items such as clothing and jewelry, as well as the location of the hearth in the home, were used to communicate group affiliation. A high degree of conformity bias (*sensu* Boyd and Richerson 2005; Richerson and Boyd 2005) was presumably one mechanism that perpetuated shared styles. In situations where there is less resource competition, one might predict a relaxed conformity bias and potentially less emphasis on imitative learning. This would lead to more diversity in the strategies used to manufacture stone tools. Resource availability is likely to have shifted through Pleistocene Africa in response to changing climatic conditions (Ambrose and Lorenz 1990; Barham and Mitchell 2008; Marean et al. 2015), potentially influencing intergroup competition and territoriality (Marean 2014, 2015).

From an evolutionary perspective, the concept of emulation has implications at both the micro and macro scales. Emulation generates diversity in technological procedures and behaviors. In other words, emulation is one process that changes the relative frequency of behavioral traits within a human population on short time scales. These different behavioral traits may have differing fitness values and be subject to selective forces. What I have suggested here is that during some periods in prehistory, emulative learning might have been emphasized over imitative learning, and that it was during those periods that new innovative behaviors were more likely to appear.

A significant branch of evolutionary archaeology is focused on explaining behavioral patterns at the macro scale and on large time scales. Recognizing convergent evolution in lithic tools relies on identifying similarity in form or technical procedures that arose independently without cultural inheritance. Because humans have the capacity for emulative learning and can maintain far-reaching social networks with variable degrees of interaction, it will sometimes be challenging to ascertain whether shared traits are truly independent or whether emulative learning between socially distant individuals played a role.

Conclusion

The stratum 4a assemblage at KP1 exhibits a high degree of intrasite diversity, consistent with the predictions presented above for emulative learning. There were diverse processes leading to the same end goal of stone-point tipped spear manufacture. Multiple *chaînes opératoires* were used to manufacture the same types of tools. These observations are consistent with a greater emphasis on emulative learning over imitative learning compared to the Acheulean and potentially the later MSA. This chapter is an early attempt at recognizing and identifying emulative learning in the archaeological record of the Middle Pleistocene and represents a departure from traditional lithic analyses rooted in an imitation-based framework. Emulative learning has a long history in the human lineage, predating modern *H. sapiens,* and generates behavioral diversity—a key component of innovation, modern human capacities (Shea 2011), and cumulative cultural change (Caldwell and Millen 2009). Somewhat paradoxically, emulative learning also generates similarity in final artifact form and behavioral end goals, with the potential to mimic convergence. Stone tools are only one part of our human behavioral repertoire, so the results presented here for KP1 are not a holistic representation of learning capacities in Middle Pleistocene hominins. However, the results suggest that differing emphases on emulative and imitative learning could explain, in part, variation in stone-tool technology through time and space in Stone Age Africa and beyond. As such, the results demonstrate that emulation is a concept warranting explicit consideration from lithic analysts.

Acknowledgments

I thank Mike O'Brien, Briggs Buchanan, and Metin Eren for the invitation to participate in the Konrad Lorenz Institute workshop and to contribute to this volume. Thank you to all the workshop participants for their insights on convergent evolution that improved this manuscript. Funding for this research was provided by the Social Science and Humanities Research Council of Canada, the University of Toronto, the University of Cape Town, and the South African Department of Science and Technology and National Research Foundation Centre of Excellence in Palaeosciences.

References

Ambrose, S. H. (2001). Paleolithic Technology and Human Evolution. *Science*, *291*, 1748–1753.

Ambrose, S. H. (2010). Coevolution of Composite-Tool Technology, Constructive Memory, and Language: Implications for the Evolution of Modern Human Behavior. *Current Anthropology*, *51*, S135–S147.

Ambrose, S. H., & Lorenz, K. G. (1990). Social and Ecological Models for the Middle Stone Age in Southern Africa. In P. Mellars (Ed.), *The Emergence of Modern Humans: An Archaeological Perspective* (pp. 3–33). Edinburgh: Edinburgh University Press.

Barham, L., & Mitchell, P. (2008). *The First Africans: African Archaeology from the Earliest Toolmakers to Most Recent Foragers*. Cambridge: Cambridge University Press.

Beaumont, P. B. (1990). Kathu Pan. In P. B. Beaumont & D. Morris (Eds.), *Guide to Archaeological Sites in the Northern Cape* (pp. 75–100). Kimberley, South Africa: McGregor Museum.

Beaumont, P. B. (2004). Kathu Pan and Kathu Townlands/Uitkoms. In D. Morris & P. B. Beaumont (Eds.), *Archaeology in the Northern Cape: Some Key Sites* (pp. 50–53). Kimberley, South Africa: McGregor Museum.

Bleed, P. (1986). The Optimal Design of Hunting Weapons: Maintainability or Reliability. *American Antiquity*, *51*, 737–747.

Boëda, E. (1995). Levallois: A Volumetric Construction, Methods, a Technique. In H. L. Dibble & O. Bar-Yosef (Eds.), *The Definition and Interpretation of Levallois Technology* (pp. 41–68). Madison, Wis.: Prehistory Press.

Bouzouggar, A., Barton, N., Vanhaeren, M., d'Errico, F., Collcutt, S., Higham, T., et al. (2007). 82,000-Year-Old Shell Beads from North Africa and Implications for the Origins of Modern Human Behavior. *Proceedings of the National Academy of Sciences of the United States of America*, *104*, 9964–9969.

Boyd, R., & Richerson, P. J. (2005). *The Origin and Evolution of Cultures*. New York: Oxford University Press.

Brown, K. S., Marean, C. W., Jacobs, Z., Schoville, B. J., Oestmo, S., Fisher, E. C., et al. (2012). An Early and Enduring Advanced Technology Originating 71,000 Years Ago in South Africa. *Nature*, *491*, 590–593.

Caldwell, C. A., & Millen, A. E. (2009). Social Learning Mechanisms and Cumulative Cultural Evolution: Is Imitation Necessary? *Psychological Science*, *20*, 1478–1483.

Clark, J. D. (1982). The Cultures of the Middle Paleolithic/Middle Stone Age. In J. D. Clark (Ed.), *The Cambridge History of Africa: From the Earliest Times to c. 500 BC* (pp. 248–340). New York: Cambridge University Press.

Clark, J. D. (1992). African and Asian Perspectives on the Origins of Modern Humans. *Philosophical Transactions of the Royal Society of London. Series B, Biological Sciences*, *337*, 201–215.

Clark, J. D., Beyene, Y., WoldeGabriel, G., Hart, W. K., Renne, P. R., Gilbert, H., et al. (2003). Stratigraphic, Chronological and Behavioural Contexts of Pleistocene *Homo sapiens* from Middle Awash, Ethiopia. *Nature*, *423*, 747–752.

Corbey, R., Jagich, A., Vaesen, K., & Collard, M. (2016). The Acheulean Handaxe: More Like a Bird's Song Than a Beatles' Tune? *Evolutionary Anthropology*, *25*, 6–19.

Fischer, A., Vemming Hansen, P., & Rasmussen, P. (1984). Macro and Micro Wear Traces on Lithic Projectile Points: Experimental Results and Prehistoric Examples. *Journal of Danish Archaeology*, *3*, 19–46.

Foley, R., & Gamble, C. (2009). The Ecology of Social Transitions in Human Evolution. *Philosophical Transactions of the Royal Society of London. Series B, Biological Sciences*, *364*, 3267–3279.

Foley, R. A., & Lahr, M. M. (1997). Mode 3 Technologies and the Evolution of Modern Humans. *Cambridge Archaeological Journal*, *7*, 3–36.

Foley, R., & Lahr, M. M. (2003). On Stony Ground: Lithic Technology, Human Evolution, and the Emergence of Culture. *Evolutionary Anthropology*, *12*, 109–122.

Gamble, C. (1999). *The Palaeolithic Societies of Europe*. Cambridge: Cambridge University Press.

Henrich, J. (2004). Demography and Cultural Evolution: How Adaptive Cultural Processes Can Produce Maladaptive Losses: The Tasmanian Case. *American Antiquity*, *69*, 197–214.

Henshilwood, C. S., d'Errico, F., Vanhaeren, M., Van Niekerk, K., & Jacobs, Z. (2004). Middle Stone Age Shell Beads from South Africa. *Science*, *304*, 404.

Henshilwood, C. S., d'Errico, F., & Watts, I. (2009). Engraved Ochres from the Middle Stone Age Levels at Blombos Cave, South Africa. *Journal of Human Evolution*, *57*, 27–47.

Henshilwood, C. S., d'Errico, F., van Niekerk, K. L., Coquinot, Y., Jacobs, Z., Lauritzen, S.-E., et al. (2011). A 100,000-Year-Old Ochre-Processing Workshop at Blombos Cave, South Africa. *Science*, *334*, 219–222.

Herries, A. I. R. (2011). A Chronological Perspective on the Acheulian and Its Transition to the Middle Stone Age in Southern Africa: The Question of the Fauresmith. *International Journal of Evolutionary Biology*, *2011*, 1–25.

Hodder, I. (1982). *Symbols in Action: Ethnoarchaeological Studies of Material Culture*. Cambridge: Cambridge University Press.

Kandler, A., & Laland, K. N. (2009). An Investigation of the Relationship between Innovation and Cultural Diversity. *Theoretical Population Biology*, *76*, 59–67.

Klein, R. G. (1988). The Archaeological Significance of Animal Bones from Acheulean Sites in Southern Africa. *African Archaeological Review*, *6*, 3–25.

Lombard, M., Parsons, I., & van der Ryst, M. M. (2004). Middle Stone Age Lithic Point Experimentation for Macro-Fracture and Residue Analyses: The Process and Preliminary Results with Reference to Sibudu Cave Points. *South African Journal of Science*, *100*, 159–166.

Marean, C. W. (2014). The Origins and Significance of Coastal Resource Use in Africa and Western Eurasia. *Journal of Human Evolution*, *77*, 17–40.

Marean, C. W. (2015). An Evolutionary Anthropological Perspective on Modern Human Origins. *Annual Review of Anthropology*, *44*, 533–556.

Marean, C. W., Anderson, R. J., Bar-Matthews, M., Braun, K., Cawthra, H. C., Cowling, R. M., et al. (2015). A New Research Strategy for Integrating Studies of Paleoclimate, Paleoenvironment, and Paleoanthropology. *Evolutionary Anthropology*, *24*, 62–72.

McBrearty, S., & Brooks, A. S. (2000). The Revolution That Wasn't: A New Interpretation of the Origin of Modern Human Behavior. *Journal of Human Evolution*, *39*, 453–563.

McDougall, I., Brown, F. H., & Fleagle, J. G. (2005). Stratigraphic Placement and Age of Modern Humans from Kibish, Ethiopia. *Nature*, *433*, 733–736.

Mesoudi, A., Chang, L., Dall, S. R. X., & Thornton, A. (2016). The Evolution of Individual and Cultural Variation in Social Learning. *Trends in Ecology & Evolution*, *31*, 215–225.

Mourre, V., Villa, P., & Henshilwood, C. S. (2010). Early Use of Pressure Flaking on Lithic Artifacts at Blombos Cave, South Africa. *Science*, *330*, 659–662.

Odell, G. H., & Cowan, F. (1986). Experiments with Spears and Arrows on Animal Targets. *Journal of Field Archaeology*, *13*, 195–212.

Oswalt, W. H. (1976). *An Anthropological Analysis of Food-Getting Technology*. New York: Wiley.

Pargeter, J. (2011). Human and Cattle Trampling Experiments in Malawi to Understand Macrofracture Formation on Stone Age Hunting Weaponry. *Antiquity*, *85*, 327.

Pleurdeau, D. (2006). Human Technical Behavior in the African Middle Stone Age: The Lithic Assemblage of Porc-Epic Cave (Dire Dawa, Ethiopia). *African Archaeological Review*, *22*, 177–197.

Porat, N., Chazan, M., Grün, R., Aubert, M., Eisenmann, V., & Horwitz, L. K. (2010). New Radiometric Ages for the Fauresmith Industry from Kathu Pan, Southern Africa: Implications for the Earlier to Middle Stone Age Transition. *Journal of Archaeological Science*, *37*, 269–283.

Potts, R. (1998). Variability Selection in Hominid Evolution. *Evolutionary Anthropology*, *7*, 81–96.

Powell, A., Shennan, S., & Thomas, M. G. (2009). Late Pleistocene Demography and the Appearance of Modern Human Behavior. *Science*, *324*, 1298–1301.

Richerson, P. J., & Boyd, R. (2005). *Not by Genes Alone: How Culture Transformed Human Evolution*. Chicago: University of Chicago Press.

Rightmire, G. P. (2001). Patterns of Hominid Evolution and Dispersal in the Middle Pleistocene. *Quaternary International*, *75*, 77–84.

Rightmire, G. P. (2004). Brain Size and Encephalization in Early to Mid-Pleistocene Homo. *American Journal of Physical Anthropology*, *124*, 109–123.

Sano, K. (2012). Functional Variability in the Magdalenian of North-western Europe: A Lithic Microwear Analysis of the Gönnersdorf K-Ii Assemblage. *Quaternary International*, *272*, 264–274.

Schoville, B. J. (2016). *Landscape Variability in Tool-Use and Edge Damage Formation in South African Middle Stone Age Lithic Assemblages*. Ph.D. dissertation, Arizona State University, Tempe.

Schoville, B. J., Brown, K. S., Harris, J. A., & Wilkins, J. (2016). New Experiments and a Model-Driven Approach for Interpreting Middle Stone Age Lithic Point Function Using the Edge Damage Distribution Method. *PLoS One*, *11*(10), e0164088.

Shea, J. J. (2006). The Origins of Lithic Projectile Point Technology: Evidence from Africa, the Levant, and Europe. *Journal of Archaeological Science*, *33*, 823–846.

Shea, J. J. (2011). *Homo sapiens* Is as *Homo sapiens* Was: Behavioral Variability versus "Behavioral Modernity" in Paleolithic Archaeology. *Current Anthropology*, *52*, 1–35.

Shea, J. J. (2013). Lithic Modes A-I: A New Framework for Describing Global-Scale Variation in Stone Tool Technology Illustrated with Evidence from the East Mediterranean Levant. *Journal of Archaeological Method and Theory*, *20*, 151–186.

Shea, J. J., Davis, Z., & Brown, K. (2001). Experimental Tests of Middle Palaeolithic Spear Points Using a Calibrated Crossbow. *Journal of Archaeological Science*, *28*, 807–816.

Shipton, C. (2010). Imitation and Shared Intentionality in the Acheulean. *Cambridge Archaeological Journal*, *20*, 197–210.

Sisk, M. L., & Shea, J. J. (2009). Experimental use and quantitative performance analysis of triangular flakes (Levallois points) used as arrowheads. *Journal of Archaeological Science*, *36*, 2039–2047.

Stringer, C. (2012). The Status of *Homo heidelbergensis* (Schoetensack 1908). *Evolutionary Anthropology*, *21*, 101–107.

Tennie, C., Call, J., & Tomasello, M. (2009). Ratcheting up the Ratchet: On the Evolution of Cumulative Culture. *Philosophical Transactions of the Royal Society of London. Series B, Biological Sciences*, *364*, 2405–2415.

Tomasello, M. (1999). *The Cultural Origins of Human Cognition*. Cambridge: Harvard University Press.

Tomasello, M., Carpenter, M., Call, J., Behne, T., & Moll, H. (2005). Understanding and Sharing Intentions: The Origins of Cultural Cognition. *Behavioral and Brain Sciences*, *28*, 675–735.

Tomasello, M., Kruger, A. C., & Ratner, H. H. (1993). Cultural Learning. *Behavioral and Brain Sciences*, *16*, 495–511.

Tostevin, G. B. (2007). Social Intimacy, Artefact Visibility, and Acculturation Models of Neanderthal—Modern Human Interaction. In P. Mellars, K. Boyle, O. Bar Yosef, & C. Stringer (Eds.), *Rethinking the Human Revolution: New Behavioural and Biological Perspectives on the Origins and Dispersal of Modern Humans* (pp. 341–357). Cambridge: McDonald Institute for Archaeological Research.

Tryon, C. A. (2006). "Early" Middle Stone Age Lithic Technology of the Kapthurin Formation (Kenya). *Current Anthropology*, *47*, 367–375.

Tryon, C. A., & McBrearty, S. (2006). Tephrostratigraphy of the Bedded Tuff Member (Kapthurin Formation, Kenya) and the Nature of Archaeological Change in the Later Middle Pleistocene. *Quaternary Research*, *65*, 492–507.

Tryon, C. A., McBrearty, S., & Texier, P.-J. (2005). Levallois Lithic Technology from the Kapthurin Formation, Kenya: Acheulian Origin and Middle Stone Age Diversity. *African Archaeological Review*, *22*, 199.

Underhill, D. (2011). The Study of the Fauresmith: A Review. *South African Archaeological Bulletin*, *66*, 15–26.

Villa, P., & Lenoir, M. (2006). Hunting Weapons of the Middle Stone Age and the Middle Palaeolithic; Spear Points from Sibudu, Rose Cottage and Bouheben. *Southern African Humanities*, *18*, 89–122.

Villa, P., Boscato, P., Ranaldo, F., & Ronchitelli, A. (2009). Stone Tools for the Hunt: Points with Impact Scars from a Middle Paleolithic Site in Southern Italy. *Journal of Archaeological Science*, *36*, 850–859.

Wadley, L. (2001). What Is Cultural Modernity? A General View and a South African Perspective from Rose Cottage Cave. *Cambridge Archaeological Journal, 11*, 201–221.

Wasielewski, H. (2014). Imitation Is Necessary for Cumulative Cultural Evolution in an Unfamiliar, Opaque Task. *Human Nature, 25*, 161–179.

Watts, I., Chazan, M., & Wilkins, J. (2016). Early Evidence for Brilliant Ritualized Display: Specularite Use in the Northern Cape (South Africa) between ~500 Ka and ~300 Ka. *Current Anthropology, 57*, 287–309.

White, T. D., Asfaw, B., DeGusta, D., Gilbert, H., Richards, G. D., Suwa, G., et al. (2003). Pleistocene *Homo sapiens* from Middle Awash, Ethiopia. *Nature, 423*, 742–747.

Wiessner, P. (1983). Style and Social Information in Kalahari San Projectile Points. *American Antiquity, 48*, 253–276.

Wilkins, J. (2010). Style, Symboling, and Interaction in Middle Stone Age Society. *Explorations in Anthropology, 10*, 102–125.

Wilkins, J. (2013). *Technological Change in the Early Middle Pleistocene: The Onset of the Middle Stone Age at Kathu Pan 1, Northern Cape, South Africa*. Ph.D. dissertation, University of Toronto, Toronto.

Wilkins, J., & Chazan, M. (2012). Blade Production ~500 Thousand Years Ago at Kathu Pan 1, South Africa: Support for a Multiple Origins Hypothesis for Early Middle Pleistocene Blade Technologies. *Journal of Archaeological Science, 39*, 1883–1900.

Wilkins, J., & Schoville, B. J. (2016). Edge Damage on 500-Thousand-Year-Old Spear Tips from Kathu Pan 1, South Africa: The Combined Effects of Spear Use and Taphonomic Processes. In R. Iovita & K. Sano (Eds.), *Multidisciplinary Approaches to the Study of Stone Age Weaponry* (pp. 101–117). New York: Springer.

Wilkins, J., Pollarolo, L., & Kuman, K. (2010). Prepared Core Reduction at the Site of Kudu Koppie in Northern South Africa: Temporal Patterns across the Earlier and Middle Stone Age Boundary. *Journal of Archaeological Science, 37*, 1279–1292.

Wilkins, J., Schoville, B. J., Brown, K. S., & Chazan, M. (2012). Evidence for Early Hafted Hunting Technology. *Science, 338*, 942–946.

Wilkins, J., Schoville, B. J., Brown, K. S., & Chazan, M. (2015). Kathu Pan 1 Points and the Assemblage-Scale, Probabilistic Approach: A Response to Rots and Plisson, "Projectiles and the Abuse of the Use-Wear Method in a Search for Impact. *Journal of Archaeological Science, 54*, 294–299.

Wynn, T., & Coolidge, F. L. (2011). The Implications of the Working Memory Model for the Evolution of Modern Cognition. *International Journal of Evolutionary Biology, 2011*, 741357.

Yaroshevich, A., Kaufman, D., Nuzhnyy, D., Bar-Yosef, O., & Weinstein-Evron, M. (2010). Design and Performance of Microlith Implemented Projectiles during the Middle and the Late Epipaleolithic of the Levant: Experimental and Archaeological Evidence. *Journal of Archaeological Science, 37*, 368–388.

10 Small, Sharp, and Standardized: Global Convergence in Backed-Microlith Technology

Chris Clarkson, Peter Hiscock, Alex Mackay, and Ceri Shipton

The question "How common is convergence?" remains unanswered and may be unanswerable. Our examples indicate that even the minimum detectable levels of convergence are often high, and we conclude that at all levels convergence has been greatly underestimated.
—Moore and Willmer (1997: 1)

Convergence in stone-tool technology, much like in biology, was likely a recurring phenomenon throughout the last three million years of human evolution, where functional and economic constraints exerted strong selection on tool size and form as well as other characteristics of technological systems. Some of the best examples of convergent stone working include the Nubian Levallois method (Will, Mackay, and Phillips 2015); overshot flaking of Solutrean and Paleoindian points (Eren, Patten, O'Brien, and Meltzer 2013b; chapter 1, this volume); fluting on Paleoindian and southern Arabian points (Crassard 2009); ground-edge axe technology in Pleistocene Australasia, Japan, and multiple Neolithic societies (Clarkson et al. 2015; Hiscock, O'Connor, Balme, and Maloney 2016; Takashi 2012); pressure blade technology in Mesoamerica and Eurasia (Crabtree 1968; Pelegrin 2003); and punch flaking on Danish and Polynesian adzes (Shipton, Weisler, Jacomb, Clarkson, and Walter 2016; Stueber 2010). Likewise, countless more or less identical tool forms appear around the globe in different times and places as the product of seemingly independent invention to meet local needs, be they burins, end scrapers, blades, or discoidal cores.

The question is not whether convergence took place, but whether it was common and widespread or took place only under exceptional circumstances. There are many reasons for thinking it was the former, but providing compelling evidence for independent origins without contact between regions, as well as deriving robust evolutionary explanations, are ongoing challenges for archeology. Multiple lines of evidence are required to test such arguments, and these might typically involve experimentation, modeling selective environments, and developing appropriate means of analyzing archeological and environmental data to determine the context of autochthonous development rather than cultural transmission from other populations.

Here we propose that the backing of microliths—applying steep, blunting retouch along one edge—is a highly evolvable trait that emerged many times in different places around the world for the specific advantages it conferred in certain contexts. This view contradicts a popular notion that the microlithic technology emerged once in Africa, then spread as a package with modern humans to neighboring regions of Europe and Southwest Asia and eventually to Asia and Australia (Mellars 2006)—an idea that replicates much earlier attempts to use microliths to track population movements across these continents (Brown 1899). To deconstruct that idea, we briefly review the record of microlith origins worldwide, illustrating that microlithic technology occurs not only on separate continents at vastly different time periods but also among different hominin species. We argue that this is because backing was both highly discoverable and advantageous, and as such the trait evolved into near-identical lithic industries in many parts of the world, despite a lack of any recent historical connections. To better frame this argument, we consider the issue of what is, in fact, advantageous and evolvable about backing. We present experimental results that show the existence of key properties shared by backed microliths that may have been selected multiple times in the past—some of which are functional constraints created by the backing itself, whereas others are desirable properties that likely conveyed certain advantages to their users.

Deconstructing the "Microlithic Package"

The creation of microliths is often depicted as a single entity, including geometric (e.g., triangles, rhomboids, trapezoids, and rectangular pieces) and nongeometric (e.g., crescents and points) pieces that are backed along one edge and supplied by a systematic microblade technology. As we noted above, this technology is often considered so distinctive that some researchers believe that its multiple appearances in Africa, Eurasia, and Australia must have been linked by a single common historical process such as migration or diffusion. Mellars (2006; Mellars, Gori, Carr, Soares, and Richards 2013), for example, argued that the spread of microlithic technology through southern and eastern Africa, and then to Southwest Asia, South Asia, Europe, and Australia, was a direct consequence of the exit from Africa of modern human populations that carried with them a microlithic technology. In fact, the microlithic package comprised three separate elements: backing, miniaturization (the systematic production of small flakes from fine-grained stone [Pargeter 2016]), and microblade production from prismatic cores. These three phenomena often occur independently of one another and cannot be uncritically treated as a package wherever one is seen to occur (Pargeter 2016). Here we focus on backed microliths.

Backed microliths have such vastly different ages and distributions worldwide that they cannot possibly be connected to a single journey out of Africa (figures 10.1 and 10.2). We have shown elsewhere (Clarkson, Harris, and Shipton 2017; Clarkson, Petraglia, Harris,

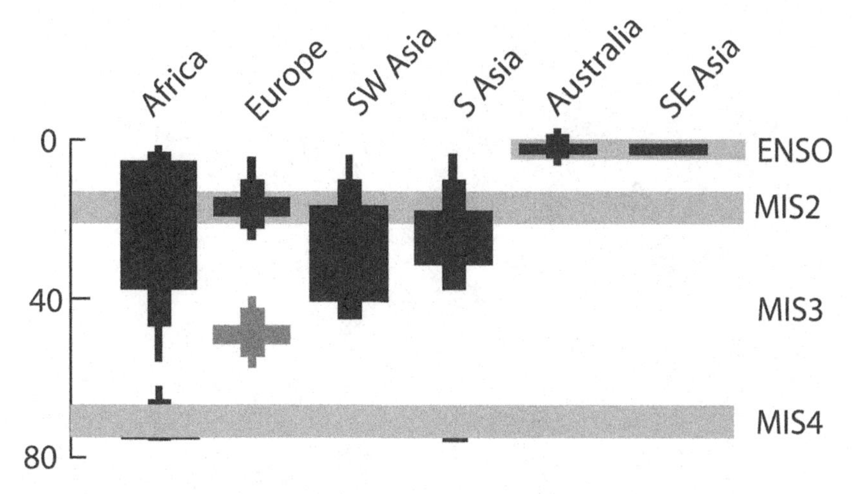

Figure 10.1
Approximate global timing and frequency distribution of backing technologies in relation to major climatic shifts (gray represents Neanderthals).

Shipton, and Norman 2017b; Hiscock, Clarkson, and Mackay 2011) that microlithic industries emerged in different times and places out of existing local industries at times of dramatic climate change that likely generated significant economic risk. The oldest known backed microliths in the world occur in the Lupemban industry at Twin Rivers and Kalambo Falls in central Africa and may date to around 300,000 years ago, made by an unknown, and possibly archaic, hominin species (Barham 2002). The next appearance of backed microliths was at 71,000 years ago in southern Africa (Brown et al. 2012), with a profusion of production from 65,000 years ago during the Howieson's Poort interval (Jacobs et al. 2008). Howieson's Poort was sustained for at least 6,000 years, but few backed pieces occur after the onset of marine isotope stage 3. Microliths appeared initially in very low numbers in Sri Lanka around 65,000–70,000 years ago (Deraniyagala 1992) and in Australia from at least the terminal Pleistocene (Slack, Fullagar, Field, and Border 2004; but see Hamm et al. 2016). In all of these regions, backed microliths reappeared in large numbers after a long absence, around 8,000 years ago in South Africa (Deacon 1982), 40,000 years ago in India (Baskak, Srivastava, Dasgupta, Kumar, and Rajaguru 2014; Clarkson et al. 2009; Mishra, Chauhan, and Singhvi 2013; Perera et al. 2011), and 3,000 years ago in southern Australia (Hiscock and Attenbrow 1998). In the latter two places they trailed off again into the late Holocene.

More damaging still for arguments in favor of a single package is the appearance of microliths and backing technology among Neanderthals in both the Mousterian of the Acheulean tradition and the Châtelperronian industries of southwest France and the Uluzzian of southern Italy, both dating before 40,000 years ago (see chapter 3, this volume).

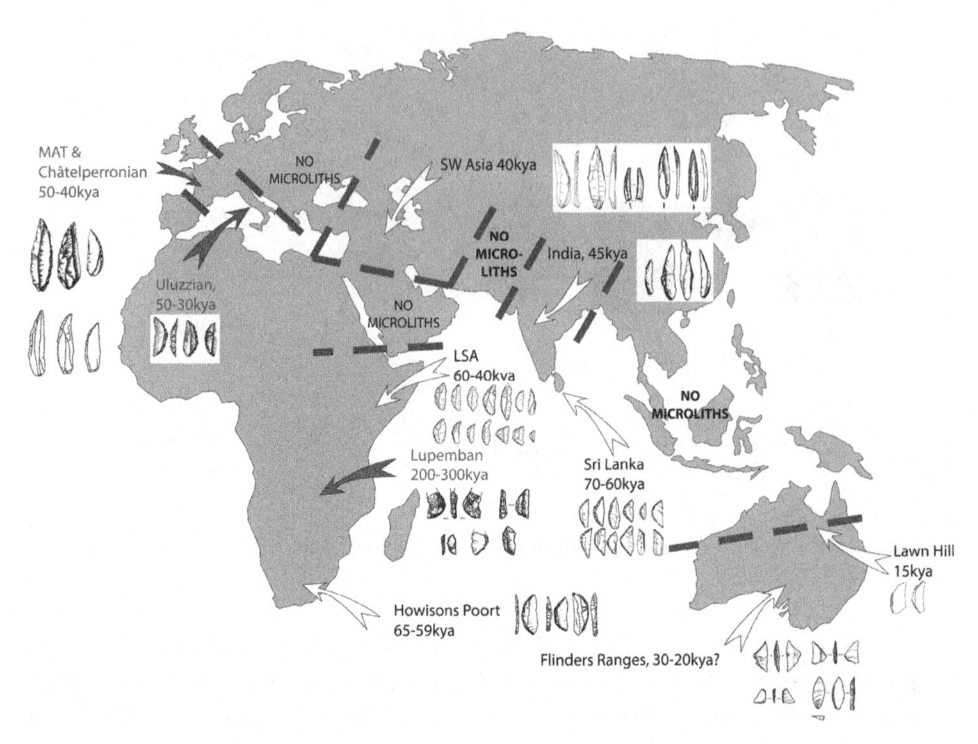

Figure 10.2
Map showing the locations and ages of the first/oldest Pleistocene occurrence of microlithic industries in various regions, as well as the locations of large geographical and temporal gaps in microlithic distribution. Gray arrows refer to industries made by Neanderthals, and white arrows refer to an unknown, potentially archaic, hominin.

Although debates are ongoing over whether the Châtelperronian and Uluzzian might have resulted from modern humans or acculturation of microlithic technology by Neanderthals (Mellars 2005), Neanderthal use of backing technology in southern France is undisputed, as this industry dates to well before modern humans entered Europe (d'Errico 2003; d'Errico, Zilhão, Julien, Baffier, and Pelegrin 1998; Koumouzelis et al. 2001; Peresani 2008).

The link between backed-microlith technology and microblade technology is also overstated. Indian and Southwest Asian backed-microlithic technology is associated with microblade technology, but this is not true for other regions. Howeisons Poort microliths in the western and southern Cape regions of South Africa are frequently made on small flakes, many from Levallois cores (Clarkson 2010; Porraz et al. 2013); the later Robberg industry (18,000–12,000 years ago) has abundant microblades but few backed artifacts, while the Holocene Wilton industry (8,000–4,000 years ago) has numerous backed pieces but few microblades (Deacon 1982; Lombard et al. 2012). Uluzzian, Sri Lankan, and

Australian microliths are commonly made on small flakes and bipolar flakes (Hiscock 2002; Lewis, Perera, and Petraglia 2014; Perera et al. 2011; Peresani 2008), and microblades are usually rare.

The varied association between microliths and microblade technology, the large gaps between time periods and regions where microliths are found, and the clear association between microliths and Neanderthals in some parts of Europe are all facts that challenge the notion of a single microlithic package connected by shared recent common ancestry or a single migration episode out of Africa. The record is, in fact, explicable as recurring, independent emergence of microliths across parts of Africa, Asia, and Australia. The patchy distribution in time and space demonstrates that microliths were invented/rediscovered many times over, most likely driven by the advantages they offered in solving common economic problems experienced by different groups in different places. Just what those advantages might have been is explored below.

Advantages of Microlithic Technology: Concepts and Experimental Tests

Archaeologists have explored a range of potential advantages offered by microlithic technology in certain contexts. These are usually framed in terms of the problems generated by high mobility and uncertainty over the scheduling of opportunities to procure stone. Here we outline a range of advantageous characteristics thought to be associated with the use of microliths in composite technologies. As few tests have ever been conducted to determine whether these characteristics hold up under conditions of manufacture and use, we also present the results of several preliminary tests that explored the performance characteristics of microliths. The concepts outlined below include transportability, efficiency of raw-material use, ease of manufacture, standardization, haftability, maintainability, and reliability.

Transportability

Put simply, microliths are very small and weigh very little, thus making them highly transportable, even in large numbers. A sample of assemblages from Middle Stone Age southern Africa, early Lower Stone Age East Africa, Upper Paleolithic India, and Holocene Australia demonstrates that microliths are on average half the size of unretouched flakes and less than a third the size of retouched flakes, making them far more transportable than alternative tools and blanks (table 10.1).

The manufacture of microliths in advance of use therefore greatly increases the transportability of the toolkit and alleviates the need to carry heavy raw material, much of which may in any case be discarded after knapping. Further, microliths are typically hafted (the backing itself is thought to be a hafting modification to aid adhesion), and, as discussed below, when used in serial configurations on multicomponent tools, the need

Table 10.1
Comparison of mean mass of microliths from the eastern and southern African MSA, Indian Upper Paleolithic, and Holocene Australia

Mean	South African MSA (71–59 kya)	East African early LSA (c.50 kya)	Indian Upper Paleolithic (45–35 kya)	Holocene Australia (4–1 kya)[3]
Microliths	2.34	1.22	1.7	1.19
Flakes	5.45	2.12	4.9	4.3
Retouched flakes	9.65	10.29	15.7	26.57

LSA, Later Stone Age; MSA, Middle Stone Age.

to carry replacement stone may be further reduced, as multiple replacement parts are already present.

Efficiency of Raw-Material Use

The manufacture of small tools also increases the efficiency of raw-material use in terms of cutting edge—total length of sharp edge around the flake margins—per gram of stone. Muller and Clarkson (2016) found that efficiency in raw-material use is most easily attained by creating small, thin flakes (see also Mackay 2008). This is exemplified in the manufacture of microblades that are elongate, thin, and narrow, which significantly increases the efficiency of raw material (Bleed 2002; Muller and Clarkson 2016).

Ease of Manufacture

Small flakes are among the easiest and most abundant products of stone knapping, requiring only small amounts of force and raw material and expedient hammerstones. This is particularly true of bipolar reduction, which can involve short, repetitive actions and limited problem solving to produce desirable tool shapes (Muller and Clarkson 2016). Bipolar knapping is easily learned and may be applied to almost any raw material of any size, often yielding microblade-like flakes. In an experiment, bipolar knapping was performed on small nodules of quartz, flint, and obsidian, reducing the stone through anvil striking until no further flakes could be detached. Following Arnold's (1987) definition of microblades as thin flakes less than 5 centimeters long with a length-to-width ratio equal to or greater than 2:1, we found that all raw materials from best (obsidian) to worst (milky quartz) quality yielded microblades well suited to backing at a rate of 5–20 per kilogram of stone per minute. A comparison of time spent preparing cores between strikes for different core technologies revealed that bipolar cores can be prepared more quickly, with only 50 percent of the total manufacture time spent preparing, turning, and inspecting bipolar cores versus 85 percent of the time for blade cores.

Standardization

One of the key advantages argued for microlithic technology is that it produces highly standardized replicates with little effort. We hypothesize that this is likely because, first, the blanks typically require little retouching to transform them from a flake or microblade into a microlith (requiring only abrupt trimming of one edge) and second, little skill or training is required to learn and complete the task. To test this proposition, we constructed an experiment involving 10 novice to intermediate knappers who were asked to copy two stone implements, a microlith and a bifacial point. The aim of the experiment was to determine the ease and accuracy with which a microlith could be copied using only a small stone hammerstone and pressure flaker, as opposed to a more challenging implement, in this case a bifacial point (figure 10.3). A tray of flakes suitable for making a small bifacial point was provided for the participants to choose from. For the microlith experiment, participants chose from both a tray of small flakes and a tray of microblades and copied

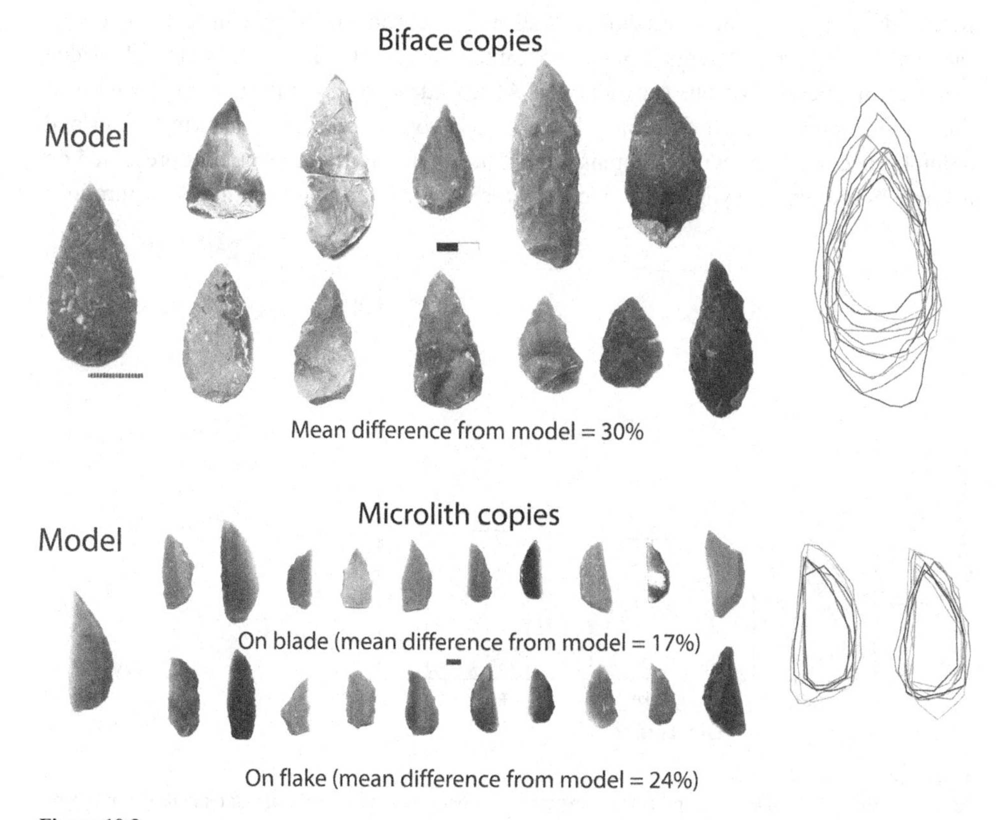

Figure 10.3
Results of the microlith- and point-copying exercise, showing models, actual results, and overlayed outlines of all copies.

the microlith on both blank types. The resulting quality of the copy was ranked by the senior author with a rank of 1, indicating a very good copy, and a rank of 5 indicating a very poor copy.

Ranking was based on the following criteria in order of importance: outline symmetry, flatness of faces, size, shape, and thinness. Length, width, and thickness of each copy were also recorded and compared to the model in order to calculate the degree of deviation. As predicted, results indicated that backed tools made from both flakes and microblades were produced with higher accuracy than points (figures 10.3 and 10.4); microliths made on microblades were the most accurate copies overall. They deviated from the metric dimensions (length, width, and thickness) of the model by only 17 percent on average, and microliths made on flakes by 24 percent, whereas points deviated by an average of 30 percent.

Functional Advantages

Microliths can be extremely versatile tool elements, capable of being employed in a wide range of hafting arrangements (figure 10.5) suited to different tasks. Use-wear and residue analyses of microliths from Australia and Africa attest to the multifunctional nature of these tools, indicating use as projectile tips and barbs, awls for skin working, plant and animal processing, and wood scraping, sometimes with multiples use traces preserved on a single microlith, implying multifunctional tools with a variety of hafting configurations

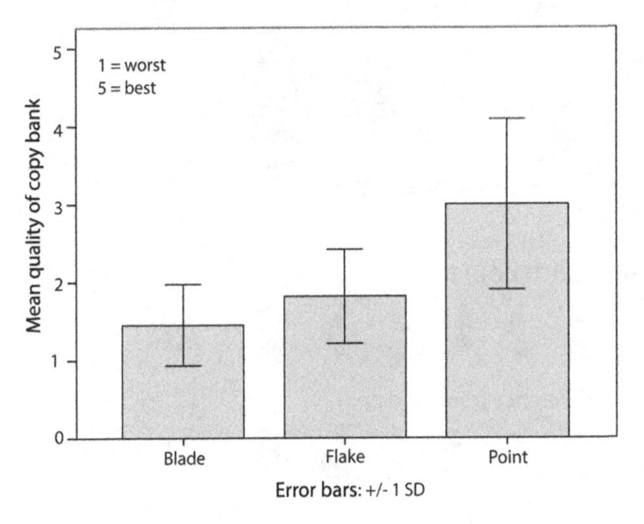

Figure 10.4
Results of the stone-implement-reproduction experiment, comparing quality of copy for microliths made on flakes and blades versus bifacial points made on larger flakes.

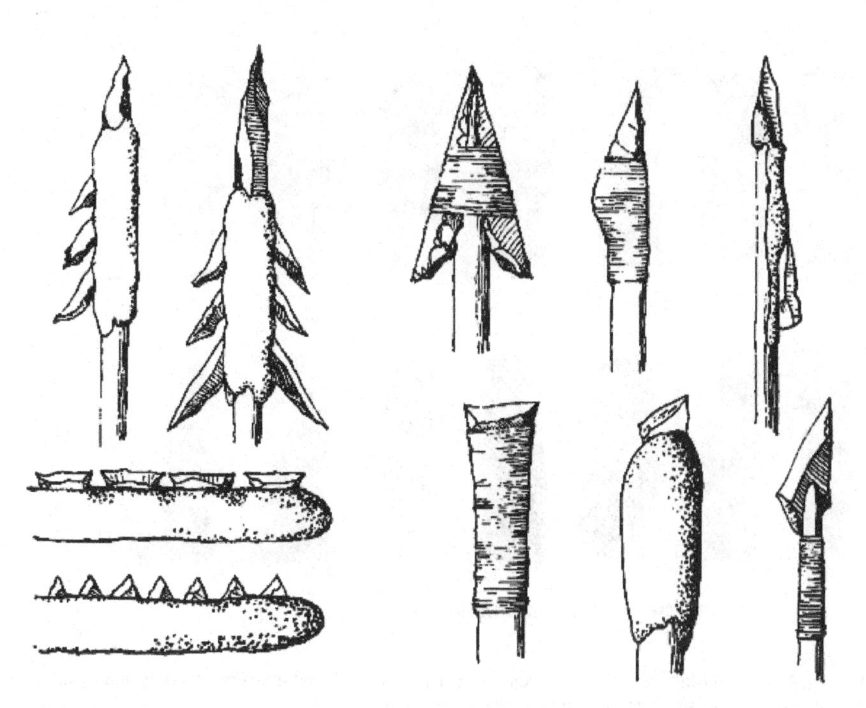

Figure 10.5
Ethnographic and archaeological examples of microlith hafting arrangements (from McCarthy 1967).

(Lombard 2005, 2008; Lombard and Phillipson 2010; McDonald et al. 2007; Robertson, Attenbrow, and Hiscock 2009).

Haftability

Microliths can be used in multicomponent weapons and tools as serial attachments of sharp and narrow inserts that sit neatly against a haft and do not overly protrude. Further, the backed edge itself may aid in adhesion to a haft by increasing surface area and "roughness," thereby aiding adhesion with gums and resins. To test whether backed microlithic inserts remain attached to a haft more frequently than unbacked flakes and microblades, we manufactured composite projectile heads in which four microliths (figure 10.6) were mounted into a shallow groove on one side of the head and four unbacked microblades into a slightly deeper groove on the other side to compensate for slightly greater width. Inserts were mounted with a composite adhesive made of beeswax, charcoal, and fiber (figure 10.7). The heads were inserted into a hollow 2-meter-long bamboo javelin and thrust into a barrel of 1:10 gravel/loam mix. The experiment revealed that microblades fell out on average two to three times more than microliths did (figure 10.8). We argue this

Figure 10.6
Examples of the ca. 3-cm-long flint microblade segments used in the adhesion and snap tests. Half the microblades were backed along one edge to serve as microliths in the experiment.

was because of the greater adhesion offered by the backed edge, given that microblades and microliths protruded an equal distance from the haft. Hence, protrusion was unlikely to have affected the results. On average, it took 50 thrusts before a microlith detached versus only 19 thrusts for a microblade to detach ($t = 2.42$, df = 23, $p = 0.02$).

Maintainability

Serial microlithic inserts represent a highly maintainable design whereby lost or broken inserts can be easily replaced with standardized duplicates (Bleed 1986; Hiscock 2002; Myers 1989). This implies that microliths are cheap and expendable items in the toolkit. It also implies that microliths can be easily produced to tight morphological specifications. Because microliths typically were joined to composite tools using adhesive, they could be reattached very quickly—in under a few minutes with gentle heating of adhesives, often using only the adhesive remaining attached to the haft or insert. Compare this to situations where notching, lashing, and adhesives are employed to attach tool bits, which takes significantly longer. Further, damage to hafts containing notched and tied implements are more severe and can be unrepairable, resulting in lengthy down times (Akerman 1978).

Figure 10.7
Composite projectile heads with microblades (M) on one side and backed microliths (B) on the other. The shaft on the left is of dense wood (gidgee), and the one on the right is of antler.

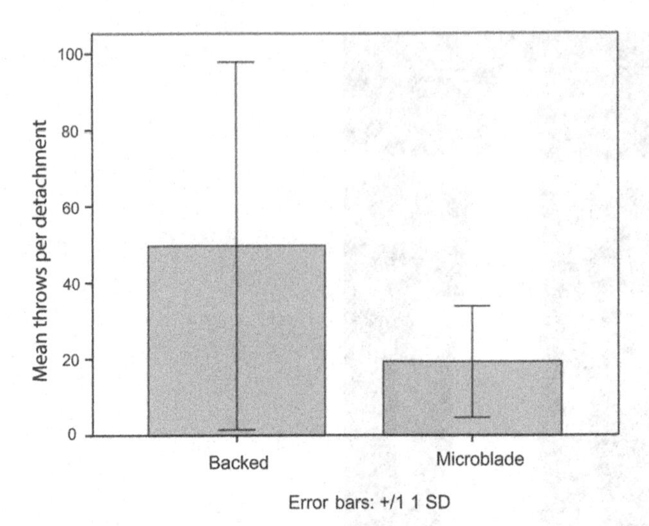

Figure 10.8
Rate of loss for microliths versus microblades from composite spear heads thrust repeatedly into a loam and gravel mix.

Reliability

A further advantage offered by the use of serial backed microlithic insert technology is the ability for tools to remain serviceable even when one or several inserts have been lost (Bleed 1986; Hiscock 2002; Myers 1989). Use of multiple cutting edges, barbs, or even points could mean that loss of some inserts does not render the tool useless. In an experiment in which sickles employing multiple backed inserts in a series were used to cut thick grass stalks for six hours (Clarkson and Shipton 2015), several inserts detached from the sickles, but the implements were still able to cut grass.

One further property that may be enhanced by backing the edge of thin flakes and microblades is durability, or, in this case, resistance to bending failure. Making a narrow, thin blade potentially weakens it as it becomes prone to bending failure. Making a thicker blade reduces bending failure but can result in increased width, thus making the flake blank undesirable for backing and use as a microlithic insert. Backing offers a solution to this problem by thickening the blade relative to width. As shown in figure 10.9, as height increases as a proportion of the base width, the elastic section modulus increases. Because thin, narrow flakes are not always easy to make, backing offers a solution to creating a thin implement that is thick relative to width.

To test whether this works in practice, we snapped 70 elongate microblades to create blades that were 3 centimeters long and 15 millimeters wide. Thirty-five of them were backed along one edge using pressure flaking, whereas the remaining edges were left unbacked. The microblades and microliths were then fastened one by one into a vice and

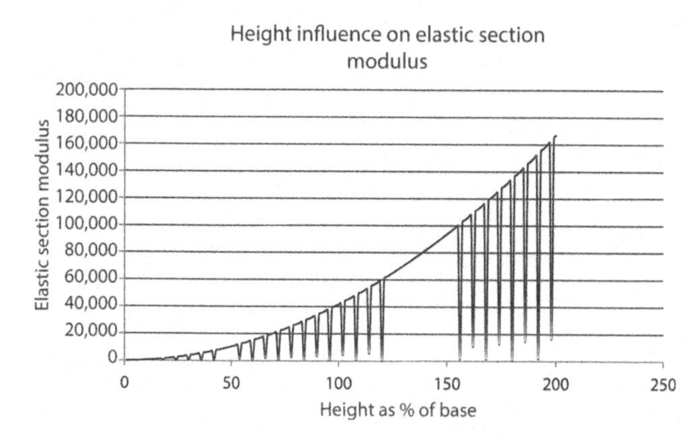

Figure 10.9
The influence of height relative to base width on elastic section modulus for brittle solids.

a pull exerted on their midpoints with a spring scale until they snapped. The spring scale registered the weight of maximum pull at the point of breakage, allowing us to compare the relative strength of microliths versus microblades. The results are shown in figure 10.10. We found that microliths required twice the force to incur breakage through bending failure than microblades (t = 4.49, df = 69, p = < 0.0005). It appears, then, that backed microblades confer an advantage in terms of strength that was likely important in tasks where tool bits experienced hefty bending stresses, ensuring greater durability and longer functioning of the tool.

The Case for Convergence in Microlithic Industries in Five Regions

The striking resemblance of small backed implements across the globe but in vastly different times and places immediately raises the possibility of convergence. But under what conditions did such industries arise? Does the archeological record of key regions support the notion of local innovations arising as a result of strong selective pressures toward small, standardized, transportable, reliable, maintainable, and multifunctional tools that made efficient use of raw material and were easy to make and replace? We briefly the review the evidence for five regions—southern Africa, East Africa, Arabia, South Asia, and Australia—where microliths arose multiple times, proliferated, and died out. We argue that such pulses in microlith use are strongly associated with changes in mobility, often as a response to dramatic climate change that forced rapid adjustments in economy.

Southern Africa

Microliths are a recurrent feature of the southern African record, occurring intermittently in low frequencies through the last 120,000 years (Mackay 2016). Within this temporal

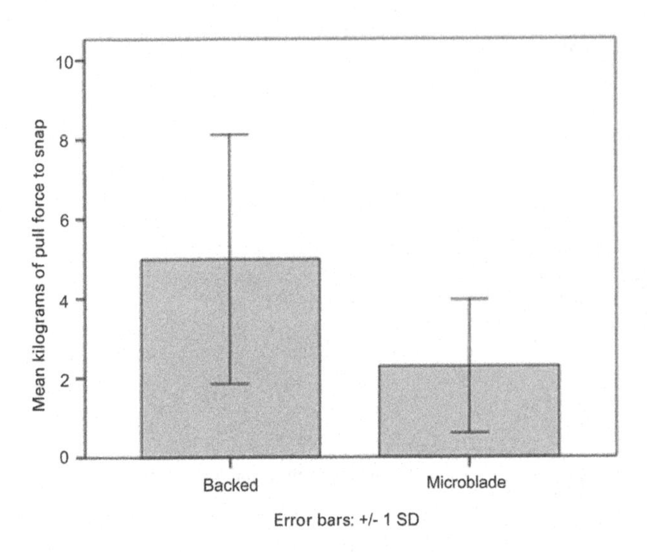

Figure 10.10
Results of snap test for force required to transversely snap microblades versus microliths.

range there were two distinct pulses of production, each associated with different climatic conditions. The oldest of these began around 71,000 years ago—for example, at the site of Pinnacle Point on the south coast (Brown et al. 2012)—although contested ages suggest that the technology may be as old or older on the west coast at Diepkloof (Jacobs et al. 2008; Tribolo et al. 2013). Throughout the latter half of MIS 4 (65,000–59,000 years ago), and broadly coeval with the latter half of this glacial period, microliths were prevalent across southern Africa in association with Howieson's Poort assemblages, which are both unusually widespread, occurring across multiple biomes (Mackay, Stewart, and Chase 2014).

Georcheological studies of site formation document greater intensity of site use in this period. Which might explain these elevated densities (Karkanas, Brown, Fisher, Jacobs, and Marean 2015; Miller, Goldberg, and Berna 2013). Paleoclimatic data suggest that late MIS 4 may have been a period of atypical humidity (Chase 2010); increased water security may thus have altered land-use systems in ways resulting in extended periods of site occupancy at key locations (Mackay and Hallinan 2017). Interestingly, however, both faunal and lithic data are consistent with economization behaviors. Flakes, cores, and nonbacked elements of the Howieson's Poort toolkit are all small (de la Pena and Wadley 2014; Mackay 2009), with elevated edge length to mass values characterizing flake assemblages (Mackay 2008). Faunal assemblages from sites on either side of the subcontinent reflect increased diet breadth and a greater emphasis on small package items during Howieson's Poort than during earlier or later periods (Clark and Kandel 2013; Steele and Klein 2013). Although predation pressure associated with increased population (Powell et al. 2009) might explain the faunal data, the spread of sites, and the density of artifacts, it doesn't readily explain the economization of technological elements. Alternatively, it may be that increased humidity facilitated a change in land-use patterns, with weak food security under glacial conditions encouraging frugal use of lithic resources and greater investment in reliable technological systems.

With diminishing humidity into early MIS 3, microliths ceased to be produced in large numbers throughout much of southern Africa (Lombard et al. 2012; Wurz 2013). From 50,000 to 25,000 years ago, the archaeological signal in the southwest of southern Africa dramatically attenuates, with little evidence of occupation (Faith 2013; Mackay et al. 2014; Mitchell 2008) and no microliths (Deacon 1979; Mackay et al. 2014; Mackay, Jacobs, and Steele 2015; Wendt 1972). The occupational signal in the southeast is stronger, but there is evidence of only intermittent production of microliths and at only a few sites (Clark 1999; Kaplan 1990; Opperman 1996; Stewart et al. 2012; Wadley 2005). Microblade production became common in the next glacial phase, MIS 2, during the Robberg industry, but, as noted earlier, microliths in the sense we use the term here remained scarce (Mackay et al. 2014).

The next sustained period of microlith production did not occur until after 8,000 years ago, some 50,000 years after the end of Howieson's Poort. During the period known as the Wilton (8,000–2,000 years ago), microliths were again the dominant tool type. As with Howieson's Poort, Wilton is particularly widespread and associated with small implements generally and shows an increase in small, often bipolar cores (Deacon 1982; Leslie Brooker 1989; Wadley 2000). Paleoclimatic conditions during Wilton were distinct from those prevailing during Howieson's Poort, the latter associated with humid glacial conditions and the former with the warmest conditions of the last 130,000 years. Rainfall appears to have been regionally variable during this period. In some regions, the Wilton period occurs in the context of a general postglacial drying trend (Chase et al. n.d.; Cowling, Cartwright, Parkington, and Allsopp 1999; Scott and Woodborne 2007), whereas in others it occurs in association with a period of relative humidity. There was a notable persistence, however, of microlith production over the last 2,000 years along the arid and semiarid west coast of South Africa (Orton, Hart, and Halkett 2004). These artifacts are also probably the smallest microliths known in southern Africa, with examples regularly less than 10 millimeters in maximum dimension (Orton 2012), reflecting the ready association between microlith production and raw-material economization under conditions of low primary productivity.

In summary, microliths in southern Africa are best characterized by two discrete phases of amplified production separated by an extremely long period of irregular, low-frequency occurrence. The conditions under which increased microlith production occurred were highly variable, including the humid, glacial MIS 4 and the arid, warm mid- to late Holocene. More than anything, these data emphasize both the range of conditions under which microliths were adaptive and their tendency for reinvention even within regions.

East Africa

In East Africa a number of sites have yielded backed crescents that date to ca. 50,000–4,000 years ago, including Mumba at the southern end of the Gregory Rift in Tanzania (Diez-Martín et al. 2009; Gliganic, Jacobs, Roberts, Domínguez-Rodrigo, and Mabulla 2012), Panga ya Saidi on the coast of Kenya (Shipton et al. 2016), Enkapune ya Muto in the central Gregory Rift in Kenya (Ambrose 1998), and Mochena Borago in southern Ethiopia (Brandt et al. 2012). The emergence of backed crescents was clearly preceded by a shift in emphasis to bipolar flaking and small flakes at Mumba and Panga ya Saidi (Diez-Martín et al. 2009; Eren, Diez-Martín, and Domínguez-Rodrigo 2013a; Shipton et al. 2016). Small-flake production also characterizes the lithic industry preceding the earliest crescents at Mochena Borago. Faunal remains from Mochena Borago indicate relatively dry environments when these early backed artifacts were being produced (Brandt et al. 2012), with paleoenvironmental records suggesting the onset of relatively dry conditions about 50,000 years ago (Blome, Cohen, Tryon, Brooks, and Russell 2012). Tryon and Faith (2016) also correlate backed microliths at Nasera Rockshelter in Tanzania with drier, more open environments.

Several Late Pleistocene East African sites either lack occupation or lack evidence for backing during later MIS3 and MIS2 (Ambrose 1998; Gliganic et al. 2012; Helm et al. 2012; Pleurdeau et al. 2014; Shipton et al. 2016). At these sites, backing then reappeared in the Holocene, where it was associated with systematic blade production. This appears to be a classic case of reinvention, with backing being technologically distinct and separated by several thousand years.

At Panga ya Saidi, the two phases where backing appears in the record follow a very similar pattern. Compared to what occurred in the preceding phases, blades have relatively high instances of retouch, there are higher percentages of cores, and there is a more formal Levallois technology (table 10.2). This suggests a pattern of high mobility (Kuhn 1995). The environmental evidence from Panga ya Saidi indicates the phases with backing were relatively drier than the preceding phases, though not dramatically so (Shipton et al. 2016).

Table 10.2
Comparison of layers with backing and the preceding layers at Panga ya Saidi, Kenya

	Late backing phase at Panga ya Saidi (Layers 5 & 6, N=4615)	Preceding phase (Layers 7 & 8, N=3756)	Early backing phase at Panga ya Saidi (Layers 11 & 12, N=4548)	Preceding phase (Layer 13 & 14, N=2446)
Percent retouch	0.6	0	0.7	0
Percent chert	34.45	2.24	31.27	1.76
Percent cores	1.28	0.43	4.22	0.86
Ratio of Levallois to bipolar cores and flakes	0.08	(No Levallois)	0.15	0.03

The evidence from Panga ya Saidi and elsewhere in East Africa suggests that backing emerged during drier periods, when there was a need for higher mobility.

Arabia

Backed microliths are extremely rare on the Arabian Peninsula (Hilbert et al. 2014; Kallweit 2004). One dated site that has yielded a relatively large assemblage is Al-Rabyah at the Jubbah oasis in northern Arabia (Hilbert et al. 2014). There, occupation beside a drying paleolake is dated to about 10,000 years ago. Most of the Al-Rabyah tools are of exotic materials that cannot be sourced in the Jubbah Basin. Thirty-seven percent of the pieces were retouched, suggesting a highly mobile population. Tools from Al-Rabyah bear close resemblance to those from the Geometric Kebaran industry of the Levant, including backed bladelets, inversely retouched bladelets, obliquely truncated bladelets, and trapezoidal backed geometrics (Macdonald, Chazan, and Janetski 2016). Geometric Kebaran sites display the widest distribution of the Levantine Epipaleolithic industries (Goring-Morris 1987; Goring-Morris, Hovers, and Belfer-Cohen 2009; Schuldenrein 1986), so their presence at Jubbah fits with a pattern of broad dispersal. It is noteworthy that Al-Rabyah represents the first evidence of occupation at the Jubbah oasis since the Middle Paleolithic (Petraglia et al. 2012). The finds from Al-Rabyah indicate backed-microliths were associated with a highly mobile and geographically expanding population.

South Asia

The current lithic evidence for major industrial transitions in two regions of India has led us to form several conclusions (Clarkson et al. 2017a; Clarkson et al. 2017b). First, technological change in the lead-up to microlith production was gradual. In two regions studied in detail (the Kurnool district and the Middle Son River district), the microlithic industry emerged gradually out of classic Middle Paleolithic industries characterized by Levallois and scraper technology. Blade technology became increasingly common after 50,000 years ago in both regions, with core technology shifting from centripetal to unidirectional and bidirectional orientations. At the same time, blades, burins, and fine-grained raw-material use became common, and blades diminished in size.

Current evidence suggests that microlithic industries were not present in India prior to 44,000 years ago and that the Thar Desert likely acted as a barrier to new populations dispersing into the region (Petraglia et al. 2009). The oldest known microlithic sites are located inland, where climatic changes would have quickly taken effect, reinforcing the idea that the microlithic may have been a response to drying. Further, Lewis et al.'s (2014) comparison of South Asian and African microliths revealed significant differences in core technology and morphology, suggesting the South Asian microlithic was probably not derived from an African dispersal. The East African microlithic was often based on bipolar technology, which was not a feature of the South Asian microlithic. The earliest microlithic industry in South Asia may also have arisen at the end of a major demographic expansion, when innovation rates were high (Petraglia et al. 2009), but became common only once climatic conditions worsened, as also seen in Australia (Attenbrow, Robertson, and Hiscock 2009). Microlithic technology was also adopted later in the south, where rainforest vegetation might have buffered local populations from deteriorating climatic conditions experienced in the northern interior. The Sri Lankan microlithic lacks microblades and also contains points, which are likely to be a local development. Finally, the gradual adoption of microlithic technology occurred alongside a weakening monsoon, with few early microliths appearing in the interior north prior to their spread across South Asia.

The appearance of the microlithic in South Asia beginning around 44,000 years ago seems to support the notion that foragers discovered and developed microlithic technology at a time of dramatic climatic and technological change. It seems this took place over many millennia, with an early origin in the central north—regions that are today highly drought-prone—when populations likely were large. They later spread across South Asia as climatic conditions deteriorated. Widespread technological changes in the late Middle Paleolithic appear to have led to microblade production, with backed microliths appearing after that.

Australia

In Australia, a clear but imprecise coincidence existed between the start of a period of intensive microlith production and the Holocene onset of an El Niño–Southern Oscillation (ENSO) climate that involved reduced effective precipitation and increased climatic variability (Hiscock 2017). An increasingly high-resolution record of climate proxies is being created by several lines of paleoenvironmental research (e.g., Black, Mooney, and Attenbrow 2008; Petherick et al. 2013), but the cultural timeline is currently of far lower resolution, limiting analysis of the covariation in temporal phases. Table 10.3, taken from Hiscock (2002), shows an inverse relationship between effective precipitation and microlith production rates in southeastern Australia. Hiscock's interpretation was that the onset

Table 10.3
Summary of trends in effective precipitation and microlith abundance for the southeastern region of Australia (from Hiscock 2002)

Years BP	Effective precipitation	Microlith production
0–2,000	Increasing but still variable	Low
2–4,000	Low and variable	High
4–5,000	Declining	Increasing
> 5,000	High	Very low

of more-variable climatic conditions created a context in which resource distribution and availability were less easily and reliably mapped or predicted. Adjustments to foraging practices and social interactions were made to reduce the increased risk attached to these circumstances, and a component of the response was the increased use of tools containing a specific insert to produce organic craft items. When effective precipitation increased in the last 2,000 years, microlith production rates declined as the technological system responded to other pressures affecting foragers.

Environmental changes that occurred 3,000–4,000 years ago were not limited to climate change. Human introduction of the dingo altered faunal structures, with serious implications for foragers. The effectiveness of this new high-order predator is revealed by its likely role in exterminating the thylacine from mainland Australia (Fillios, Crowther, and Letnic 2012; Letnic, Fillios, and Crowther 2012), but the really dramatic change was wrought by its impact on large and small fauna. Comparisons between modern regions with and without dingoes indicate that the dingo would have suppressed kangaroo to a small fraction of their previous numbers. The onset of this direct competition with human foragers created resource reductions at almost the same time that the ENSO amplification magnified the climatically driven resource decrease. Although long-term climatic fluctuations of greater size had occurred in the Pleistocene, this conjunction of both climate-derived and competition-derived depression of terrestrial resources had few, if any, parallels in Australian prehistory.

Reduction of terrestrial fauna required altered foraging strategies. A shift in hunting emphasis toward smaller game may have occurred, although elements of the small-mammal fauna may also have been depleted. The economic balance between hunting and gathering probably shifted, as we have evidence of increased exploitation of lower-ranked plant resources as ENSO intensification increased subsistence risks and lowered productivity (Asmussen and McInnes 2013). Additionally, as terrestrial game availability was reduced and coastal ecosystems stabilized following sea-level stabilization, marine resources probably took on a significant role in human diets. In concert, these multiple resource shifts would have required new economic systems that pursued the different resource structures and available biomass that were in place by around 3,500 years ago.

Microlith production was emphasized during the proliferation event, approximately 4,000–2,000 years ago in eastern Australia, in new economic and social contexts. Australian microliths normally were not used directly as extractive tools, but rather as processing tools. Although some of the processing may have been involved in food preparation, the primary use of hafted microliths was in making objects from hides, wood, bones, and feathers. The objects could have been extractive tools such as wooden spears, boomerangs, and throwing and digging sticks. Craft objects could have been made for storage and transportation or for shelter, clothing, and other practical purposes. Such objects may all have had value in negotiating economic actions during the onset and intensification of conditions of lower and less-predictable resource availability around 3,000–4,000 years ago. It is possible that the craft items provided social signals that acted to mediate human actions.

David and Lourandos (1998) suggested that stone tools such as microliths acted as social mechanisms in the negotiation of territory and group composition. Hiscock (2017), however, argues that this proposition can be only partially correct, at best, because it is now clear that although microliths may themselves have had social meaning, their low visibility limited their value as signals. Hence it was the craft items, perhaps elaborately decorated, being manufactured with the microliths that were used as part of social negotiations in processes of cultural change. Those changes were situated in a context of resource scarcity and unpredictability and may therefore have involved patterns of migration, altered territorial boundedness, and new social arrangements for resource access. It is likely that in such contexts new tools were not only adding security through reduction of foraging risk, but were being used to express identity and to construct and manage rules about access to resources.

Conclusion

Convergence in microlithic technology is not only well supported by the archeological evidence, it is the only viable explanation for the widespread but temporally disparate appearance of backing technology in far-flung regions with no recent common ancestry or direct contact. As the chapters in this volume demonstrate, convergence is a common evolutionary process in cultural and biological systems, and accepting this proposition and adopting appropriate methods to test for its existence will become a major goal in future cultural-evolutionary explanations. Experimentation to understand selective advantages of certain properties of stone tools, phylogenetic analysis, and careful attention to archeological context will be critical in determining the role convergence has played in shaping human diversity and the evolutionary unfolding of the human past.

References

Akerman, K. (1978). Notes on the Kimberley Stone-Tipped Spear 544 Focusing on the Point Hafting Mechanism. *Mankind, 11*, 486–490.

Ambrose, S. H. (1998). Chronology of the Later Stone Age and Food Production in East Africa. *Journal of Archaeological Science, 25*, 377–392.

Arnold, J. E. (1987). Technology and Economy: Microblade Core Production from the Channel Islands. In J. K. Johnson & C. A. Morrow (Eds.), *The Organization of Core Technology* (pp. 207–237). Boulder, Colo.: Westview Press.

Asmussen, B., & McInnes, P. (2013). Assessing the Impact of Mid-to-Late Holocene ENSO-driven Climate Change on Toxic Macrozamia Seed Use: A 5000 Year Record from Eastern Australia. *Journal of Archaeological Science, 40*, 471–480.

Attenbrow, V., Robertson, G., & Hiscock, P. (2009). The Changing Abundance of Backed Artefacts in South-Eastern Australia: A Response to Holocene Climate Change? *Journal of Archaeological Science*, *36*, 2765–2770.

Barham, L. (2002). Backed Tools in Middle Pleistocene Central Africa and Their Evolutionary Significance. *Journal of Human Evolution*, *43*, 585–603.

Baskak, B., Srivastava, P., Dasgupta, S., Kumar, A., & Rajaguru, S. N. (2014). Earliest Dates and Implications of Microlithic Industries of Late Pleistocene from Mahadebbera and Kana, Purulia District, West Bengal. *Current Science*, *107*, 1167–1171.

Black, M. P., Mooney, S. D., & Attenbrow, V. (2008). Implications of a 14,200 Year Contiguous Fire Record for Understanding Human-Climate Relationships at Goochs Swamp, New South Wales, Australia. *Holocene*, *18*, 437–447.

Bleed, P. (1986). The Optimal Design of Hunting Weapons: Maintainability or Reliability. *American Antiquity*, *51*, 737–747.

Bleed, P. (2002). Cheap, Regular, and Reliable: Implications of Design Variation in Late Pleistocene Japanese Microblade Technology. In R. G. Elston & S. L. Kuhn (Eds.), *Thinking Small: Global Perspectives on Microlithization* (pp. 95–102). Archeological Papers, no. 12. Washington, D.C.: American Anthropological Association.

Blome, M. W., Cohen, A. S., Tryon, C. A., Brooks, A. S., & Russell, J. (2012). The Environmental Context for the Origins of Modern Human Diversity: A Synthesis of Regional Variability in African Climate 150,000–30,000 Years Ago. *Journal of Human Evolution*, *62*, 563–592.

Brandt, S. A., Fisher, E. C., Hildebrand, E. A., Vogelsang, R., Ambrose, S. H., Lesur, J., et al. (2012). Early MIS 3 Occupation of Mochena Borago Rockshelter, Southwest Ethiopian Highlands: Implications for Late Pleistocene Archaeology, Paleoenvironments and Modern Human Dispersals. *Quaternary International*, *274*, 38–54.

Brown, J. A. (1899). On Some Small Highly Specialized Forms of Stone Implements, Found in Asia, North Africa, and Europe. *Journal of the Anthropological Institute of Great Britain and Ireland*, *18*, 134–139.

Brown, K. S., Marean, C. W., Jacobs, Z., Schoville, B. J., Oestmo, S., Fisher, E. C., et al. (2012). An Early and Enduring Advanced Technology Originating 71,000 Years Ago in South Africa. *Nature*, *491*, 590–593.

Chase, B. M. (2010). South African Palaeoenvironments during Marine Oxygen Isotope Stage 4: A Context for the Howiesons Poort and Still Bay Industries. *Journal of Archaeological Science*, *37*, 1359–1366.

Chase, B. M., J. T. Faith, A. Mackay, M. Chevalier, A. S. Carr, A. Boom, S. Lim, and P. J. Reimer, n.d. Climatic Controls on Later Stone Age Human Adaptation in Africa's Southern Cape. Unpublished manuscript.

Clark, A. M. B. (1999). Late Pleistocene Technology at Rose Cottage Cave: A Search for Modern Behavior in an MSA Context. *African Archaeological Review*, *16*, 93–119.

Clark, J. L., & Kandel, A. W. (2013). The Evolutionary Implications of Variation in Human Hunting Strategies and Diet Breadth during the Middle Stone Age of Southern Africa. *Current Anthropology*, *54*, S269–S287.

Clarkson, C. (2010). Regional Diversity within the Core Technology of the Howiesons Poort Techno-Complex. In S. Lycett & P. Chauhan (Eds.), *New Perspectives on Old Stones: Analytical Approaches to Paleolithic Technologies* (pp. 43–59). New York: Springer.

Clarkson, C., & Shipton, C. (2015). Teaching Ancient Technology Using "Hands-on" Learning and Experimental Archaeology. *Ethnoarchaeology*, *7*, 157–172.

Clarkson, C., Petraglia, M., Korisettar, R., Haslam, M., Boivin, N., Crowther, A., et al. (2009). The Oldest and Longest Enduring Microlithic Sequence in India: 35 000 Years of Modern Human Occupation and Change at the Jwalapuram Locality 9 Rockshelter. *Antiquity*, *83*, 1–23.

Clarkson, C., Smith, M., Marwick, B., Fullagar, R., Wallis, L. A., Faulkner, P., et al. (2015). The Archaeology, Chronology and Stratigraphy of Madjedbebe (Malakunanja II): A Site in Northern Australia with Early Occupation. *Journal of Human Evolution, 83*, 46–64.

Clarkson, C., Harris, C., & Shipton, C. (2017a, in press). Lithics at the Crossroads: A Review of Technological Transitions in South Asia. In R. Korisettar (Ed.), *Archaeology of India*. Delhi: Primus.

Clarkson, C., Petraglia, M., Harris, C., Shipton, C., & Norman, K. (2017b, in press). The South Asian Microlithic: *Homo sapiens* Dispersal or Adaptive Response? In E. Robinson & F. Sellet (Eds.), *Lithic Technological Organization and Paleoenvironmental Change*. New York: Springer.

Cowling, R. M., Cartwright, C. R., Parkington, J. E., & Allsopp, J. C. (1999). Fossil Wood Charcoal Assemblages from Elands Bay Cave, South Africa: Implications for Late Quaternary Vegetation and Climates in the Winter-Rainfall Fynbos Biome. *Journal of Biogeography, 26*, 367–378.

Crabtree, D. E. (1968). Mesoamerican Polyhedral Cores and Prismatic Blades. *American Antiquity, 1968*, 446–478.

Crassard, R. (2009). Modalities and Characteristics of Human Occupations in Yemen during the Early/Mid-Holocene. *Comptes Rendus Geoscience, 341*, 713–725.

d'Errico, F. (2003). The Invisible Frontier. A Multiple Species Model for the Origin of Behavioral Modernity. *Evolutionary Anthropology, 12*, 188–202.

d'Errico, F., Zilhão, J., Julien, M., Baffier, D., & Pelegrin, J. (1998). Neanderthal Acculturation in Western Europe? A Critical Review of the Evidence and Its Interpretation. *Current Anthropology, 39*, S1–S44.

David, B., & Lourandos, H. (1998). Rock Art and Socio-Demography in Northeast Australian Prehistory. *World Archaeology, 30*, 193–219.

de la Pena, P., & Wadley, L. (2014). Quartz Knapping Strategies in the Howiesons Poort at Sibudu. *PLoS One, 9*(7), e101534.

Deacon, H. J. (1979). Excavations at Boomplaas Cave—A Sequence through the Upper Pleistocene and Holocene in South Africa. *World Archaeology, 10*, 241–257.

Deacon, J. (1982). *The Later Stone Age in the Southern Cape, South Africa*. Ph.D. dissertation, University of Stellenbosch, Stellenbosch, South Africa.

Deraniyagala, S. U. (1992). *The Prehistory of Sri Lanka: An Ecological Perspective*, vol. 2. Department of Archaeological Survey, Government of Sri Lanka, Colombo.

Diez-Martín, F., Domínguez-Rodrigo, M., Sánchez, P., Mabulla, A. Z., Tarriño, A., Barba, R., et al. (2009). The Middle to Later Stone Age Technological Transition in East Africa: New Data from Mumba Rockshelter Bed V (Tanzania) and Their Implications for the Origin of Modern Human Behavior. *Journal of African Archaeology, 2009*, 147–173.

Eren, M. I., Diez-Martín, F., & Domínguez-Rodrigo, M. (2013a). An Empirical Test of the Relative Frequency of Bipolar Reduction in Beds VI, V, and III at Mumba Rockshelter, Tanzania: Implications for the East African Middle to Late Stone Age Transition. *Journal of Archaeological Science, 40*, 248–256.

Eren, M. I., Patten, R. J., O'Brien, M. J., & Meltzer, D. J. (2013b). Refuting the Technological Cornerstone of the Ice-Age Atlantic Crossing Hypothesis. *Journal of Archaeological Science, 40*, 2934–2941.

Faith, J. T. (2013). Taphonomic and Paleoecological Change in the Large Mammal Sequence from Boomplaas Cave, Western Cape, South Africa. *Journal of Human Evolution, 65*, 715–730.

Fillios, M., Crowther, M., & Letnic, M. (2012). The Impact of the Dingo on the Thylacine in Holocene Australia. *World Archaeology, 44*, 118–134.

Gliganic, L. A., Jacobs, Z., Roberts, R. G., Domínguez-Rodrigo, M., & Mabulla, A. Z. P. (2012). New Ages for Middle and Later Stone Age Deposits at Mumba Rockshelter, Tanzania: Optically Stimulated Luminescence Dating of Quartz and Feldspar Grains. *Journal of Human Evolution, 62*, 533–547.

Goring-Morris, A. N. (1987). *At the Edge: Terminal Pleistocene Hunter—Gatherers in the Negev and Sinai.* British Archaeological Reports, International Series, no. 361, Oxford.

Goring-Morris, A. N., Hovers, E., & Belfer-Cohen, A. (2009). The Dynamics of Pleistocene and Early Holocene Settlement Patterns and Human Adaptations in the Levant: An Overview. In J. J. Shea & D. E. Lieberman (Eds.), *Transitions in Prehistory: Essays in Honor of Ofer Bar-Yosef* (pp. 185–252). Oxford: Oxbow.

Hamm, G., Mitchell, P., Arnold, L. J., Prideaux, G. J., Questiaux, D., Spooner, N. A., et al. (2016). Cultural Innovation and Megafauna Interaction in the Early Settlement of Arid Australia. *Nature, 539*, 280–283.

Helm, R., Crowther, A., Shipton, C., Tengeza, A., Fuller, D., & Boivin, N. (2012). Exploring Agriculture, Interaction and Trade on the Eastern African Littoral: Preliminary Results from Kenya. *Azania, 47*, 39–63.

Hilbert, Y. H., White, T. S., Parton, A., Clark-Balzan, L., Crassard, R., Groucutt, H. S., et al. (2014). Epipalaeolithic Occupation and Palaeoenvironments of the Southern Nefud Desert, Saudi Arabia, during the Terminal Pleistocene and Early Holocene. *Journal of Archaeological Science, 50*, 460–474.

Hiscock, P. (2002). Pattern and Context in the Holocene Proliferation of Backed Artifacts in Australia. In R. G. Elston & S. L. Kuhn (Eds.) *Thinking Small: Global Perspectives on Microlithization* (pp. 163–177). Archeological Papers, no. 12. Washington, D.C.: American Anthropological Association.

Hiscock, P. (2017, in press). Horizons of Change: Entanglement of Palaeoenvironment and Cultural Dynamic in Australian Lithic Technology. In E. Robinson & F. Sellet (Eds.), *Lithic Technological Organisation and Paleoenvironmental Change*. New York: Springer.

Hiscock, P., & Attenbrow, V. (1998). Early Holocene Backed Artefacts from Australia. *Archaeology in Oceania, 33*, 49–62.

Hiscock, P., Clarkson, C., & Mackay, A. (2011). Big Debates over Little Tools: Ongoing Disputes over Microliths on Three Continents. *World Archaeology, 43*, 653–664.

Hiscock, P., O'Connor, S., Balme, J., & Maloney, T. (2016). World's Earliest Ground-Edge Axe Production Coincides with Human Colonisation of Australia. *Australian Archaeology, 82*, 2–11.

Jacobs, Z., Roberts, R. G., Galbraith, R. F., Deacon, H. J., Grun, R., Mackay, A., et al. (2008). Ages for the Middle Stone Age of Southern Africa: Implications for Human Behavior and Dispersal. *Science, 322*, 733–735.

Kallweit, H. (2004). Lithics from the Emirates: The Abu Dhabi Airport Sites. *Proceedings of the Seminar for Arabian Studies, 34*, 139–145.

Kaplan, J. (1990). The Umhlatuzana Rock Shelter Sequence: 100 000 Years of Stone Age History. *Natal Museum Journal of Humanities, 2*, 1–94.

Karkanas, P., Brown, K. S., Fisher, E. C., Jacobs, Z., & Marean, C. W. (2015). Interpreting Human Behavior from Depositional Rates and Combustion Features through the Study of Sedimentary Microfacies at Site Pinnacle Point 5–6, South Africa. *Journal of Human Evolution, 85*, 1–21.

Koumouzelis, M., Ginter, B., Kozlowski, J. K., Pawlikowski, M., Bar-Yosef, O., Albert, R. M., et al. (2001). The Early Upper Palaeolithic in Greece: The Excavations in Klisoura Cave. *Journal of Archaeological Science, 28*, 515–539.

Kuhn, S. L. (1995). *Mousterian Lithic Technology*. Princeton, N.J.: Princeton University Press.

Leslie Brooker, M. (1989). The Holocene Sequence from Uniondale Rock Shelter in the Eastern Cape. *South African Archaeological Society Goodwin Series, 6*, 17–32.

Letnic, M., Fillios, M., & Crowther, M. (2012). Could Direct Killing by Larger Dingoes Have Caused the Extinction of the Thylacine from Mainland Australia? *PLoS One*, *7*(5), e34877.

Lewis, L., Perera, N., & Petraglia, M. (2014). First Technological Comparison of Southern African Howiesons Poort and South Asian Microlithic Industries: An Exploration of Inter-Regional Variability in Microlithic Assemblages. *Quaternary International*, *350*, 7–25.

Lombard, M. (2005). Evidence of Hunting and Hafting during the Middle Stone Age at Sibidu Cave, KwaZulu-Natal, South Africa: A Multianalytical Approach. *Journal of Human Evolution*, *48*, 279–300.

Lombard, M. (2008). Finding Resolution for the Howiesons Poort through the Microscope: Micro-residue Analysis of Segments from Sibudu Cave, South Africa. *Journal of Archaeological Science*, *35*, 26–41.

Lombard, M., & Phillipson, L. (2010). Indications of Bow and Stone-Tipped Arrow Use 64 000 Years Ago in KwaZulu-Natal, South Africa. *Antiquity*, *84*, 635–648.

Lombard, M., Wadley, L., Deacon, J., Wurz, S., Parsons, I., Mohapi, M., et al. (2012). South African and Lesotho Stone Age Sequence Updated. *South African Archaeological Bulletin*, *67*, 120–144.

Macdonald, D. A., Chazan, M., & Janetski, J. C. (2016). The Geometric Kebaran Occupation and Lithic Assemblage of Wadi Mataha, Southern Jordan. *Quaternary International*, *396*, 105–120.

Mackay, A. (2008). A Method for Estimating Edge Length from Flake Dimensions: Use and Implications for Technological Change in the Southern African MSA. *Journal of Archaeological Science*, *35*, 614–622.

Mackay, A. (2009). *History and Selection in the Late Pleistocene Archaeology of the Western Cape, South Africa.* Ph.D. dissertation, Australian National University, Canberra.

Mackay, A. (2016). Technological Change and the Importance of Variability: The Western Cape of South Africa from MIS 5 to MIS 2. In S. Jones & B. A. Stewart (Eds.), *Africa from MIS 6–2: Population Dynamics and Paleoenvironments* (pp. 49–63). Dordrecht, Netherlands: Springer.

Mackay, A., & Hallinan, E. (2017, in press). Provisioning Responses to Environmental Variation in the Late Pleistocene of Southern Africa. In E. Robinson & F. Sellet (Eds.), *Lithic Technological Organisation and Paleoenvironmental Change*. New York: Springer.

Mackay, A., Stewart, B. A., & Chase, B. M. (2014). Coalescence and Fragmentation in the Late Pleistocene Archaeology of Southernmost Africa. *Journal of Human Evolution*, *72*, 26–51.

Mackay, A., Jacobs, Z., & Steele, T. E. (2015). Pleistocene Archaeology and Chronology of Putslaagte 8 (PL8) Rockshelter, Western Cape, South Africa. *Journal of African Archaeology*, *13*, 71–98.

McCarthy, F. D. (1967). *Australian Aboriginal Stone Implements: Including Bone, Shell and Teeth Implements.* Sydney: Australian Museum.

McDonald, J. J., Donlon, D., Field, J. H., Fullagar, R. L., Coltrain, J. B., Mitchell, P., et al. (2007). The First Archaeological Evidence for Death by Spearing in Australia. *Antiquity*, *81*, 877–885.

Mellars, P. (2005). The Impossible Coincidence: A Single-Species Model for the Origins of Modern Human Behavior in Europe. *Evolutionary Anthropology*, *14*, 12–27.

Mellars, P. (2006). Going East: New Genetic and Archaeological Perspectives on the Modern Human Colonization of Eurasia. *Science*, *313*, 796–800.

Mellars, P., Gori, K. C., Carr, M., Soares, P. A., & Richards, M. B. (2013). Genetic and Archaeological Perspectives on the Initial Modern Human Colonization of Southern Asia. *Proceedings of the National Academy of Sciences of the United States of America*, *110*, 10699–10704.

Miller, C. E., Goldberg, P., & Berna, F. (2013). Geoarchaeological Investigations at Diepkloof Rock Shelter, Western Cape, South Africa. *Journal of Archaeological Science*, *40*, 3432–3452.

Mishra, S., Chauhan, N., & Singhvi, A. K. (2013). Continuity of Microblade Technology in the Indian Subcontinent since 45 Ka: Implications for the Dispersal of Modern Humans. *PLoS One*, *8*(7), e69280.

Mitchell, P. J. (2008). Developing the Archaeology of Marine Isotope Stage 3. *South African Archaeological Society Goodwin Series*, *10*, 52–65.

Moore, J., & Willmer, P. (1997). Convergent evolution in invertebrates. *Biological Reviews*, *72*, 1–60.

Muller, A., & Clarkson, C. (2016). Identifying Major Transitions in the Evolution of Lithic Cutting Edge Production Rates. *PLoS One*, *11*(12), e0167244.

Myers, A. (1989). Reliable and Maintainable Technological Strategies in the Mesolithic of Mainland Britain. In R. Torrence (Ed.), *Time, Energy and Stone Tools* (pp. 78–91). Cambridge: Cambridge University Press.

Opperman, H. (1996). Strathalan Cave B, North-Eastern Cape Province, South Africa: Evidence for Human Behaviour 29,000–26,000 years ago. *Quaternary International*, *33*, 45–53.

Orton, J. (2012). *Late Holocene Archaeology in Namaqualand, South Africa: Hunter-Gatherers and Herders in a Semi-arid Environment*. Oxford: University of Oxford Press.

Orton, J., Hart, T., & Halkett, D. (2004). Shell Middens in Namaqualand: Two Later Stone Age Sites at Rooiwalbaai, Northern Cape. *South African Archaeological Bulletin*, *60*, 24–30.

Pargeter, J. (2016). Lithic Miniaturization in Late Pleistocene Southern Africa. *Journal of Archaeological Science: Reports*, *10*, 221–236.

Pelegrin, J. (2003). Blade Making Techniques from the Old World. In K. Hirth (Ed.), *Experimentation and Interpretation in Mesoamerican Lithic Technology* (pp. 55–71). Salt Lake City: University of Utah Press.

Perera, N., Kourampas, N., Simpson, I. A., Deraniyagala, S. U., Bulbeck, D., Kamminga, J., et al. (2011). People of the Ancient Rainforest: Late Pleistocene Foragers at the Batadomba-Lena Rockshelter, Sri Lanka. *Journal of Human Evolution*, *61*, 254–269.

Peresani, M. (2008). A New Cultural Frontier for the Last Neanderthals: The Uluzzian in Northern Italy. *Current Anthropology*, *49*, 725–731.

Petherick, L., Bostock, H., Cohen, T. J., Fitzsimmons, K., Tibby, J., Fletcher, M.-S., et al. (2013). Climatic Records over the Past 30 Ka from Temperate Australia? A Synthesis from the Oz-INTIMATE Workgroup. *Quaternary Science Reviews*, *74*, 58–77.

Petraglia, M., Clarkson, C., Boivin, N., Haslam, M., Korisettar, R., Chaubey, G., et al. (2009). Population Increase and Environmental Deterioration Correspond with Microlithic Innovations in South Asia ca. 35,000 Years Ago. *Proceedings of the National Academy of Sciences of the United States of America*, *106*, 12261–12266.

Petraglia, M. D., Alsharekh, A., Breeze, P., Clarkson, C., Crassard, R., Drake, N. A., et al. (2012). Hominin Dispersal into the Nefud Desert and Middle Palaeolithic Settlement along the Jubbah Palaeolake, Northern Arabia. *PLoS One*, *7*(11), e49840.

Pleurdeau, D., Hovers, E., Assefa, Z., Asrat, A., Pearson, O., Bahain, J. J., et al. (2014). Cultural Change or Continuity in the Late MSA/Early LSA of Southeastern Ethiopia? The Site of Goda Buticha, Dire Dawa Area. *Quaternary International*, *343*, 117–135.

Porraz, G., Texier, P.-J., Archer, W., Piboule, M., Rigaud, J.-P., & Tribolo, C. (2013). Technological Successions in the Middle Stone Age Sequence of Diepkloof Rock Shelter, Western Cape, South Africa. *Journal of Archaeological Science*, *40*, 3376–3400.

Powell, A., Shennan, S., & Thomas, M. G. (2009). Late Pleistocene Demography and the Appearance of Modern Human Behavior. *Science*, *324*, 1298–1301.

Robertson, G., Attenbrow, V., & Hiscock, P. (2009). Multiple Uses for Australian Backed Artefacts. *Antiquity*, *83*, 296–308.

Schuldenrein, J. (1986). Paleoenvironment, Prehistory, and Accelerated Slope Erosion along the Central Israeli Coastal Plain (Palmahim): A Geoarchaeological Case Study. *Geoarchaeology*, *1*, 61–81.

Scott, L., & Woodborne, S. (2007). Pollen Analysis and Dating of Late Quaternary Faecal Deposits (Hyraceum) in the Cederberg, Western Cape, South Africa. *Review of Palaeobotany and Palynology*, *144*, 123–134.

Shipton, C., Weisler, M., Jacomb, C., Clarkson, C., & Walter, R. (2016). A Morphometric Reassessment of Roger Duff's Polynesian Adze Typology. *Journal of Archaeological Science: Reports*, *6*, 361–375.

Slack, M. J., Fullagar, R. L., Field, J. H., & Border, A. (2004). New Pleistocene Ages for Backed Artefact Technology in Australia. *Archaeology in Oceania*, *39*, 131–137.

Steele, T. E., & Klein, R. G. (2013). The Middle and Later Stone Age Faunal Remains from Diepkloof Rock Shelter, Western Cape, South Africa. *Journal of Archaeological Science*, *40*, 3453–3462.

Stewart, B. A., Dewar, G. I., Morley, M. W., Inglis, R. H., Wheeler, M., Jacobs, Z., et al. (2012). Afromontane Foragers of the Late Pleistocene: Site Formation, Chronology and Occupational Pulsing at Melikane Rockshelter, Lesotho. *Quaternary International*, *270*, 40–60.

Stueber, D. O. (2010). The Use of Indirect Percussion with Stone Punches for Manufacturing Rectangular Cross Section Type 1 Adzes during the Moa-Hunter Period of the Maori Culture. In H. G. Nami (Ed.), *New Zealand Experiments and Interpretation of Traditional Technologies: Essays in Honor of Errett Callahan* (pp. 325–343). Buenos Aires: Ediciones de Arqueología Contemporánea.

Takashi, T. (2012). MIS3 Edge-Ground Axes and the Arrival of the First *Homo sapiens* in the Japanese Archipelago. *Quaternary International*, *248*, 70–78.

Tribolo, C., Mercier, N., Douville, E., Joron, J. L., Reyss, J. L., Rufer, D., et al. (2013). OSL and TL Dating of the Middle Stone Age Sequence at Diepkloof Rock Shelter (South Africa): A Clarification. *Journal of Archaeological Science*, *40*, 3401–3411.

Tryon, C. A., & Faith, J. T. (2016). A Demographic Perspective on the Middle to Later Stone Age Transition from Nasera Rockshelter, Tanzania. *Philosophical Transactions of the Royal Society of London. Series B, Biological Sciences*, *371*, 20150238.

Wadley, L. (2000). The Wilton and Pre-Ceramic Post-classic Wilton Industries at Rose Cottage Cave and their Context in the South African Sequence. *South African Archaeological Bulletin*, *55*, 90–106.

Wadley, L. (2005). A Typological Study of the Final Middle Stone Age Stone Tools from Sibudu Cave, Kwazulu-Natal. *South African Archaeological Bulletin*, *60*, 51–63.

Wendt, W. E. (1972). Preliminary Report on an Archaeological Research Programme in South West Africa. *Cimbebasia B*, *2*, 1–61.

Will, M., Mackay, A., & Phillips, N. (2015). Implications of Nubian-like Core Reduction Systems in Southern Africa for the Identification of Early Modern Human Dispersals. *PLoS One*, *10*(6), e0131824.

Wurz, S. (2013). Technological Trends in the Middle Stone Age of South Africa between MIS 7 and MIS 3. *Current Anthropology*, *54*, S305–S319.

IV PATTERNS IN THE ARCHAEOLOGICAL RECORD

11 The Convergent Evolution of Serrated Points on the Southern Plains–Woodland Border of Central North America

Ashley M. Smallwood, Heather L. Smith, Charlotte D. Pevny, and Thomas A. Jennings

Here we present analyses of serrated projectile-point blades to identify evidence of evolutionary convergence in stone tools using an empirical case study from the prehistoric record of the Southern Plains–Woodland border in central North America. Point-edge serration was introduced in this region during the Late Paleoindian period (12,850–11,700 B.P.), variably used throughout the Archaic period, abandoned by the Middle Woodland period (2,100–1,500 B.P.), and introduced again in the Late Woodland to Early Mississippian/Late Prehistoric period (1,300–650 B.P.). We focus on the evolutionary relationship between the earliest serrated Late Paleoindian Dalton and the Late Prehistoric Scallorn point types and explore how a cultural-evolutionary approach can help identify convergent evolution in the stone-tool record and consider the behavioral situations in which convergence arose. First, we use cladistic analysis to generate a phylogenetic tree that shows hypotheses of relatedness, or, in this case, change within lineages that does not reflect ancestry. Second, we evaluate morphological similarities and differences using geometric morphometric analysis. Third, we consider point evolutionary trajectory and morphology as a response to adaptive challenges that caused two populations with at best dim ancestral connections to converge on the same tool design.

Background

Serrations are a series of isolated tooth-like projections on a tool margin (figure 11.1). In prehistoric stone-tool production, serrations were applied during the final steps, requiring the flintknapper to intentionally modify the edge to isolate portions of the margin. Because serration is an obvious feature of artifact form, it is often used as a diagnostic trait to define tool types.

The Study Area

The Southern Plains–Woodland border encompasses several physiographic provinces (Fenneman and Johnson 1946; see figure 11.2). The area was occupied repeatedly

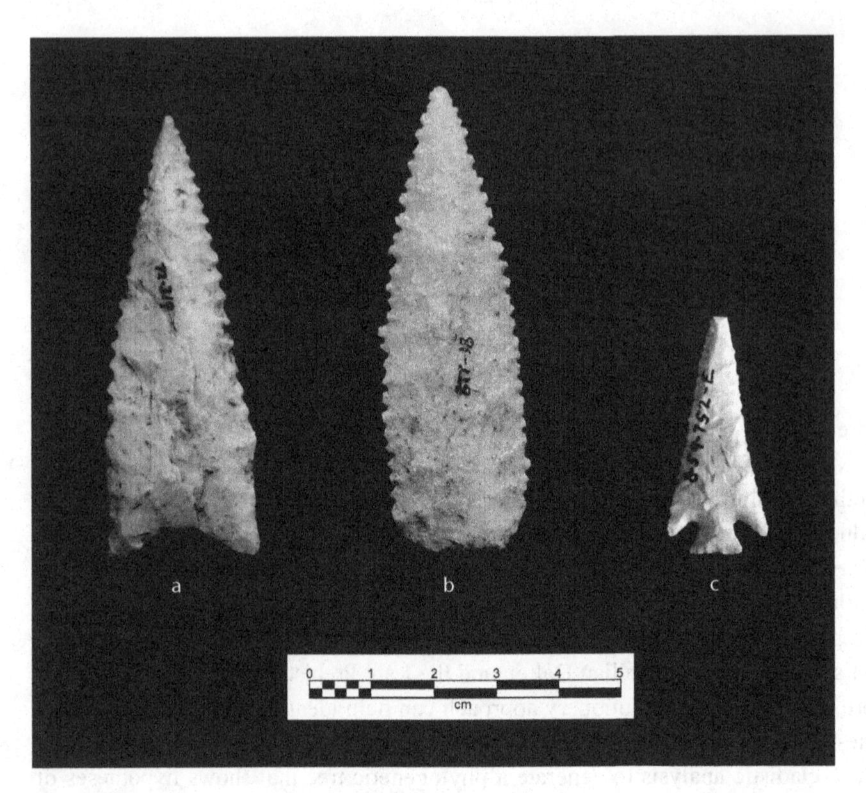

Figure 11.1
Examples of serrated-point types compared in this study: (a) Dalton, (b) Kimberley, and (c) Scallorn.

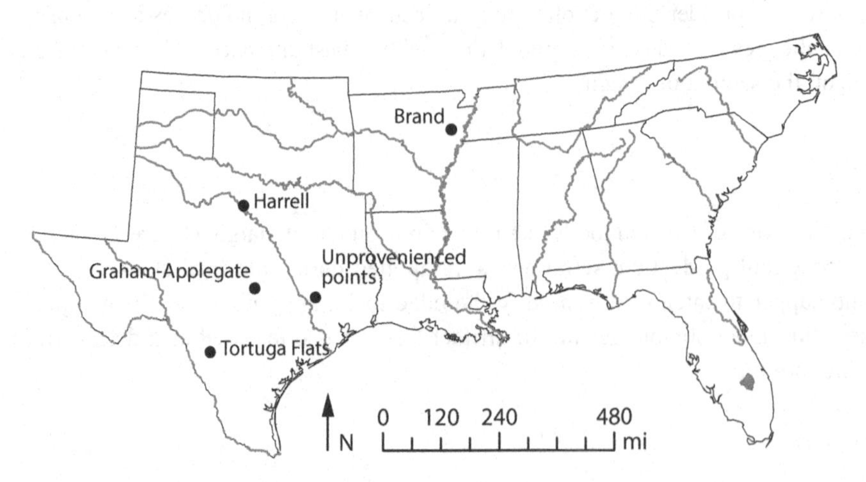

Figure 11.2
Map of the study area showing the location of sites that yielded serrated points used in this analysis.

throughout prehistory, from ca. 15,500 B.P. on, but most extensively after about 13,000 B.P., beginning with the well-documented Clovis complex (Waters and Stafford 2007; Waters et al. 2011). The region includes Dalton and Scallorn "heartlands," where the earliest and latest dated serrated-point types have been found, and it encompasses an area where Dalton and Scallorn points geographically co-occur.

The Evolution of Serrated Point Technology: Late Paleoindian Dalton Points

In North America, serrated bifacial points first occur in the portion of the archaeological record associated with the Late Paleoindian Dalton period (Anderson and Sassaman 2012). Dalton is associated with a date range of ca. 12,500–11,300 B.P. (Goodyear 1982). Components with Dalton points have been radiocarbon dated at sites throughout the Eastern Woodlands (e.g., Dust Cave and Stanfield-Worley, both in Alabama), and dates consistently demonstrate that Dalton postdates the Clovis complex and predates or co-occurs with Early Archaic side-notched complexes (Anderson, Smallwood, and Miller 2015; Miller and Gingerich 2013; Sherwood, Driskell, Randall, and Meeks 2004).

The Dalton complex spans the latter part of the Younger Dryas, a climatic episode dated to ca. 12,850–11,700 B.P. and marked by a reversal of general warming trends and a return to glacial-like conditions (Anderson et al. 2015). During the Younger Dryas, biotic communities underwent a major reorganization, but the impacts of such changes on Paleoindian populations are regular topics of debate (e.g., Anderson and Bissett 2015; Eren 2012; Meltzer and Holliday 2010). In a continental-scale extinction event, more than 30 genera of mammals disappeared by the first part of the Younger Dryas (Grayson and Meltzer 2015; Haynes 2009), including many species of megafauna on which Clovis hunters relied. The extinction of these large herbivores allowed for the competitive release of medium-bodied mammals (e.g., deer), requiring an adaptive shift in hunting strategies among later populations, including Dalton (Koldehoff and Walthall 2009). It is in this context that serrations first entered the archaeological record.

Dalton flintknappers crafted lanceolate points with concave bases, often with basal thinning (Bradley 1997; Morse 1971). Dalton-point blades are variable in form. Some are serrated along the lateral margins above the hafted area; others are serrated and beveled; and still others have no obvious modification beyond sharp edges. Less frequently, Dalton points exhibit burins or are blunted and rounded at the distal end. Functional hypotheses explain the variation in blade morphology as changes produced from knife use and progressive dulling, resharpening, and reworking (Goodyear 1974), or as forms deliberately crafted for varied hunting needs (O'Brien and Wood 1998). Both approaches recognize Dalton points functioned, either intermittently or exclusively, as dart points, likely thrown with an atlatl (Goodyear 1974; O'Brien and Wood 1998).

Dalton points occur throughout the Eastern Woodlands, from the Atlantic Coast in the East, south along the Gulf Coastal Plain, north to the Upper Mississippi River valley, and west into parts of the Central Lowlands (Anderson et al. 2015). Some of the best-documented Dalton sites are located along the Southern Plains–Woodland border. Because of the density and nature of Dalton sites in northeastern Arkansas, that area is referred to as the "Dalton Heartland" (Koldehoff and Walthall 2009). Variation in Dalton assemblages from sites such as Sloan, identified by Morse (1997) as a Dalton cemetery, and Lace Place, a possible long-term camp, have been used to model Dalton settlement (Ballenger 2001; Gillam 1996; Morse 1971; Schiffer 1975). Although these models disagree on the degree of logistical versus residential mobility, as well as drainage versus cross-drainage landscape use, they elude to underlying shifts in Paleoindian lifeways that emerged during Dalton times. As post-Pleistocene hunters, Dalton populations exploited a variety of resources, with an inferred emphasis on deer, and by the Late Paleoindian period, Dalton territory size apparently decreased, which likely required more-intensive use of local resources.

After the disappearance of Dalton, point types with serrations occurred variably throughout the Archaic period in the study area. However, during the subsequent Woodland period in the Southeast and the Late Archaic period in Texas, serrations decreased in frequency and became essentially absent by the Middle Woodland period, 2,100–1,500 B.P. That would soon change with the introduction of the bow and arrow.

Late Woodland and Mississippian Period Technological Transitions and Scallorn Points

The emergence of small, narrow-stemmed triangular and triangular notched points is often attributed to the technological shift to the bow and arrow (Thomas 1978). Although the timing of the adoption and nature of the spread of the bow and arrow varied regionally (see Nassaney and Pyle 1999 for a panregional overview), most researchers agree that by the Late Woodland period, ca. 1,300 B.P., bow-and-arrow technology was the dominant weapon system in the Eastern Woodlands (Blitz and Porth 2013; Nassaney and Pyle 1999).

Investigations of the transition from dart to arrow points consider changes in point variation and point size as a means to understand the change in projectile technology. Blitz and Porth (2013), who date the adoption of the bow in the Eastern Woodlands to approximately 1,700–1,600 B.P., identify a significant size-reduction threshold associated with Lowe-cluster points dated to the Middle Woodland–Late Woodland transition. This size reduction is attributed to the alteration of dart-point design for initial bow technology. Further, the authors conclude that a later reduction in thickness among smaller, lightweight triangular arrow points—forms such as Jack's Reef and Hamilton/Madison, which date to the Late Woodland–Mississippian transition, ca. 1,400–1,000 B.P.—represents a technological refinement.

Roughly 11,000 years after the introduction of serrated Dalton points, a new type of finely serrated projectile emerged coincident with the spread of the bow and arrow. Originally typed on the basis of specimens from Texas and named "Scallorn Stemmed" by Kelley (1947), Scallorn points are among the earliest dated arrow points in the region and have been identified at sites throughout the Mississippi Valley and the Midwest (Anderson and Smith 2003; McGahey 2000; O'Brien and Wood 1998). In Central Texas and along the Texas coast, Scallorn points have been radiometrically dated to between 1,270 B.P. and 650 B.P. (Lohse, Black, and Cholak 2014; Ricklis 2004a).

Scallorn points, considered temporally diagnostic of the Late Prehistoric Austin interval (Collins 2004; Prewitt 1981), are small, triangular, corner-notched points with straight to convex lateral edges and pronounced barbs (Turner, Hester, and McReynolds 2011). Stems can be straight or expanding, and bases are straight, convex, or concave. Fine serrations extend along the blade edges. Asphaltum has been noted on the stems of Scallorn points (Huebner and Comuzzie 1992), and its presence is inferred to be an adhesive to haft the points to shafts. Abraded and grooved "shaft straighteners" are also indications of this new technology (Hester 2004).

Scallorn points have been recovered from archaeological sites throughout the study area. The eastern half shares environmental and cultural affinities with the rest of the Southeastern Woodlands. In this area, Scallorn points were contemporaneous with the rise of the Caddo culture ca. 1,200 B.P. and associated with the use of pottery, increased sedentism, population growth, organizational complexity, intense reliance on agriculture, and a panregional, ceremonially based interaction network (Anderson and Mainfort 2002). The western portion of the study area is environmentally heterogeneous and includes savannah, prairies, plains, and desert. Annual rainfall steadily decreases to the west. Agriculture, if present at all, came late; varied depending on precipitation; and was not very important compared to what occurred in the Eastern Woodlands and on the High Plains. In Central Texas, Late Prehistoric Scallorn-point makers continued to live a hunter-gather lifestyle, and settlement and subsistence patterns were a continuation of adaptations of the preceding Late Archaic period (Collins 2004). Thus, the distribution of Scallorn points does not appear to coincide with a particular ecological niche or subsistence strategy.

Materials

The samples we examined include Dalton and Scallorn points recovered from buried contexts at sites that contributed to defining the characteristics of the Late Paleoindian Dalton period and the Late Prehistoric Austin interval. These samples were limited to points that have complete bases, serrated edges, and a minimum of 7 millimeters of blade length.

Goodyear (1974) recovered a Dalton assemblage from the Brand site (30PO139) located in northeastern Arkansas. A total of 305 Dalton points were reported, and 77 of these have serrated blades. Goodyear explored variation in Dalton-point morphology from Brand and the functionality that may have created the varied point forms. Based on the high incidence of serrated points and the occurrence of other tools suggested to be used in bone and antler working, Goodyear (1974) interpreted Brand as a hunting camp used for butchering and processing deer. The large number of serrated points makes the Dalton assemblage from Brand an excellent sample for the earliest evidence of serrated points. For this paper, we reanalyzed and digitized 30 serrated points to compare blade-edge shape.

A sample of Scallorn points was analyzed from Graham-Applegate Rancheria (41LL419), located in central Texas (Hixson 2003). The Austin-interval component contains large burned-rock features and five houses, each of which has a large central hearth and stone circles and pavements representing foundations and floors. Although the lithic assemblage has not been fully analyzed, it contains dozens of Scallorn points (Hixson, personal communication, 2016). For this study, we used 13 of the serrated points.

The Harrell site (NT-5), located in Young County in north-central Texas, was an Austin-interval habitation site with a well-defined cemetery, suggesting an extended period of occupation (Fox 1939; Hughes 1942). Numerous hearths were found in the upper few feet of the refuse midden. Digging tools suggest Harrell inhabitants practiced farming, but very few macrobotanical remains were recovered. Of the 555 points recovered, we used six serrated Scallorn points.

Tortuga Flats (41ZV155) is located in south-central Texas (Hester and Hill 1975; Hill and Hester 1973; Inman, Hill, and Hester 1998). Small mobile groups seasonally used the site. Archaeological materials, including hammerstones, exhausted cores, biface thinning flakes, other flakes and flake fragments, and preforms indicate tool production and resharpening activities took place at the locale. Food processing occurred in at least one area where a metate fragment was identified, and a refuse area contained the remains of bison, antelope, deer, coyote, and rabbit as well as discarded tools. Thirteen Scallorn points were collected from surface and excavation contexts, and we used a sample of four points.

We supplemented our sample of point types for the cladistics analyses to ensure a robust phylogenetic tree with point types representing temporal and geographic variation. The additional points were found in published sources that covered sites throughout the study area—Arkansas, Louisiana, Oklahoma, and Texas—and represent types ranging in time from the Early Paleoindian to the Mississippian/Late Prehistoric periods.

The geometric morphometric analyses focused on a subsample of points, including those from Brand, Graham-Applegate, Harrell, and Tortuga Flats. Points from these sites were recorded, analyzed, and photographed firsthand by the authors. To highlight the similarities and differences in the two point types, a sample of unequivocally unrelated serrated points was also included for comparison—seven serrated Kimberley points collected in 1902 from the region East Kimberley, Australia. Kimberley points are found in archaeological and ethnographic contexts in northwest Australia and are suggested to date from 1,500 to 1,000 B.P., with production continuing into the nineteenth century (Akerman, Fullagar, and van Gijn 2002). These pressure-flaked bifacial points have very fine serrations along the blade margin (Akerman et al. 2002). The ethnographic record shows Kimberley points tipped dart spears used for hunting (Akerman et al. 2002); residues on these points suggest they were also used as knives (Lommel 1997). Akerman et al. (2002) suggest that they were also important commodities in trade and exchange networks.

Methods I: Cladistic Analysis

The dataset included 349 points spread over 50 previously defined point types (chapter 12, this volume). Rather than evaluating the classifications of all the point types in the dataset, our focus is on understanding the evolutionary relationship between Dalton and Scallorn points and demonstrating evolutionary distance between these serrated point types. We incorporated 12 Dalton points from the Brand site, three Scallorn points from Tortuga Flats, and five Scallorn points from Graham-Applegate. The remaining sample of point measurements came from published sources collected from images using the program ImageJ (Schneider, Rasband, and Eliceiri 2012).

The character set contained two descriptive traits and six continuous measures. Following O'Brien and Lyman (2003), continuous data were condensed into categorical groups. The divisions were arbitrary and relative to the data spread for each variable. Because size variation can potentially distort cladistic comparisons, continuous data were size-adjusted following Lycett, von Cramon-Taubadel, and Foley (2006). The geometric mean was calculated for each point using four measures: maximum length, base length, maximum base width, and maximum blade width. The geometric mean is the fourth root of the product of these four measures. For each point, all six continuous character measurement values were then divided by the points' specific geometric mean to yield the size-adjusted values (see chapter 12, this volume). These calculated values became characters 3–8 in the cladistic analyses.

Table 11.1 and figure 11.3 show the eight characters and character states. Although point-edge characteristics, including serrations, were likely influenced by evolutionary pressures related to tool design and intended use or other factors, the presence or absence of serrations was not included as a character in the cladistic analyses. This ensured a degree of independence between the trait of interest—serrations—and the reconstructed phylogenetic tree(s).

The dataset of 349 points and eight characters was analyzed using PAUP* 4.0 (Swofford 1998), following methods described by O'Brien et al. (2014). Clovis points, the oldest dated points in the dataset, were selected as the outgroup (see arguments for this approach

Table 11.1
Point characters and character states or description of values used in the cladistic analysis

Variable type	Character	Character state or description of value calculation
Descriptive	Base type	Lanceolate; side notched; corner notched; straight stemmed; contracting stemmed; expanding stemmed; basal notched
	Proximal base shape	Concave; flat; convex
Continuous*	SA base length	Maximum length of the point base (i.e., the portion of the point that would have been in the haft)
	SA base width	Maximum width of the point base
	SA concavity	Calculated using the depth of basal indentation from the proximal towards the distal tip (flat or convex based points received values of zero)
	SA length/blade width	Calculated using the shape ratio of maximum point length divided by the maximum blade width
	SA basal constriction	Calculated by dividing the maximum basal width by the minimum basal width
	SA blade width/base width	Calculated using the shape ratio of maximum blade width divided by maximum base width

*Continuous characters were size-adjusted (SA) using the geometric mean of each individual point in the database.

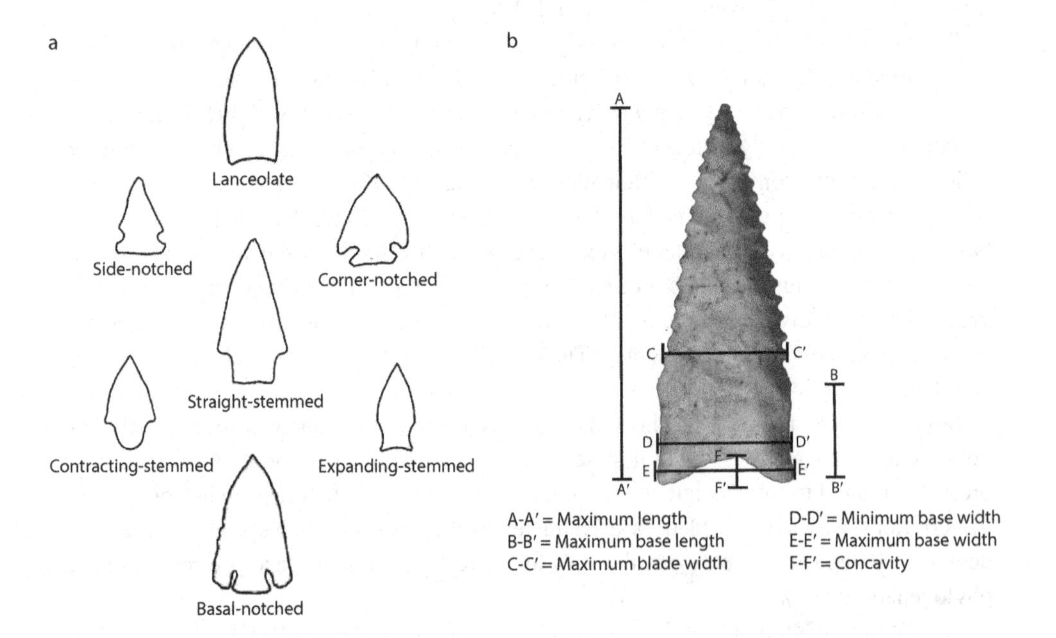

Figure 11.3
Schematics showing point characters and character states or description of values used in the cladistic analysis: (a) diagram illustrating base-type character states and (b) diagram illustrating how continuous variables were measured.

in chapter 12, this volume). A heuristic search using the principle of parsimony was then performed using 1,000 replicates. From these, a single, 50-percent majority-rule consensus tree was produced. Because the consistency index can be affected by the number of taxa (Sanderson and Donoghue 1989), we used the retention index (RI) as a goodness-of-fit measure to assess tree support (Collard, Shennan, and Tehrani 2006).

Methods II: Geometric Morphometric Analyses

Geometric morphometric analyses were performed on two datasets combining Dalton, Scallorn, and Kimberley points. First, an analysis was conducted on complete serrated points (n = 50): 21 serrated Dalton points from Brand; 23 serrated Scallorn, including points from Harrell, Tortuga Flats, and 15 unprovenienced points donated to Texas A&M University; and six Kimberley points. Second, an analysis was performed on 60 fragmented serrated projectile points: 30 serrated Dalton points from Brand; 23 serrated Scallorn points from Graham-Applegate, Harrell, and Tortuga Flats; and 7 Kimberley points. The first dataset was analyzed to compare blade morphology, assessing the entire serrated distal-end shape. The second dataset was used to analyze an isolated segment of serrations along the blade margin, magnifying the serrations to compare the morphology on a finer scale.

Dataset One

A landmark approach to geometric morphometric shape analysis was used to assess variation of serrated-point blade morphology in plan view (figure 11.4a). The three point types considered here were complete specimens; however, their base shapes varied and prevented identification of landmarks that could represent uniform morphological features across the sample. Such corresponding landmarks are often used as homologous landmarks in Procrustes superimposition to align specimens horizontally along the X-axis in a Cartesian coordinate system (Bookstein 1991; Rohlf and Slice 1990; Zelditch, Swiderski, Sheets, and Fink 2004).

To compare three different point types with heterogeneous base shapes, a different approach was used to align the data clouds along the X-axis, averaging the slopes of the distal lateral margins around the X-axis (following Smith and DeWitt 2016). To digitize artifact images, tools were positioned horizontally with basal margin to the left in digital photographs, and a constellation of type II semilandmarks was placed along each artifact's perimeter in tpsDig2 (v. 2.12; Rohlf 2008a). Landmark constellations were reduced to a suite of 400 semilandmarks that consisted of outlines made of 200 equidistant semilandmarks assigned to each lateral margin, defined as the length of the blade edge from the distal tip to its confluence with the basal portion. This position was identified by the highest and lowest coordinate along the Y-axis, with points positioned horizontally along

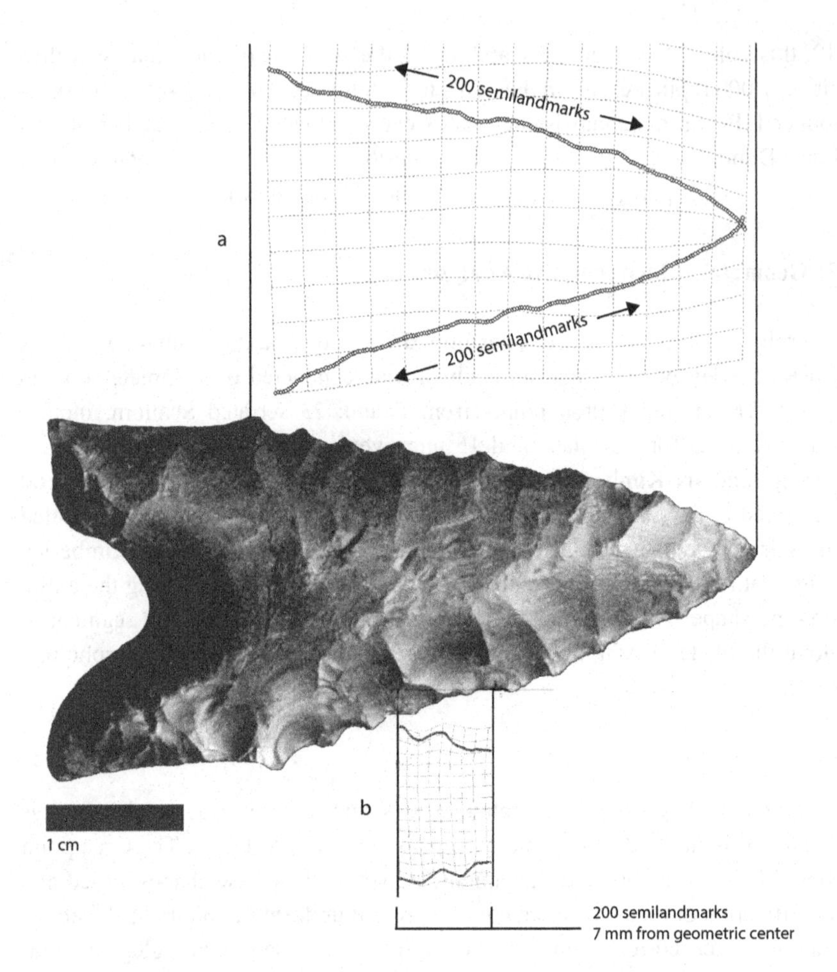

Figure 11.4
Illustration of landmark approach to geometric morphometric shape analysis in plan view, showing landmark distribution for assessing variation in serrated-point blade morphology (a) and for assessing variation in isolated serrated edges with 7 millimeters of lateral blade margins (b).

the X-axis. Inadvertently, this position was also found to represent the location where serrated edges ceased in most specimens. Outlines representing the proximal portion of the artifacts—the proximal portion of each lateral margin as defined above—were deleted to focus analyses on blade shape. The resulting semilandmark density was more than sufficiently saturated to capture serrated-edge shape differences.

Generalized least-squares Procrustes superimposition (generalized Procrustes analysis) was conducted in tpsRelw (v. 1.45; Rohlf 2008b) to superimpose the constellations of corresponding semilandmarks (Rohlf and Slice 1990), translating each constellation to the same centroid location, scaling each constellation to the same centroid size, and iteratively rotating each constellation until the summed squared distances between the semilandmarks and mean semilandmark position was minimized (Bookstein 1991; Mitteroecker, Gunz, Windhager, and Schaefer 2013; Rohlf 1999). Superimposed semilandmark constellations (Procrustes shape coordinates) were subjected to principal component (PC) analysis (Adams, Rohlf, and Slice 2004, 2013; Bookstein 1991; Mitteroecker et al. 2013). Centroid size, the square root of the summed squared distances between all semilandmarks to their common centroid, served as an unbiased size variable (Bookstein 1991) in analyses of shape and form (e.g., Smith, Smallwood, and DeWitt 2014).

Multivariate analysis of variance (MANOVA) was used to test models of morphological homogeneity by testing variance in shape among artifacts organized by point type. In MANOVA, shape analysis used size as a covariate to characterize and statistically control for linear allometry (Smith et al. 2014). Principal components of shape variation were also used to visualize shape characteristics that represent major factors of variability in the samples. Statistical analyses were conducted using JMP software version 10 (SAS Inst. Inc., Cary, NC).

Dataset Two

To identify variation in the shape of serrated edges, 7 millimeters of lateral blade margins were isolated on 59 specimens of Dalton, Scallorn, and Kimberley points (figure 11.4b). Isolating a section of the blade margins allowed fragmented specimens to be included; the same semilandmark density (200 semilandmarks) was used to saturate the small section. Specimens in the dataset were digitized using the same procedure described above except that, before landmark constellations were reduced, points in excess of 7 millimeters (determined by the length of the longest fragment) from the geometric center of each artifact were deleted to standardize the length of the serrated edges and to isolate lateral-edge shape from the morphology present at the transition from base to blade. Landmarks placed along the lateral-edge segments were reduced to 200 type II semilandmarks, and generalized least-squares Procrustes superimposition and PC analysis were conducted as described above.

Results

Cladistic Analysis

The heuristic search returned 1,000 equally parsimonious trees. The 50-percent majority-rule consensus tree has a retention index of 0.86, which falls well within the range of values for cladograms produced using cultural datasets (Collard et al. 2006), indicating strong tree support (chapter 1, this volume).

All Dalton points in the study sample fall within a single clade (figure 11.5), the ancestral node of which is characterized by lanceolate points with concave bases. Points in this clade have relatively low values (character-state group values of 1 or 2) for all metric size-adjusted shape characters.

Scallorn points occur in multiple clades. Eleven of the 18 Scallorn points fall in clades that also contain Ellis points, which are transitional, terminal Archaic/Woodland forms that date ca. 2,150–1,270 B.P. (Lohse et al. 2014) and are often classified as dart points. Other Scallorn specimens fall occur in clades with Motley points, a dart point dated to ca. 3,650–2750 B.P. (Anderson and Smith 2003), and other later arrow points, including Alba, Bonham, Cuney, and Friley. No Scallorn points fall within clades that also contain Dalton points, which we would expect, given the millennia between them.

Geometric Morphometric Analysis

Multivariate Shape Analysis for Dataset One: Complete Serrated Projectile Points

The first four PCs explain 95.09 percent of variability in the dataset (n = 50). Figure 11.6a illustrates shape characteristics expressed at the positive and negative ends of the PC axes. Models were organized by point type, and tests of blade shape found strong differences among Dalton, Scallorn, and Kimberley points ($p = 0.001$). Results illustrated as canonical centroid plots show that Scallorn and Kimberley samples overlap in shape space, demonstrating that blade shape in these two point types have more in common with each other than either has with Dalton (figure 11.7). Dalton points are described by the negative loading of PC1 and positive loadings of PC3 and PC4, which describe blade shapes that are long relative to width and have incurvate lateral margins.

Multivariate Shape Analysis for Dataset Two: Isolated Serrated Margin

Although variability in blade shape supports a typological difference between Dalton and Scallorn, or perhaps different blade functions, the shapes of serrations on the isolated blade margins are indistinguishable. The first six PCs explain 92.64 percent of variability in the dataset (n = 60). Figure 11.6b illustrates shape characteristics expressed at the positive and negative ends of the PC axes, demonstrating variation in margin shape. The shapes of serrated edges are not significantly different among point types ($p = 0.22$), suggesting that the variability observable in each PC axis is present within each type.

Discussion

Results of the cladistic analysis provide support for the hypothesis that Dalton and Scallorn point serrations reflect an example of convergence among separate projectile-point lineages. The consensus tree shows that serrated points do not form a single evolutionary

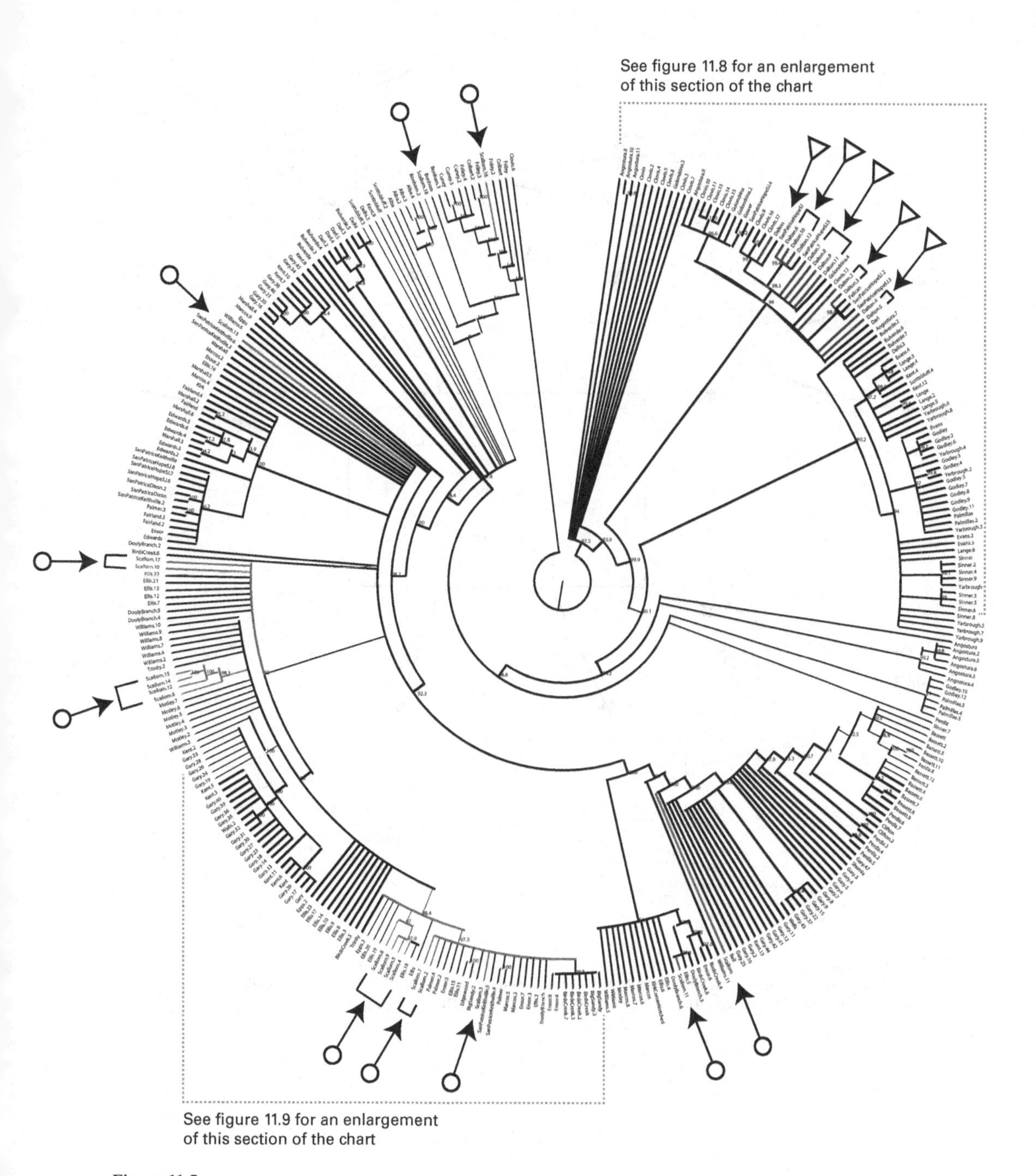

Figure 11.5
Majority-rule consensus tree, demonstrating all Scallorn clades (arrows with circles) are phylogenetically distinct and distant from the Dalton clade (arrows with triangles). Note that this figure is included at this scale simply to illustrate evolutionary distance between the point types.

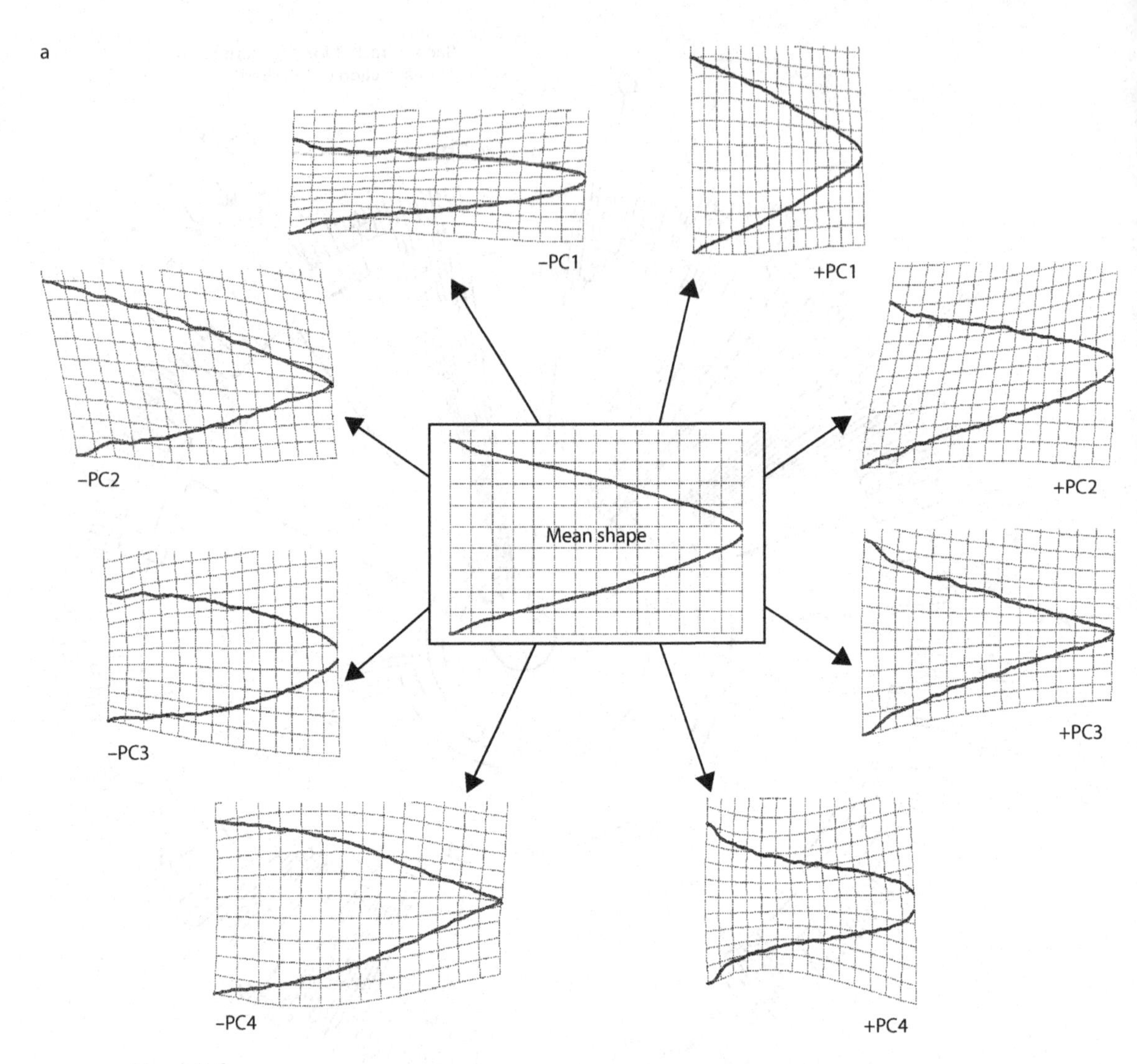

Figure 11.6
Multivariate analysis of shape characteristics expressed at the positive and negative ends of the principal component axes for (a) complete serrated projectile points and (b) isolated serrated margin.

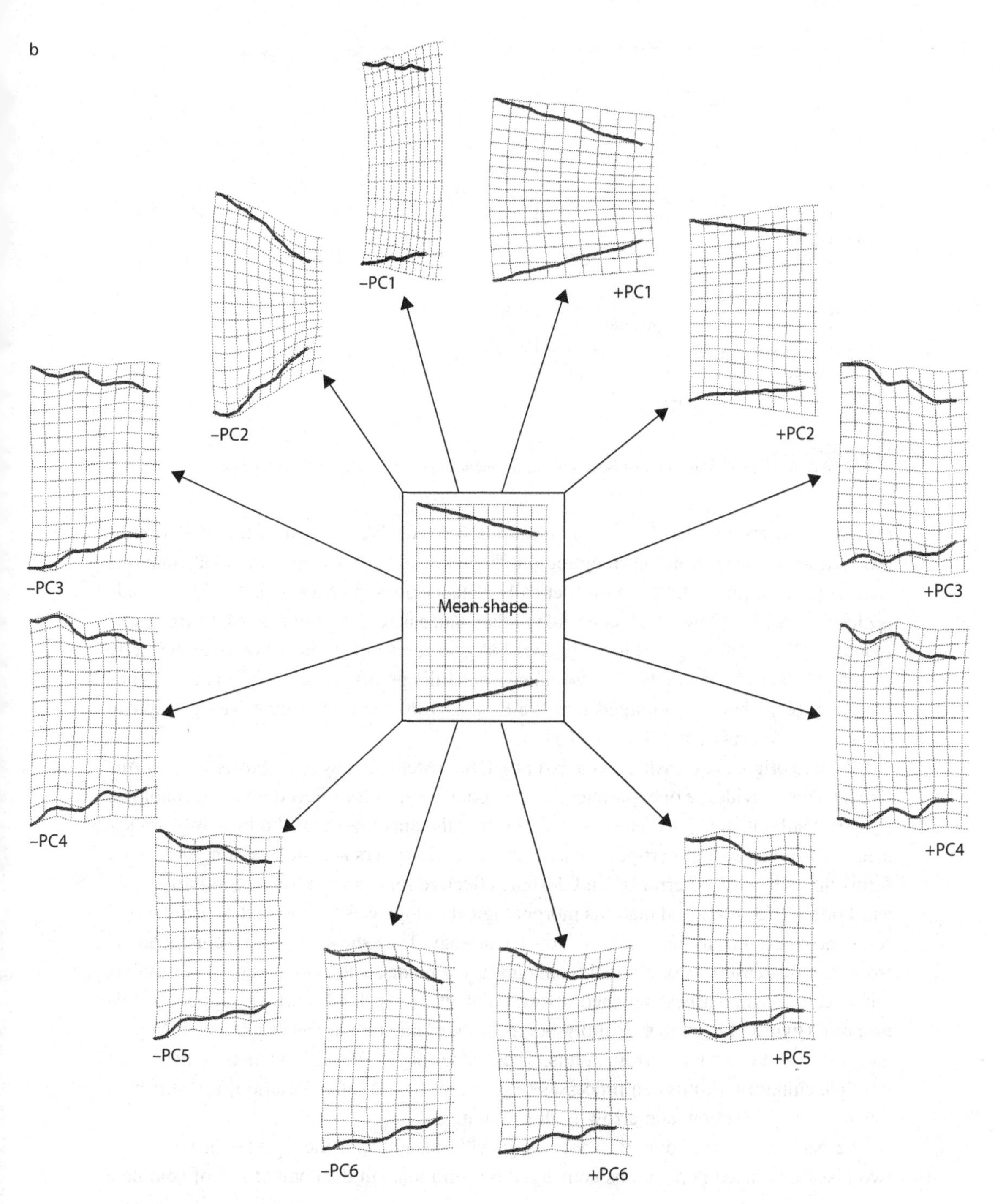

Figure 11.6 (continued)

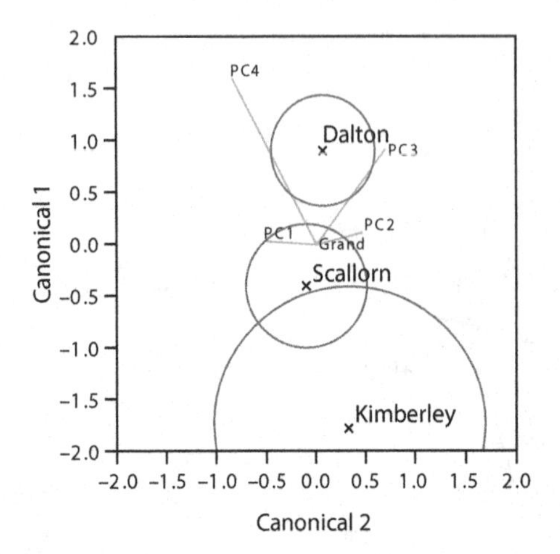

Figure 11.7
Canonical centroid plots, showing that Scallorn and Kimberley samples overlap in shape space.

clade, demonstrating that the use of serrations was not directly tied to the evolution of projectile-point shapes. In the study region, the application of serrations to blade margins was adopted within multiple phylogenetically distant clades. As expected, the Dalton clade is closely linked to Clovis and other Paleoindian lanceolates (see figure 11.8; O'Brien and Lyman 2003). Surprisingly, this clade does not give rise to many later points in the study region. Most Scallorn points are associated with Ellis points, which are found predominately in Texas but also grouped into the broader Lowe cluster, which has widespread occurrence (Anderson and Smith 2003).

Ellis was originally classified as a dart point, but recent reanalysis suggests it and forms like it could be evidence of experimentation in point designs for a new delivery technology. Lyman, VanPool, and O'Brien (2008) found that the appearance of the bow was associated with an increase in dart-point variation, as flintknappers adjusted existing dart-point forms through trial and error to find designs effective for tipping arrows. Similarly, Blitz and Porth (2013) proposed that this morphological change was the product of arrow-point refinement designed to function with bow technology. They show that Lowe-cluster points from the Southeast represent transitional dart-arrow points. The association of Ellis points with Scallorn, the earliest notched arrow point in Texas, lends further support to this proposal (figure 11.9). Other Scallorn points in the cladogram are associated with Motley dart points and other later arrow points. These other phylogenetic links may indicate that multiple cultural-transmission processes (e.g., direct bias, guided variation) accompanied the adoption of the bow and arrow in this region.

The geometric morphometric analyses provide insight into the context in which the two distantly related populations converged on serrations. In the comparison of complete

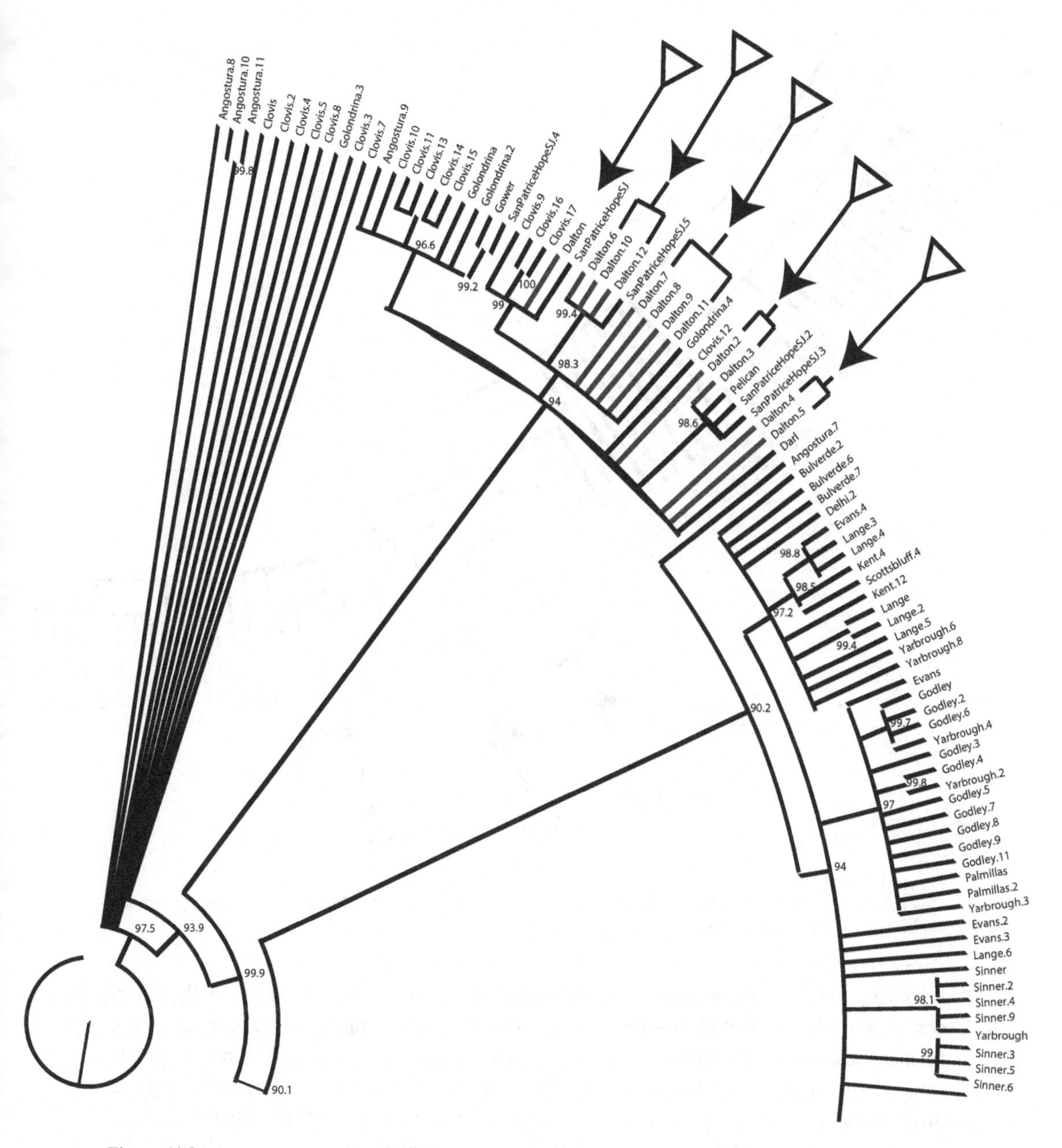

Figure 11.8
Majority-rule consensus tree, highlighting the Dalton clade closely linked to Clovis and other Paleoindian lanceolates. Dalton points are marked with arrows with triangles.

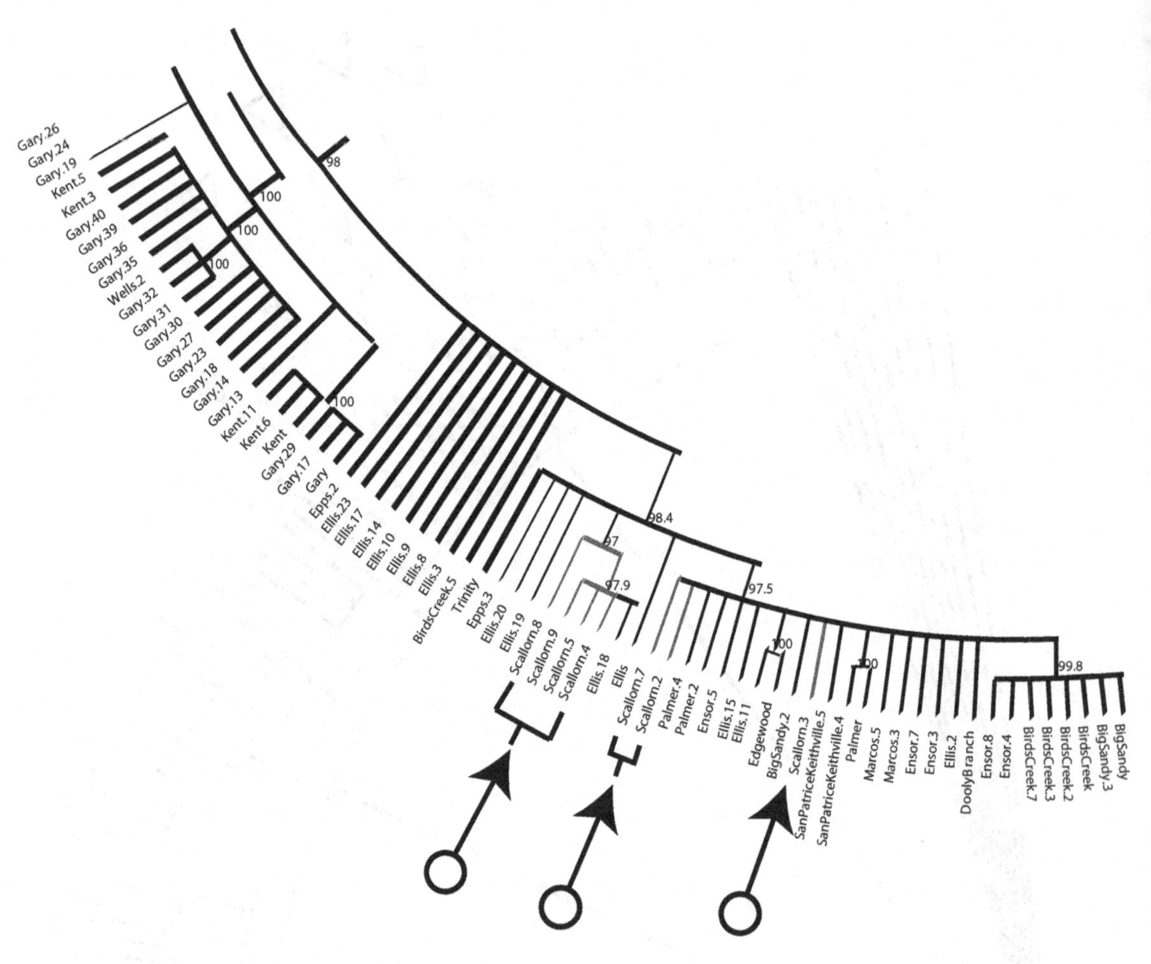

Figure 11.9
Majority-rule consensus tree, highlighting the association of Scallorn with Ellis points. Scallorn points are marked with arrows with circles.

serrated-point blade shape, Dalton, Scallorn, and Kimberley points are significantly different; thus, the blades that Dalton and Scallorn flintknappers serrated are different in shape. In fact, Scallorn point blades share more shape similarities with the Kimberley points from Australia than with Dalton points. We suspect these differences are reflections of differences in design for point function. However, in the analysis of the isolated serrated margin, the shape of Dalton, Scallorn, and Kimberley serrations are not significantly different. The lack of significant difference in the shape suggests the serrations themselves were designed for the same function. Thus, Dalton and Scallorn point makers converged on the use of morphologically indistinguishable serrations.

Why Serrate?

Why did these distantly related (if at all) populations, with distinct blade-shape designs, converge on the application of serrations along their point margins? Experimental studies help answer this question. Wilkins, Schoville, and Brown (2014) compared wound tracks between untipped and stone-tipped spears and found no significant difference in penetration depth. Similarly, Loendorf et al. (2015) found no difference in penetration for serrated versus nonserrated stone tips. These results suggest that penetration depth was likely not a key characteristic influencing the use of serrated projectiles. An alternative functional effect might relate to wound size and associated internal shredding. Human hunters are usually slower than their prey and pursue prey for long periods of time (chapter 6, this volume); poison or increased bleeding decreases capture time. A spear or arrow point rarely kills large animals right away, but following a blood trail is an effective technique to track animals as they weaken (Kelly 1995). Unfortunately, no experimental studies of wound size have been conducted comparing serrated to nonserrated projectiles. However, Wilkins et al. (2014) found that, compared to untipped spears, stone-tipped spears produced significantly larger wounds, with a widening of the inner wound track. Forensic studies comparing serrated to nonserrated-knife stab wounds show serrated knives produce damage striations to skin (Pounder, Bhatt, Cormack, and Hunt 2011), cartilage (Pounder, Cormack, Broadbent, and Millar 2011), and soft tissues, including arteries (Jacques, Kogon, and Shkrum 2014) and that these striations are not produced by nonserrated knives.

These experimental results show that the shape of the weapon tip has a significant impact on wound size and shape and that serrations cause additional damage to multiple internal tissues. Serrations increase damage caused by tearing, which could be produced either through projectile or knife use. Explaining the adoption of serrations by Dalton and Scallorn populations must take into account this functional advantage.

Serrations for Tearing in Dalton

Not all Dalton points are serrated, and serrated forms vary, which has led researchers to propose a variety of functional hypotheses. Morse (1971) proposed that variation in body shapes result from resharpening and changes in point use over time (see also Goodyear 1974). According to Morse, early in their use-lives Dalton points were socketed into foreshafts, allowing them to be used both as dart points and as hafted knives. He suggests that the jagged edges of serrated Dalton points, with fine serrations at the tip and longer serrations at the shoulder, could help perform several functions related to the butchering of deer.

O'Brien and Wood (1998) propose an alternative hypothesis, suggesting that the variation in Dalton shape was the result of primary engineering design. Thus, Dalton point makers initially manufactured points with distinct blade characteristics (e.g., beveling, serrations). They reference variation in modern archery equipment to argue all Dalton point shape variants were designed to be projectiles (O'Brien and Wood 1998).

Both scenarios rely on a common fundamental function—the teethlike serrations on Dalton points are ultimately used for tearing. In Morse's (1971) explanation, the fine serrations on a Dalton point tip could tear through deer skin for an initial cut, assist in gutting, and, when used in a sawing-like motion, tear apart the front quarters from the hindquarters. Similarly, in O'Brien and Wood's (1998) hypothesis, serrated Dalton points could have been particularly lethal because the teeth-like serrations tear open wounds, causing internal damage, perhaps even encouraging a prominent blood trail for tracking.

Both scenarios also help explain adaptive responses to environmental changes at the end of the Pleistocene. After the extinction of the megafauna and the competitive release of deer populations, serrated point edges offered Dalton populations the selective functional advantage of increased wound tearing to track blood trails of smaller, quicker prey and to process that prey during butchering.

Serrations for Tearing in Scallorn?

Scallorn points are associated with a major shift in projectile technology—the adoption of the bow and arrow. The adoption and social impacts of the spread of bow-and-arrow technology in North America varied from region to region with unique environmental, social, and historical conditions (e.g., Blitz and Porth 2013; Shott 1996; VanPool and O'Brien 2013). In our study area, adoption of the technology has been associated with bow-based warfare and enforced cooperation. Nassaney and Pyle (1999) conclude that bow-and-arrow technology was abruptly adopted in central Arkansas during the Late Woodland period, approximately 1,400 B.P., and that social factors such as warfare played a role. Populations to the south and west, in Texas and Oklahoma, were refining their use of bow technology and, with the advantages of this new technology, expanding their territories.

While Nassaney and Pyle (1999) argue that warfare was an important incentive for the adoption of the technology in Arkansas, Blitz and Porth (2013) conclude that social change took a different trajectory in the lower Southeast, an area including East Texas and Louisiana. There, wild-plant foods were abundant, making food production less of a subsistence focus. Populations aggregated in large civic-ceremonial centers, and the bow may have helped enforce communal cooperation among large, nonkin-related populations. In the western portion of the area discussed here, the adoption of the bow and arrow seems to have coincided with the onset of widespread drought conditions, and bow-based warfare was perhaps driven by population pressure and territorial disputes (Prewitt 1981). Thus, throughout the ecologically diverse study area, Scallorn points emerged in the midst of complex technological and social changes.

Although no studies have directly addressed the question of why Scallorn points are serrated, a closer examination of archaeological contexts shows a possible selective advantage: a weapon for killing not just animals but humans as well. Across large parts of the study area, including in Austin-interval burial contexts, Scallorn points are associated with evidence of widespread violence, with numerous incidents of death caused by arrow wounds (Boyd 1997, 2004; Collins 2004; Greer and Benfer 1975; Hall, Hester, and Black 1986; Hester 2004; Hester and Collins 1969; Hester, Wilson, and Headrick 1993; Huebner and Comuzzie 1992; Prewitt 1974; Prikryl 1990; Ricklis 2004b). In this context, serrating Scallorn points potentially provided an important functional advantage in a social, as opposed to an ecological, adaptation. Function remained the same, as the increased internal tearing damage caused by serrations would have intensified impact shock and made it more likely that points would remain embedded to cause additional damage, regardless of the intended target.

Converging on Serrations

The tearing function of serrated point margins offered a selective advantage that explains their adoption in the Late Paleoindian and Woodland–Mississippian/Late Prehistoric periods. For Dalton populations at the end of the Ice Age, serrations would have improved bloodletting during a deer hunt and could also have served a useful role as a hafted knife for butchery. For Scallorn populations engaged in warfare, serrations would have improved the shock, awe, and lethalness of arrows.

Conclusions

The study discussed here is an exercise in understanding how to detect cases of convergence in the stone-tool record. Rather than using the entire scale of reduction technologies or artifact types, which might lend themselves to more intuitively obvious evolutionary and adaptive explanations, we have focused on convergence at the scale of a single tool attribute. We used cladistics to support the interpretation that Dalton and Scallorn serrated points are evolutionarily unrelated and support the hypothesis that both populations converged on the use of serrations. We used geometric morphometric analysis to demonstrate differences in overall point-blade designs and show that, despite these differences, both populations converged on the same shape of serrations.

Serrations are often noted in point-type descriptions, but the functional role serrations played through time is rarely considered in detail. We show that two unrelated populations, Dalton and Scallorn, experiencing distinct environmental and social conditions, converged on the application of serrations to point margins to address a shared adaptive functional need for increased tearing.

References

Adams, D. C., Rohlf, F. J., & Slice, D. E. (2004). Geometric Morphometrics: Ten Years of Progress Following the "Revolution." *Italian Journal of Zoology*, *71*, 5–16.

Adams, D. C., Rohlf, F. J., & Slice, D. E. (2013). A Field Comes of Age: Geometric Morphometrics in the 21st Century. *Hystrix, 24*, 7–14.

Akerman, K., Fullagar, R., & van Gijn, A. (2002). Weapons and Wunan: Production, Function and Exchange of Kimberley Points. *Australian Aboriginal Studies Issue, 1*, 13–42.

Anderson, D. G., & Bissett, T. G. (2015). The Initial Colonization of North America: Sea Level Change, Shoreline Movement, and Great Migrations. In M. Frachetti & R. Spengler (Eds.), *Mobility and Ancient Society in Asia and the Americas: Proceedings of the Second International Conference on "Great Migrations"* (pp. 59–88). New York: Springer.

Anderson, D. G., & Mainfort, R. C., Jr. (2002). An Introduction to Woodland Archaeology in the Southeast. In D. G. Anderson & R. C. Mainfort, Jr., (Eds.), *The Woodland Southeast* (pp. 1–19). Tuscaloosa: University of Alabama Press.

Anderson, D. G., & Sassaman, K. E. (2012). *Recent Developments in Southeastern Archaeology: From Colonization to Complexity*. Washington, D.C.: Society for American Archaeology Press.

Anderson, D. G., & Smith, S. D. (2003). *Archaeology, History, and Predictive Modeling: Research on Fort Polk 1972–2002*. Tuscaloosa: University of Alabama Press.

Anderson, D. G., Smallwood, A. M., & Miller, D. S. (2015). Early Human Settlement in the Southeastern United States: Current Evidence and Future Directions. *PaleoAmerica, 1*, 1–45.

Ballenger, J. A. M. (2001). *Dalton Settlement in the Arkoma Basin of Eastern Oklahoma. Robert E. Bell Monographs in Anthropology, no. 2, Sam Noble Museum of Natural History*. Norman: University of Oklahoma.

Blitz, J. H., & Porth, E. S. (2013). Social Complexity and the Bow in the Eastern Woodlands. *Evolutionary Anthropology, 22*, 89–95.

Bookstein, F. L. (1991). *Morphometric Tools for Landmark Data: Geometry and Biology*. New York: Cambridge University Press.

Boyd, D. K. (1997). *Caprock Canyonlands Archeology: A Synthesis of the Late Prehistory and History of Lake Alan Henry and the Texas Panhandle Plains* , vol. II. Reports of Investigations, no. 110. Austin, Texas: Prewitt and Associates.

Boyd, D. K. (2004). The Palo Duro Complex. In T. K. Perttula (Ed.), *The Prehistory of Texas* (pp. 296–330). College Station: Texas A&M University Press.

Bradley, B. A. (1997). Sloan Site Biface and Projectile Point Technology. In D. F. Morse (Ed.), *Sloan: A Paleoindian Dalton Cemetery in Arkansas* (pp. 53–57). Washington, D.C.: Smithsonian Institution Press.

Collard, M., Shennan, S. J., & Tehrani, J. J. (2006). Branching, Blending, and the Evolution of Cultural Similarities and Differences among Human Populations. *Evolution and Human Behavior, 27*, 169–184.

Collins, M. B. (2004). Archeology in Central Texas. In T. K. Perttula (Ed.), *The Prehistory of Texas* (pp. 101–126). College Station: Texas A&M University Press.

Eren, M. I. (2012). On Younger Dryas Climate Change as a Causal Determinate of Prehistoric Hunter-Gatherer Culture Change. In M. I. Eren (Ed.), *Hunter-Gatherer Behavior: Human Response during the Younger Dryas* (pp. 11–23). Walnut Creek, Calif.: Left Coast Press.

Fenneman, N. M., & Johnson, D. W. (1946). *Physiographic Divisions of the Conterminous U.S.* Washington, D.C.: U.S. Geological Survey.

Fox, G. (1939). *Field Data and Reports on the Harrell Site (NT-5)*. Austin: Texas Archeological Research Laboratory.

Gillam, J. C. (1996). A View of Paleoindian Settlement from Crowley's Ridge. *Plains Anthropologist, 157*, 273–286.

Goodyear, A. C. (1974). *The Brand Site: A Techno-functional Study of a Dalton Site in Northeast Arkansas.* Research Series no. 7. Fayetteville: Arkansas Archeological Survey.

Goodyear, A. C. (1982). The Chronological Position of the Dalton Horizon in Southeastern United States. *American Antiquity, 47*, 382–395.

Grayson, D. K., & Meltzer, D. J. (2015). Revisiting Paleoindian Exploitation of Extinct North American Mammals. *Journal of Archaeological Science, 56*, 177–193.

Greer, J., & Benfer, R. A. (1975). Austin Phase Burials at the Pat Parker Site, Travis County, Texas. *Bulletin of the Texas Archeological Society, 46*, 189–216.

Hall, G. D., Hester, T. R., & Black, S. L. (1986). *The Prehistoric Sites at Choke Canyon Reservoir, Southern Texas: Results of Phase II Archaeological Investigations. Choke Canyon Series, no. 10.* Center for Archaeological Research, University of Texas at San Antonio.

Haynes, G. (2009). *American Megafaunal Extinctions at the End of the Pleistocene.* New York: Springer.

Hester, T. R. (2004). The Prehistory of South Texas. In T. K. Perttula (Ed.), *The Prehistory of Texas* (pp. 127–151). College Station: Texas A&M University Press.

Hester, T. R., & Collins, M. B. (1969). Burials from the Frisch Auf! Site: 41FY42. *Texas Journal of Science, 20*, 261–272.

Hester, T. R., & Hill, T. C., Jr. (1975). *Some Aspects of Late Prehistoric and Protohistoric Archaeology in Southern Texas. Special Report, no. 1.* Center for Archaeological Research, University of Texas at San Antonio.

Hester, T. R., Wilson, D., & Headrick, P. (1993). An Austin Phase Burial from Frio County, Southern Texas: Archaeology and Physical Anthropology. *Terra (Helsinki, Finland), 20*, 5–8.

Hill, T. C., Jr., & Hester, T. R. (1973). A Preliminary Report on the Tortuga Flat Site: A Protohistoric Campsite in Southern Texas. *Newsletter of the Dallas Archeological Society, 17*, 10–14.

Hixson, C. A. (2003). Horizontal Stratigraphy at the Graham/Applegate Site (41LL419). *Newsletter of the Council of Texas Archeologists, 27*, 12–16.

Huebner, J. A., & Comuzzie, A. G. (1992). *The Archeology and Bioarcheology of Blue Bayou: A Late Archaic and Late Prehistoric Mortuary Locality in Victoria County, Texas. Studies in Archeology, no. 9.* Austin: Texas Archeological Research Laboratory, University of Texas.

Hughes, J. T. (1942). *An Archeological Report on the Harrell Site of North-central Texas.* M.A. thesis, University of Texas. Austin.

Inman, B., Hill, T. C., Jr., & Hester, T. R. (1998). Archeological Investigations at the Tortuga Flat Site (41ZV155), Zavala County, South Texas. *Bulletin of the Texas Archeological Society, 69*, 11–33.

Jacques, R., Kogon, S., & Shkrum, M. (2014). An Experimental Model of Tool Mark Striations by a Serrated Blade in Human Soft Tissues. *American Journal of Forensic Medicine and Pathology, 35*, 59–61.

Kelley, C. J. (1947). The Lehmann Rock Shelter: A Stratified Site of the Toyah, Uvalde, and Round Rock Foci. *Bulletin of the Texas Archeological Society, 18*, 115–128.

Kelly, R. L. (1995). *The Foraging Spectrum.* Washington, D.C.: Smithsonian Institution Press.

Koldehoff, B., & Walthall, J. A. (2009). Dalton and the Early Holocene Midcontinent: Setting the Stage. In T. E. Emerson, D. L. McElrath, & A. C. Fortier (Eds.), *Archaic Societies: Diversity and Complexity across the Midcontinent* (pp. 137–151). Albany: State University of New York.

Loendorf, C., Oliver, T. J., Tiedens, S., Plumlee, R. S., Woodson, M. K., & Simon, L. (2015). Flaked-Stone Projectile Point Serration: A Controlled Experimental Study of Blade Margin Design. *Journal of Archaeological Science: Reports, 3*, 437–443.

Lohse, J. C., Black, S. L., & Cholak, L. M. (2014). Toward an Improved Archaic Radiocarbon Chronology for Central Texas. *Bulletin of the Texas Archeological Society, 52*, 65–89.

Lommel, A. (1997). *The Unambal: A Tribe in Northwest Australia*. Sydney: Takarakka Publications.

Lycett, S. J., von Cramon-Taubadel, N., & Foley, R. A. (2006). A Crossbeam Co-ordinate Caliper for the Morphometric Analysis of Lithic Nuclei: A Description, Test and Empirical Examples of Application. *Journal of Archaeological Science, 33*, 847–861.

Lyman, R. L., VanPool, T. L., & O'Brien, M. J. (2008). Variation in North American Dart Points and Arrow Points When One or Both Are Present. *Journal of Archaeological Science, 35*, 2805–2812.

McGahey, S. O. (2000). *Mississippi Projectile Point Guide*. Jackson: Mississippi Department of Archives and History.

Meltzer, D. J., & Holliday, V. T. (2010). Would North American Paleoindians Have Noticed Younger Dryas Age Climate Changes? *Journal of World Prehistory, 23*, 1–41.

Miller, D. S., & Gingerich, J. A. M. (2013). Regional Variation in the Terminal Pleistocene and Early Holocene Radiocarbon Record of Eastern North America. *Quaternary Research, 79*, 175–188.

Mitteroecker, P., Gunz, P., Windhager, S., & Schaefer, K. (2013). A Brief Review of Shape, Form, and Allometry in Geometric Morphometrics, with Applications to Human Facial Morphology. *Hystrix, 24*, 59–66.

Morse, D. F. (1971). Recent Indications of Dalton Settlement Pattern in Northeast Arkansas. *Southeastern Archaeological Conference Bulletin, 13*, 5–10.

Morse, D. F. (Ed.). (1997). *Sloan: A Paleoindian Dalton Cemetery in Arkansas*. Washington, D.C.: Smithsonian Institution.

Nassaney, M. S., & Pyle, K. (1999). The Adoption of the Bow and Arrow in Eastern North America: A View from Central Arkansas. *American Antiquity, 64*, 243–263.

O'Brien, M. J., & Lyman, R. L. (2003). *Cladistics and Archaeology*. Salt Lake City: University of Utah Press.

O'Brien, M. J., & Wood, W. R. (1998). *The Prehistory of Missouri*. Columbia: University of Missouri Press.

O'Brien, M. J., Boulanger, M. T., Buchanan, B., Collard, M., Lyman, R. L., & Darwent, J. (2014). Innovation and Cultural Transmission in the American Paleolithic: Phylogenetic Analysis of Eastern Paleoindian Projectile-Point Classes. *Journal of Anthropological Archaeology, 34*, 100–119.

Pounder, D. J., Bhatt, S., Cormack, L., & Hunt, B. (2011). Tool Mark Striations in Pig Skin Produced by Stabs from a Serrated Blade. *American Journal of Forensic Medicine and Pathology, 32*, 93–95.

Pounder, D. J., Cormack, L., Broadbent, E., & Millar, J. (2011). Class Characteristics of Serrated Knife Stabs to Cartilage. *American Journal of Forensic Medicine and Pathology, 32*, 157–160.

Prewitt, E. (1974). *Archeological Investigations at the Loeve-Fox Site, Williamson County, Texas. Texas Archeological Survey Research Report, no. 49*. Austin: University of Texas.

Prewitt, E. (1981). Cultural Chronology in Central Texas. *Bulletin of the Texas Archeological Society, 85*, 259–287.

Prikryl, D. J. (1990). *Lower Elm Fork Prehistory: A Redefinition of Cultural Concepts and Chronologies along the Trinity River, North-central Texas. Report, no. 37*. Austin: Office of the State Archeologist, Texas Historical Commission.

Ricklis, R. A. (2004a). Prehistoric Occupation of the Central and Lower Texas Coast. In T. K. Perttula (Ed.), *The Prehistory of Texas* (pp. 155–180). College Station: Texas A&M University Press.

Ricklis, R. A. (2004b). The Archeology of the Native American Occupation of Southeast Texas. In T. K. Perttula (Ed.), *The Prehistory of Texas* (pp. 181–202). College Station: Texas A&M University Press.

Rohlf, F. J. (1999). Shape Statistics: Procrustes Superimpositions and Tangent Spaces. *Journal of Classification, 16*, 197–223.

Rohlf, F. J. (2008a). TpsDig (version 2.12). Department of Ecology and Evolution, State University of New York, Stony Brook.

Rohlf, F. J. (2008b). *Relative Warps (version 1.45)*. Stony Brook: Department of Ecology and Evolution, State University of New York.

Rohlf, F. J., & Slice, D. E. (1990). Extensions of the Procrustes Method for the Optimal Superimposition of Landmarks. *Systematic Zoology, 39*, 40–59.

Sanderson, M. J., & Donoghue, M. J. (1989). Patterns of Variation in Levels of Homoplasy. *Evolution; International Journal of Organic Evolution, 43*, 1781–1795.

Schiffer, M. B. (1975). An Alternative to Morse's Dalton Settlement Pattern Hypothesis. *Plains Anthropologist, 20*, 253–266.

Schneider, C. A., Rasband, W. S., & Eliceiri, K. W. (2012). NIH Image to ImageJ: 25 Years of Image Analysis. *Nature Methods, 9*, 671–675.

Sherwood, S. C., Driskell, B. N., Randall, A. R., & Meeks, S. C. (2004). Chronology and Stratigraphy at Dust Cave, Alabama. *American Antiquity, 69*, 533–554.

Shott, M. J. (1996). Innovation and Selection in Prehistory: A Case Study from the American Bottom. In G. H. Odell (Ed.), *Stone Tools: Theoretical Insights into Human Prehistory* (pp. 279–313). New York: Plenum.

Smith, H. L., & DeWitt, T. J. (2016). The Northern Fluted Point Complex: Technological and Morphological Evidence of Adaptation and Risk in the Late Pleistocene Arctic. *Archaeological and Anthropological Sciences*. doi: 10.1007/s12520-016-0335-y.

Smith, H. L., Smallwood, A. M., & DeWitt, T. (2014). A Geometric Morphometric Exploration of Clovis Fluted-point Shape Variability. In A. M. Smallwood & T. A. Jennings (Eds.), *Clovis: On the Edge of a New Understanding* (pp. 161–180). College Station: Texas A&M University Press.

Swofford, D. (1998). *PAUP*: Phylogenetic Analysis Using Parsimony (*and Other Methods) (version 4)*. Sunderland, Mass.: Sinauer.

Thomas, D. H. (1978). Arrowheads and Atlatl Darts: How the Stone Got the Shaft. *American Antiquity, 43*, 461–472.

Turner, E. S., Hester, T. R., & McReynolds, R. L. (2011). *Stone Artifacts of Texas Indians*. Lanham, Md.: Taylor.

VanPool, T. L., & O'Brien, M. J. (2013). Sociopolitical Complexity and the Bow and Arrow in the American Southwest. *Evolutionary Anthropology, 22*, 111–117.

Waters, M. R., & Stafford, T. W. (2007). Redefining the Age of Clovis: Implications for Peopling of the Americas. *Science, 23*, 1122–1126.

Waters, M. R., Stafford, T., McDonald, G., Gustafson, C., Rasmussen, M., Cappellini, E., et al. (2011). Pre-Clovis Mastodon Hunting 13,800 Years Ago at the Manis Site, Washington. *Science, 334*, 351–353.

Wilkins, J., Schoville, B. J., & Brown, K. S. (2014). An Experimental Investigation of the Functional Hypothesis and Evolutionary Advantage of Stone-Tipped Spears. *PLoS One, 9*(8), e104514.

Zelditch, M. L., Swiderski, D., Sheets, H. D., & Fink, W. (2004). *Geometric Morphometrics for Biologists: A Primer*. Amsterdam: Elsevier.

12 Clovis and Toyah: Convergent Blade Technologies on the Southern Plains Periphery of North America

Thomas A. Jennings and Ashley M. Smallwood

Evolutionary approaches to archaeological data apply biological principles to understand artifactual variation and cultural change through time (chapter 1, this volume). Stone tools are a particularly rich dataset because their morphology represents intentional designs—adaptations to specific environmental and social contexts—transmitted through social learning. Finished tools, especially those forms found in the archaeological record to be chronologically or spatially diagnostic of a culture, can be ordered and compared morphologically to understand evolutionary relationships (O'Brien and Lyman 2003). Because stone-tool manufacture is reductive, the process of production, or the behavioral recipe, can also reflect significant information about past populations (Schiffer and Skibo 1987; Schillinger, Mesoudi, and Lycett 2016; Smallwood 2012).

An artifact is more than its final form; it is a reflection of the behavioral recipe that guided the technological process. This recipe includes the raw materials used, the tools employed, the sequence of actions followed in production, and the rules used to solve problems that arise in production (Schiffer and Skibo 1987). Behavioral recipes are technological knowledge that is influenced by environmental conditions, and because recipes are transmitted through social learning, they are also affected by transmission biases (Schillinger et al. 2016; Shennan 2008). For lithic assemblages, evaluating core-reduction strategies—in other words, the recipe for reducing a core to produce tool blanks—is key to comparing the production sequences of past populations (Eren and Buchanan 2016; Smallwood 2012). Similarities in behavioral recipes can be the product of inheritance of information—that is, one lineage member learns technological knowledge from another—or of populations developing convergent adaptations to similar environmental conditions (Lipo, O'Brien, Collard, and Shennan 2006; Lycett 2015; O'Brien and Lyman 2003). Although many studies have investigated evidence of heritable continuity of technological knowledge, few have made convergent evolution the focus of technological comparisons (for examples, see Clark and Riel-Salvatore 2006; Eren, Patten, O'Brien, and Meltzer 2013; Lycett 2009; O'Brien et al. 2014; Straus, Meltzer, and Goebel 2005; Wang, Lycett, von Cramon-Taubadel, Jin, and Bae 2012; Will, Mackay, and Phillips 2015).

Here we explore the chronological occurrence of blades and blade-core reduction in the Southern Plains and periphery of North America (figure 12.1) as a case study for understanding convergence in the stone-tool record. Blades are specialized flakes that are long and parallel-sided and at least twice as long as they are wide. Interior blades have dorsal scars that run parallel to the trajectory of removal (Bordes 1967; Bordes and Crabtree 1969; Collins 1999). Several blades can be struck from a stone core and modified into a variety of tools (e.g., knives, scrapers, and points). Hints of blade technology have been documented in the pre-Clovis record of North America (Adovasio and Pedler 2004; Jennings and Waters 2014), but widespread evidence of blade-core reduction—that is, actual blade cores—and blade products are most clearly associated with the Clovis record, dated to approximately 13,100–12,700 calendar years B.P. (Waters and Stafford 2007).

Evidence of blade production is especially common at Clovis sites on the southern Plains and Plains periphery (Kilby 2015). Interestingly, blade technology is not found in the post-Clovis Paleoindian record, and in the southern Plains, blades and blade cores do not reenter the record until the Late Prehistoric Toyah period (Collins 1999, 2004), dating to approximately 650–300 B.P. (Lohse, Black, and Cholak 2014a). Thus, Clovis and Toyah blade technologies appear to represent an example of convergence. We use phylogenetic and phenotypic analysis to demonstrate evolutionary distance between Clovis and Toyah to support a discussion of why these disparate populations converged on the same reduction technology. We then examine how similarities in this part of the behavioral recipe reflect similar adaptations to similar environmental conditions.

Background

Blade production is best understood as a sequence of actions (Collins 1999). The first step involves the selection of a homogenous nodule of stone—a predictable material such as chert or obsidian—that accommodates the size of the intended blades to be produced. Before blades can be removed, a nodule must be shaped into a formally designed core, removing cortical material and surface irregularities. Although blade-core design can vary (e.g., conical vs. wedge-shaped cores), the goal of reduction is similar—the production of long, narrow, parallel-sided blades. The initial preparation of the formal core to create a single platform surface is key to the removal of a series of blades. The removal of the first blade, either by following the natural angular surface of a nodule or by creating a crested ridge through flaking, creates flake-scar ridges to guide subsequent blade removals. In other words, each blade detachment forms a new ridge that becomes the dorsal axis of a later blade removal. The unmodified acute edges of blades can be used for cutting or scraping, or the edge can be subsequently flaked to create curated tools such as side scrapers, end scrapers, and points. The diversity of decision options from nodule selection to tool formation offers opportunities for a diversity of blade-technology recipes.

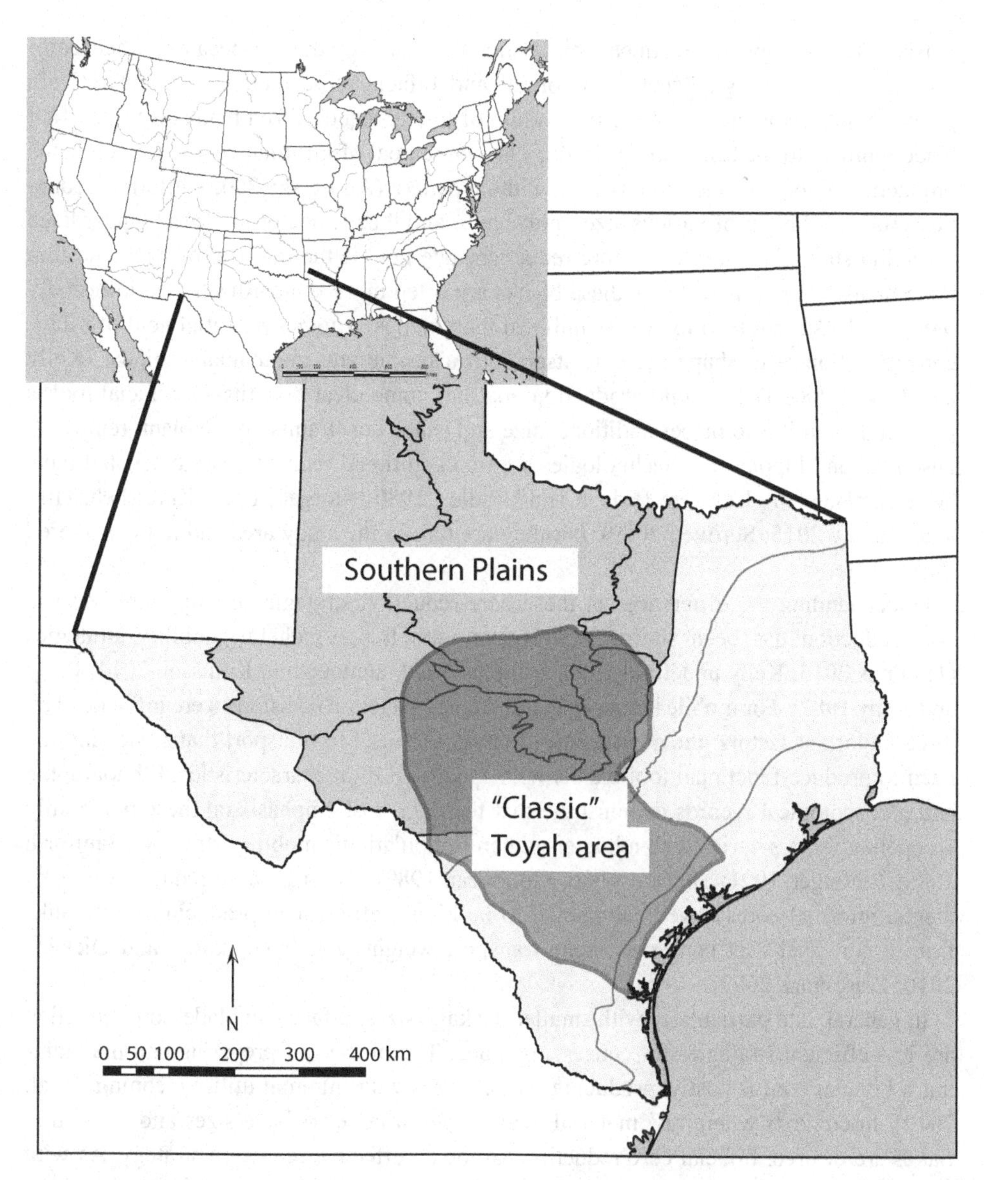

Figure 12.1
Map of the study area showing the Southern Plains and the "classic" Toyah area.

Blade technology is a distinct, unidirectional reductive strategy because, when compared to other strategies such as informal and bifacial core reduction, comparatively greater emphasis is placed on the production of highly standardized blanks from the first blade removal to the last (Collins 1999). The single goal of informal core reduction is the production of tool blanks, and flaking is without pattern (Andrefsky 2005). Informal reduction produces blanks of various sizes and shapes and thus is one of the least-standardized reduction strategies. In bifacial core reduction, one goal is the removal of wide and thin flake blanks. In size and shape, these blanks are often more standardized than informally reduced blanks, but tend to be less uniform than blades. Another potential goal of biface core reduction is to shape the core itself into a usable and maintainable biface (Kelly and Todd 1988). This second production goal has some clear benefits—a bifacial tool is produced—but it also places additional size and shape constraints on the blank removals. Discoidal and bipolar core technologies are two additional reduction strategies that have been documented elsewhere (Frison and Bradley 1980; Morgan, Eren, Khreisheh, Hill, and Bradley 2015; Surovell 2009), but they are rare in the study area and not considered further here.

Understanding the differences in these core-reduction strategies is important because core reduction has been shown to correlate with hunter-gatherer mobility strategies (Jennings 2015; Kelly and Todd 1988; Kuhn, 1994; Manninen and Knutsson 2014; Parry and Kelly 1987). For mobile hunter-gatherers, core-reduction decisions were influenced by three important factors: the availability of stone, the need to transport that stone, and the need to produce functional tool blanks with specific design characteristics. Ethnographic and archaeological records demonstrate that the degree of emphasis on these potentially competing factors varies with a hunter-gatherer population's mobility strategy (Bamforth 1986; Bettinger 1991; Binford 1980; Goodyear 1989). Through core-reduction experiments, informal cores have been shown to be highly efficient in producing tool blanks from a core that best conserves stone transport weight (Jennings, Pevny, and Dickens 2010; Prasciunas 2007).

In general, and particularly with smaller package sizes, biface and blade-core reduction are less-efficient strategies for conserving stone. The process of preparing and maintaining a bifacial core is costly, producing waste flakes with minimal utility (Jennings et al. 2010). In contexts where raw material is available in large package sizes and wide, thin flakes are desired, bifacial core reduction can be an effective reduction strategy. As with biface reduction, blade production is not an efficient strategy for conserving stone (Eren, Greenspan, and Sampson 2008). Preparing and maintaining the blade platform surface is costly because potentially usable stone is lost. However, and in contrast to other forms of core reduction, blade reduction has the added advantage of producing highly standardized blanks (Rasic and Andrefsky 2001). Thus, in blade production the emphasis is on the production of functional tool blanks with specific design characteristics: long, parallel-sided blanks. Further, once these blanks are removed from the core, they are easily transportable and can be chipped into a variety of tools. In contexts where large stone-material packages are available, blade knappers could emphasize these other advantages (Jennings et al. 2010).

Clovis Blade Technology

The Clovis Paleoindian complex is characterized by a suite of technological traits, including the presence of bifacially flaked, fluted projectile points, bifacial-point preforms, and bifacial cores; the rare presence of bone and ivory tools; and large blades and blade cores (Eren and Buchanan 2016; Haynes 2002; Jennings and Waters 2014). The distinctive Clovis point, a lanceolate-shaped form with lenticular cross sections and flutes that originate at the base and usually extend half the length of a point face, occurs at sites throughout North America that were south of the Wisconsin ice sheets (Smallwood and Jennings 2015). Of particular importance here, large blades, often called macroblades, found at many sites in the Plains, Southwest, and Southeast have become similarly diagnostic indications of Clovis occupations (Collins 1999; Kilby 2008; Waters et al. 2011). Clovis flintknappers struck prismatic blades from formally prepared conical and wedge-shaped cores. Blades from wedge-shaped cores were removed along multiple facets, or core platforms. An important characteristic of wedge-shaped blade cores is that the blade-removal face intersects the platform surface at an acute angle. Blades struck from a conical core are centripetally removed from a single platform surface. The platform angle on these cores approaches 90 degrees. Blades struck from both core types often exhibit a degree of ventral curvature, and both cores produce blades of similar sizes and shapes (Collins 1999; Waters et al. 2011).

Clovis blades, because of their sharp edges, could have been used in their unmodified form, or they could have served as blanks that were subsequently modified into a variety of tools. Clovis people chipped blades along one margin for use as side scrapers and endscrapers, the latter suggestive of hide working (Waters et al. 2011), and some blades were modified with serrations and gravers for cutting and incising. Blades with evidence of use have been found at quarry-camp locations, where they were produced for domestic use on site (Bradley, Collins, and Hemmings 2010; Waters et al. 2011). They have been recovered from kill sites and hunting camps in contexts with mammoth bones, where they were likely used for animal processing (Haynes and Huckell 2007; Sanchez et al. 2014). Blades were also cached in assemblages in the central and southern Plains, deposited as utilitarian caches in areas distant from stone sources (Huckell and Kilby 2014; Kilby 2008, 2015).

Toyah Blade Technology

The Toyah horizon of central Texas is most commonly associated with Perdiz arrow points (Black 1986; Jelks 1993: 12; Kenmotsu and Boyd 2012; Turner, Hester, and McReynolds

2011) and bone-tempered, plain pottery. The Toyah lithic toolkit also includes endscrapers, thin, beveled bifacial knives, perforators or drills, and blades and blade cores (Green and Hester 1973; Kelley 1947; Hester 1975, 1980; Hester and Parker 1970; Hester and Shafer 1975; Jelks 1962; Suhm and Jelks 1962). Assemblages with Toyah material culture occur throughout Texas, but Johnson (1994) notes regional variation among Toyah assemblages (also see Black 1986). The Classic Toyah area, including portions of the Southern Plains and Gulf Coastal Plain of Central Texas (Mandel 2000; Perttula 2004), is a region that includes archaeological sites where all elements of Toyah lithic and ceramic technology occur with no mixture of non-Toyah artifacts. It is in this region that blade technology has been most extensively documented (Collins 1999; Hester and Shafer 1975). In a description of Toyah blade technology from sites along the central and southern Texas coast, Hester and Shafer (1975) describe assemblages that include prismatic blades and conical polyhedral and bifacial (with a description suggestive of a wedge-shaped-core technology) blade cores.

Toyah blades were shaped into arrow points, end scrapers, and perforators, and some blades were unifacially trimmed, with striations indicating use as backed knives (Green and Hester 1973; Hester and Shafer 1975). Toyah lithic assemblages have been recovered from open-air campsites that include large animal-processing areas, where abundant quantities of fragmented deer and bison bone were discarded (Green and Hester 1973; Hester 1995). Toyah blades and blade segments were also cached among other lithic artifacts, including a core, bifaces, flakes, and other unifacial tools, at a site in north-central Texas along the Brazos River (Collins 1999; Tunnell 1989). This assemblage provides evidence of utilitarian caching.

Exploring Clovis and Toyah Blade Technology: A Case of Convergence?

Blade technology is a defining characteristic of both Clovis and Toyah in part because in the thousands of years between them, many populations did not reduce cores with this technique. How then did blade technology evolve in these chronologically distant populations? Does blade technology among Clovis and Toyah represent an example of convergence? In other words, did both populations adopt identical reduction strategies as a result of similar environmental conditions and adaptive paths? What were the environmental and social contexts in which the blade technologies arose? We examine these questions below.

Methods

Cladistic analysis was used to demonstrate evolutionary distance between Clovis and Toyah, an important step for identifying convergence (chapter 1, this volume). The analyses most closely follow the methods used by O'Brien et al. (2014). The dataset consists of 349 points representing 50 previously defined point types (chapter 11, this volume). It includes points from all pre-contact archaeological periods from Clovis to the Late Prehistoric. All points were recovered from the same geographic region encompassed by the states of Texas, Oklahoma, Arkansas, and Louisiana. This sample includes 17 Clovis points and eight Perdiz (Toyah) points. Point measurements were either directly recorded or measured from to-scale publication images using the program ImageJ (Schneider, Rasband, and Eliceiri 2012). The character dataset comprises two categorical traits and six variables calculated from point measurements (table 12.1). The two categorical variables, base type (seven character states) and proximal base shape (three character states), refer to characteristics of point bases related to hafting technique. The six shape variables have all been size-adjusted using the geometric mean (Lycett, von Cramon-Taubadel, and Foley 2006). We calculated the geometric mean for each specimen using maximum length, base length, maximum base width, and maximum blade width. Size adjustment then proceeded by dividing variables by the geometric mean. These six size-adjusted variables were then converted to categorical groups divided based on the data range for each variable.

Cladistic analyses commonly rely on selecting an outgroup to root the tree and serve as an evolutionary character state starting point (chapter 1, this volume). The difficulty here, however, lies in outgroup selection. Although pre-Clovis sites have been documented across North America, and some have provided hints of stone-projectile technology (Adovasio and Pedler 2004; Jennings and Waters 2014; Halligan et al. 2016; Waters et al.

Table 12.1
Point characters and character states or description of values used in cladistic analysis

Variable type	Character	Character state or description of value calculation
Descriptive	Base type	Lanceolate, side-notched, corner-notched, straight-stemmed, contracting-stemmed, expanding-stemmed, basal-notched
	Proximal base shape	Concave, flat, convex
Continuous*	SA base length	Maximum length of the point base (i.e., the portion of the point that would have been in the haft)
	SA base width	Maximum width of the point base
	SA concavity	Calculated using the depth of basal indentation from the proximal toward the distal tip (flat or convex based points received values of zero)
	SA length/blade width	Calculated using the shape ratio of maximum point length divided by the maximum blade width
	SA basal constriction	Calculated by dividing the maximum basal width by the minimum basal width
	SA blade width/base width	Calculated using the shape ratio of maximum blade width divided by maximum base width

* Continuous characters were size-adjusted (SA) using the geometric mean of each individual point in the database.

2011), the assemblages are still too few and widespread to comprise a suitable outgroup. Likewise, Upper Paleolithic assemblages in Siberia and Beringia (Goebel and Buvit 2011) have not produced an outgroup candidate. Given these limitations, we used Clovis as the outgroup, which means that all cladograms have Clovis as their root. Therefore, Toyah Perdiz points will be grouped as a descendant of Clovis, and this cladistic analysis cannot be used to demonstrate that Clovis and Perdiz developed within entirely separate evolutionary histories. Convergence, if identified, will necessarily be convergence by reversion (McGhee 2011). Despite this limitation, cladistic analysis is useful for investigating evolutionary distances between Clovis and Perdiz points.

We used PAUP* 4.0 (Swofford 1998) for analyzing the 349-taxa, 8-character dataset. A parsimony-based heuristic search was performed with 1,000 tree replicates to create a 50-percent majority-rule consensus cladogram (see below). We then compared the morphological shapes of Clovis and Perdiz points. Phenotypic distance can be visualized through principal component analysis (PCA; Louys and Faith 2015; Schillinger et al. 2016; Wang et al. 2012), which orthogonally transforms multiple, possibly correlated, variables into linearly uncorrelated components (Buchanan, Kilby, Huckell, O'Brien, and Collard 2012; Jennings, Smallwood, and Waters 2015; VanPool and Leonard 2011). Components with eigenvalues greater than 1 account for a relatively large proportion of the variation within a dataset and are retained for further comparisons. In this study, only the six continuous, size-adjusted variables were included in the PCA, and raw metric values prior to character-state grouping were used. A *t*-test was used to compare component values for the Clovis and Perdiz points.

To further explore evolutionary differences between Clovis and Toyah blade technologies and provide support for the argument that they arose through convergence, we compared the sizes and shapes of blades from Clovis and Toyah site assemblages. The data come from Collins (1999). In his foundational work describing and defining Clovis blade technology, Collins presented a preliminary comparison between Clovis blades and other, non-Clovis blades, including Toyah. In his comparisons, Collins used visual scatter plots to highlight potential differences in blade sizes and shapes. We expanded on this work in three ways. First, we employed Kruskal-Wallis tests to add statistical weight to potential differences (the data are not normally distributed). Second, we used a different method for calculating shape differences. Collins (1999) divided each blade measurement by the sum of length, width, and thickness (a general method repeated by others, including Bradley et al. 2010; Lohse, Hemmings, Collins, and Yelacic 2014b; and Waters et al. 2011), but this method biases the comparison in favor of the largest measure—length—and against differences in the smallest measure—thickness. To equalize volumetric differences between individual blades and compare allometric shape differences, we instead divided each variable by the geometric mean (Lycett et al. 2006), which is the cube root of the product of length, width, and thickness. Third, we used the coefficient of variation (CV) to compare relative differences in blade standardization (Eerkens and Bettinger 2001, 2008; Jennings 2016; Lycett et al. 2006; Smallwood 2012).

Results

The cladistic analysis demonstrated that considerable evolutionary changes separate Clovis and Perdiz points (figure 12.2). The 50-percent majority-rule consensus tree has a retention index of 0.86 (chapter 1, this volume), a value at the high end of values for other cultural datasets (Collard, Shennan, and Tehrani 2006). Because Clovis was selected as the outgroup, all other points, nodes, and branches in the tree are linked to this root. Eighty-eight percent (seven of eight) of the Perdiz points in the sample are 18 or more internal evolutionary nodes separated from Clovis. Numerous morphological character-state changes distinguish these two point styles, and many other point styles and types emerged between them.

PCA lends statistical weight to the evolutionary differences hypothesized in the consensus tree and helps quantify the significant morphological (phenotypic) distance between Clovis and Perdiz points. Three PCA components have eigenvalues greater than 1, and together they explain 77.4 percent of the variation (table 12.2). Component 1 is characterized by high, positive loadings for size-adjusted (SA) length/blade width and SA blade width/base width. Both are shape ratios involving blade width. Component 2 is characterized by high, positive loadings for SA base width and SA basal constriction. Both reflect basal shape. Component 3 is characterized by high, positive loadings in SA length and SA concavity. Clovis and Perdiz points significantly differ for all three component scores, demonstrating that these two point types are morphologically distinct in all aspects of shape (figure 12.3).

The blade comparison also reveals significant differences in shape and size. Clovis and Toyah blade sizes significantly differ in all three major dimensions (table 12.3); Clovis blades are longer ($p < 0.001$), wider ($p < 0.001$), and thicker ($p < 0.001$). Length-to-width ratios, argued to be a unique characteristic of Clovis blades (Bradley et al. 2010), also differ significantly. Clovis blades are longer relative to width than Toyah blades. After adjusting for size (table 12.4), four of the five blade allometric shape measures also differ significantly. Clovis blade shapes are long and narrow but are thicker relative to blade width, and wider relative to blade length, than Toyah blades (figure 12.4). The only shape that does not differ is thickness. Relative to their sizes, Clovis and Toyah blade shapes are equally thick.

Comparisons of blade CV values (figure 12.5), alternatively, reveal important similarities. In terms of blade sizes, both Clovis and Toyah blades are relatively unstandardized, and all variables have CVs greater than 20. Size-adjusted shape CVs for length and width are low (ranging from 14 to 16) for both Clovis and Toyah blades. These variables display

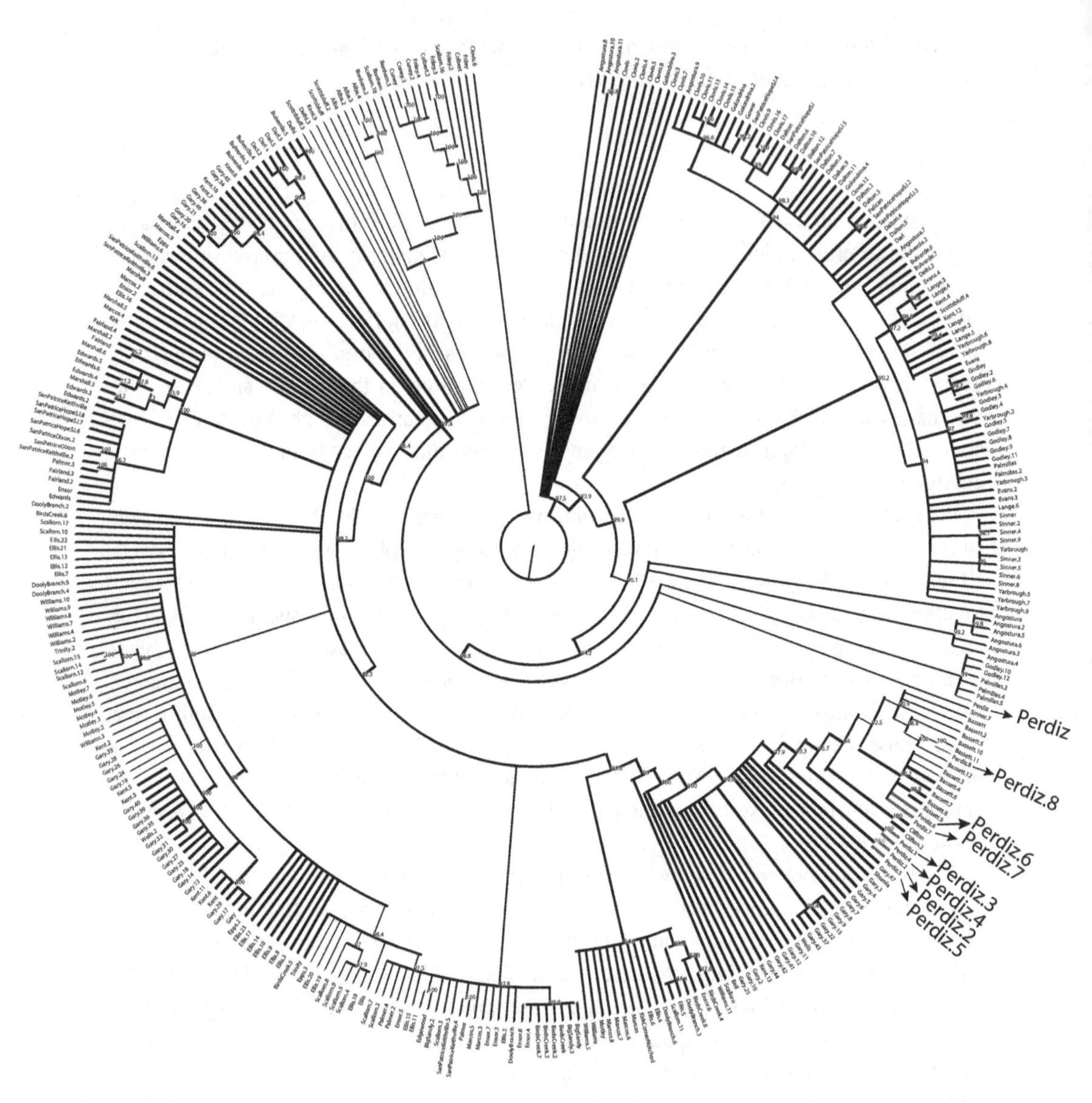

Figure 12.2
Fifty-percent majority-rule cladogram highlighting the locations of the eight Perdiz points.

Table 12.2
Principal component analysis (PCA) score comparisons for Clovis and Perdiz points with *t*-test *p*-values

PCA	Point type	Mean score	Significance (*p*)
Component 1	Clovis	–0.02	0.005
(Eigen = 2.058)	Perdiz	2.00	
Component 2	Clovis	–0.31	0.001
(Eigen = 1.393)	Perdiz	–1.37	
Component 3	Clovis	2.67	0.001
(Eigen = 1.186)	Perdiz	0.35	

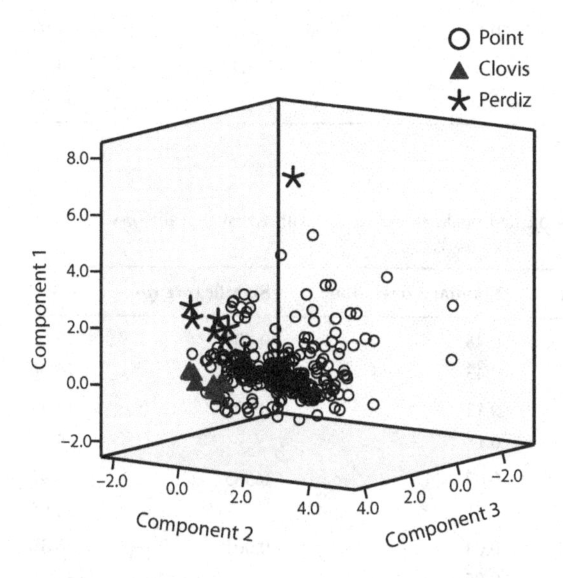

Figure 12.3
Scatterplot comparing Clovis and Perdiz point principal component analysis component scores.

the highest degree of standardization. The other size-adjusted variables all have higher CVs, reflecting minimal standardization.

Discussion

Cladistic and morphological analyses of stone points provided empirical evidence of the differences between Clovis and Perdiz points. The consensus tree, for example, shows extensive evolutionary distance between the two point types. The differences are also evident in the PCA analysis, which compared the point shapes. Clovis and Perdiz point

Table 12.3
Comparison of Clovis and Toyah blade sizes with Kruskal-Wallis *p*-values and coefficient of variation (CV) values

Size	Culture	Mean	Standard deviation	Significance (*p*)	CV
Length (mm)	Clovis	113.45	30.90	0.001	27.23
	Toyah	47.79	18.11		37.89
Width (mm)	Clovis	31.37	10.20	0.001	32.52
	Toyah	18.29	5.04		27.56
Thickness (mm)	Clovis	13.46	4.64	0.001	34.47
	Toyah	6.97	3.87		55.52
Length/width	Clovis	3.73	0.80	0.001	21.45
	Toyah	2.60	0.57		21.92
Width/thickness	Clovis	2.49	0.85	0.018	34.14
	Toyah	3.06	1.12		36.60

Table 12.4
Comparison of Clovis and Toyah average size-adjusted blade shape values with Kruskal-Wallis *p*-values and coefficient of variation (CV) values

Shape	Culture	Mean	Standard deviation	Significance (*p*)	CV
SA length	Clovis	3.19	0.48	0.001	15.05
	Toyah	2.69	0.43		16.00
SA width	Clovis	0.88	0.13	0.001	14.77
	Toyah	1.05	0.15		14.29
SA thickness	Clovis	0.37	0.07	0.565	18.92
	Toyah	0.38	0.11		28.95
SA length/width	Clovis	0.11	0.04	0.001	36.36
	Toyah	0.16	0.05		31.25
SA width/thickness	Clovis	0.07	0.04	0.018	57.14
	Toyah	0.20	0.13		65.00

differences are obvious when visually comparing the large, fluted, lanceolate Clovis dart point to the small, stemmed Perdiz arrow point, and PCA analyses show that the points significantly differ in every shape comparison.

If Clovis and Toyah populations separately converged on blade core reduction—and there is every reason to suspect this is the case, given the 12,000-year time span separating the makers of each point type—then the recipes of each technology should also significantly differ as reflected in the reduction products, the blades. The size of Clovis and Toyah blades significantly differ, as does blade shape. These differences provide strong evidence that Clovis and Toyah flintknappers employed different and unique processes to reduce

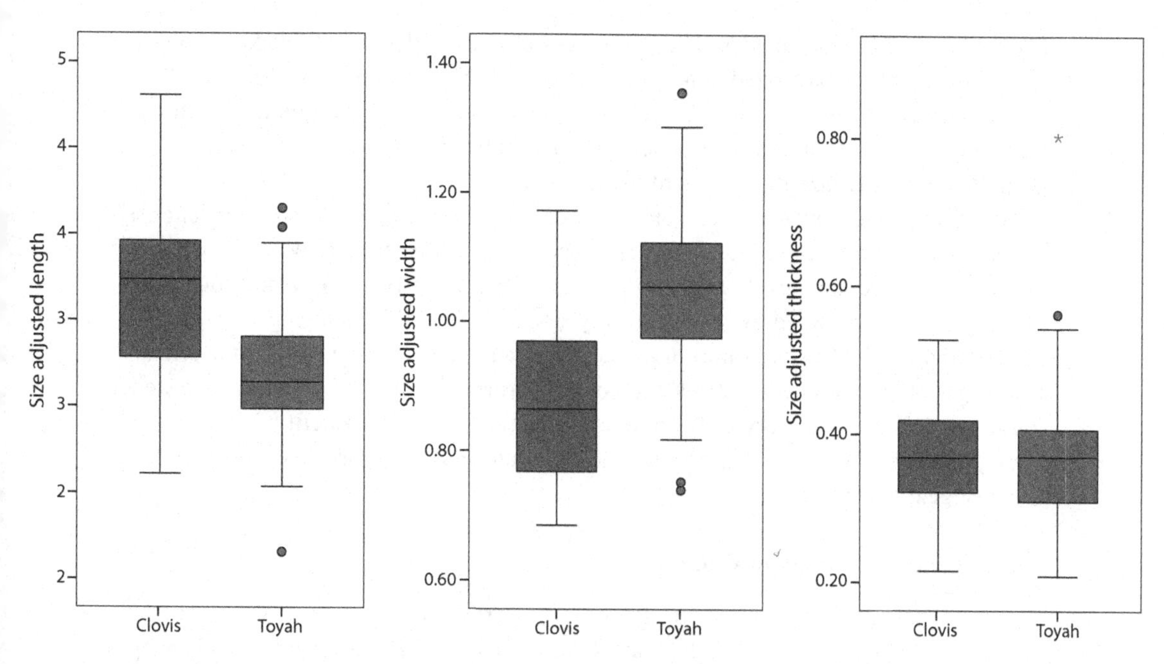

Figure 12.4
Box plots comparing Clovis and Toyah blade shapes for three variables: size-adjusted length, size-adjusted width, and size-adjusted thickness.

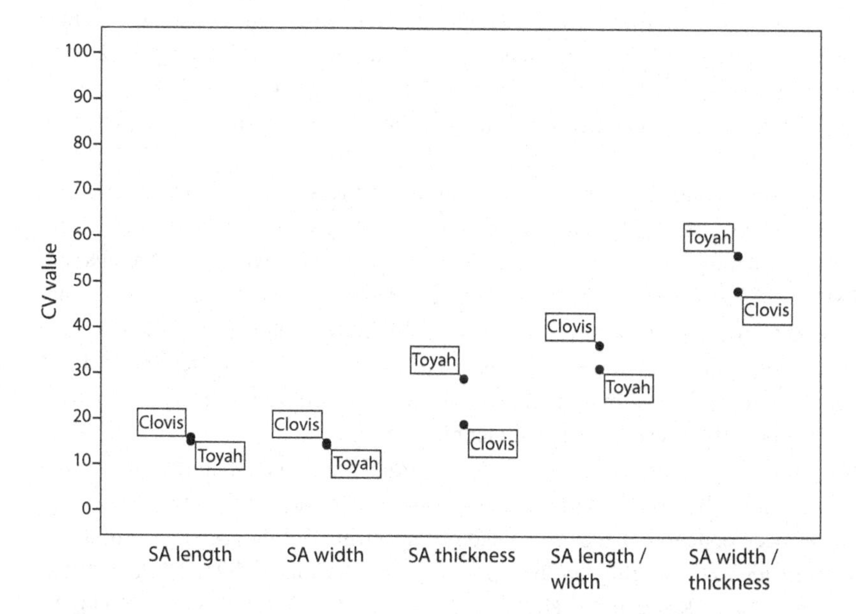

Figure 12.5
Comparison of coefficients of variation of Clovis and Toyah blades for all size-adjusted shape variables.

blade cores and produce blades. The differences also show that Toyah blade technology cannot simply be characterized as a reversion to Clovis blade technology. Whereas Clovis and Toyah blade sizes and shapes differ, reflecting unique, convergent production recipes, CV comparisons reveal shared production goals: both Clovis and Toyah blade makers attempted to standardize the shapes of blades.

That Clovis and Toyah groups converged on a technique of blade-core reduction is important, because blade reduction has been shown to be a potentially wasteful and inefficient use of stone supplies (Eren et al. 2008; Jennings et al. 2010). Why, then, would these two populations, separated by thousands of years, have used this knapping strategy? The answer might lie in the similarities in shape CV because they highlight one of the primary advantages of blade reduction: production of standardized blanks. In contexts where stone conservation is not a primary concern, blade reduction offers the benefit of standardized blank production. What might have been the environmental and adaptive contexts that led to convergent blade technologies?

Clovis Blade Technology in Context

The environment of the Southern Plains and periphery during the Clovis period (13,100–12,700 calendar years B.P.; Waters and Stafford 2007) was complex and changing. As the Ice Age came to a close, the North American continent witnessed new temperature, precipitation, and seasonality patterns, which led plant communities to undergo a dramatic reorganization (Meltzer and Holliday 2010). The extinction event that accompanied this biotic reorganization and involved approximately 35 genera of animals, including mammoth, was nearing its end. As mammoth population densities dwindled, they became separated into dispersed refugia (Ballenger 2015; Haynes 2011), which, for human hunters, would have been a spatially predictable concentration of highly ranked prey animals.

Amid these environmental changes, Clovis groups along the Southern Plains and periphery adopted subsistence, mobility, and technological strategies designed to cope with shifting resource distributions. The Clovis diet in this region was diverse (DeAngelis and Lyman 2016), and Clovis camp sites have yielded a variety of large and small faunal remains as well as some plant remains (Ferring 2001; Hill 2007; Waters et al. 2011). Applying Hill's (2007) large-fauna abundance index (LFAI) to Clovis camp sites (table 12.5) shows that large game, including mammoth, were clearly an important dietary component but not the sole focus of Clovis subsistence (Jennings 2015).

To facilitate a broad-spectrum diet that included seasonal hunting of spatially predictable mammoth herds, Clovis bands adopted a logistical mobility collecting strategy (Jennings 2015), which has been conceptualized as one end of a spectrum of hunter-gather settlement strategies, with residential mobility on the opposite end (Bettinger 1991; Binford 1980; Bleed 1986; Grove 2010; Kelly 1995; Kuhn 1995; Perreault and Brantingham 2011). As

Table 12.5
Comparison of Clovis (Jennings 2015) and Toyah (calculated from Dering 2008) large-fauna abundance indices (LFAI) at camp sites

Site	Complex	Class 4 NISP	Total NISP	LFAI (Hill 2007)
Aubrey	Clovis	401	1172	0.34
Gault	Clovis	1051	2185	0.48
Lewisville	Clovis	22	442	0.05
Camp Bowie	Toyah	0	4	0
Mustang Branch	Toyah	853	6282	0.14
Honey Creek	Toyah	3	162	0.02
Rush	Toyah	10844	10997	0.98

hypothesized by Jennings (2015; see also Collins 2007 and Meltzer 2004), Clovis collectors established relatively large and intensively occupied base camps on the edge of the Southern Plains. From these sites, logistical task groups made seasonal hunting forays into the grasslands, where they left behind kill sites from successful hunts. Because mammoth refugia and, therefore, hunting localities were spatially predictable, Clovis logistical task groups also relied on caching. Caches of high-utility cores, blanks, and tools served as supplies for future hunts in material-poor locations as well as load exchanges to offset the transport weight of meat packages (Kilby 2008).

Clovis technological organization and in particular blade-core reduction, were directly tied to Clovis mobility. The strategies of establishing relatively long-term base camps, gearing up for logistical forays, and caching stone for future needs all eased stone-transport concerns. Because stone conservation was not a primary concern for populations occupying the Southern Plains and periphery, Clovis flintknappers prioritized other core-reduction benefits and employed multiple reduction techniques, including heavy reliance on blade reduction (Jennings 2015). Blades are bulky and transport-inefficient (Eren et al. 2008; Jennings et al. 2010), but the blanks produced through blade reduction allow for a much higher degree of standardization (Rasic and Andrefsky 2001). Clovis flintknappers, unconcerned with the efficiency drawback, employed blade reduction to produce standardized blanks for multiple tool types. Blade cores, blades, and blade tools have been recovered from camp sites, kill sites, and caches, demonstrating the direct link between Clovis blade technology and subsistence and mobility strategies.

Toyah Blade Technology in Context

The Toyah period in central Texas dates to ca. 650–300 B.P. (Collins 2004; Lohse et al. 2014a) and coincided with a period of environmental change. Reconstructed Palmer Drought Severity Indices (PDSI; Cook and Krusic 2004) show that this period was much wetter than the preceding Austin phase. Further, PDSI value differences between 25-year temporal groups suggest that fluctuations in moisture increased during the Toyah period (Mauldin, Thompson, and Kemp 2012). In response to increased moisture variability, expansion of woodlands began and progressed rapidly (Bousman 1998), and grasslands became increasingly patchy (Mauldin et al. 2012). Vegetation and moisture changes coincided with a major shift in the regional availability of bison, which were virtually absent from the area for approximately 1,500 years, until they reentered the region around 650 B.P. (Dillehay 1974; Lohse, Culleton, Black, and Kennett 2014c; Mauldin et al. 2012). This fluctuating environment also likely led to patchy concentrations of bison populations, but the location of herds varied with annual moisture and grassland variability. The return of bison to the region would have provided an important subsistence opportunity, but hunters would have needed to adapt to both herd spatial predictability and temporal unpredictability.

Mauldin et al. (2012) have thoroughly outlined how these environmental changes impacted Toyah subsistence, mobility, and technological strategies. Toyah populations adopted a broad-spectrum diet that included many large and small animals and intensively processed plant foods (Dering 2008; Mauldin et al. 2013). Bison were clearly an important element of Toyah subsistence, as evidenced by the presence of bison in 44 of 53 (83 percent) archaeological components (Mauldin et al. 2012), although the degree of emphasis on bison may have varied subregionally and temporally with local shifts in bison availability (see above discussion). Applying the LFAI to data presented by Dering (2008) provides further evidence that bison were of variable importance in Toyah diets. The limited available seasonality evidence indicates that bison hunting took place in the summer and fall (Dering 2008). Toyah hunters actively and frequently pursued bison, but it was not always a primary subsistence focus.

As hypothesized by Mauldin et al. (2012), Toyah mobility was organized to take advantage of a wide array of resources and to respond rapidly to bison availability. This was accomplished by adopting a collecting settlement system emphasizing logistical mobility. Logistical task groups left base camps in pursuit of game, and bison were transported back to base camps for final butchering and processing (Lohse et al. 2014c). Toyah populations also cached tools and blanks (Collins 1999; Tunnell 1989), which would have supplied and supported logistical forays.

Toyah lithic technological organization fits the adaptive shift of logistical mobility (Mauldin et al. 2012). Operating from relatively long-term base camps allowed knappers to pre-plan for logistical hunts and "gear up" (Binford 1978) on stone supplies in advance. Therefore, stone-supply conservation was not a primary concern. This in turn allowed populations to invent and rely on the bulky and inefficient strategy of blade reduction. Freed from concerns about reduction efficiency, Toyah flintknappers took advantage of the ability to standardize blank shapes through blade reduction, and Toyah blades served as blanks for Perdiz points, knives, and scrapers. Toyah blade technology emerged as a specialized toolkit component designed for the pursuit of bison (Collins 2004; Mauldin et al. 2012).

Summary

Similarities between the contexts of the convergent evolution of blade technology in Clovis and Toyah are striking. Both populations experienced a period of environmental variability that created spatially predictable but temporally unpredictable patches of large game. In response, both populations adopted a broad-spectrum subsistence strategy that incorporated large-game hunting as a core dietary strategy. To target predictable hunting grounds, both populations adopted a logistical mobility strategy that included stone caching. This mobility strategy reduced potential concerns about stone-supply conservation and facilitated the use of blade technology which, though less efficient, offered the advantage of standardized blanks for important components of Clovis and Toyah toolkits.

Conclusion

Convergent, or homoplastic, traits have been defined as those that are similar but are not inherited from a common ancestor (McGhee 2011) and as similarities resulting from processes other than descent from a common ancestor (O'Brien et al. 2014). These definitions explicitly state the importance of knowing each trait's ancestry. In archaeology, the most widely used method for reconstructing ancestries and identifying potential instances of convergent homoplasies is outgroup-based cladistic analysis.

The case study we discuss here is important because it might not always be possible to identify homoplastic convergence through cladistics. In cases where an archaeological complex of interest is also one of the oldest expressions in a region, we need additional ways to explore the possibility of convergence. We used cladistics to calculate the phylogenetic distance between Clovis and Toyah point types and showed that the products of blade production also significantly differ. Each line of evidence shows that Clovis and Toyah are significantly different. To us, they provide strong support for the hypothesis that Clovis and Toyah blade technologies are unique, unrelated, and an example of cultural convergence.

An important difference between biological and cultural evolution is the relative stability of evolutionary building blocks. Although portions of DNA can be lost or gained, these kinds of substantial changes are rare, and the relative stability of DNA and regulatory genes places developmental constraints on biological evolution (McGhee 2011). Cultural knowledge does not necessarily require such structural permanence, and blocks of cultural knowledge can be lost or gained. We propose that in the 12,000 years of cultural evolution that separated Clovis and Toyah, the cultural knowledge of Clovis blade making and even elements of the Clovis lithic-reduction framework were lost, abandoned, or significantly altered, and that the blade technologies are, therefore, convergent. The Clovis and Toyah case also provides insights into the contexts in which we should expect convergence in stone-tool technologies. Major environmental changes that alter landscapes and the availability of plant and animal resources in similar ways can lead to major, convergent changes in hunter-gatherer lithic technological organization.

Acknowledgments

Thanks to Mike O'Brien, Briggs Buchanan, and Metin Eren for organizing and inviting us to participate in the workshop that led to this paper. We very much enjoyed the opportunity to share and discuss current approaches to the study of convergent stone technologies at the Konrad Lorenz Institute. We also thank the College of Sciences, University of West Georgia, for supporting this research.

References

Adovasio, J. M., & Pedler, D. R. (2004). Pre-Clovis Sites and Their Implications for Human Occupation before the Last Glacial Maximum. In D. B. Madsen (Ed.), *Entering America: Northeast Asia and Beringia before the Last Glacial Maximum* (pp. 139–158). Salt Lake City: University of Utah Press.

Andrefsky, W. (2005). *Lithics: Macroscopic Approaches to Analysis* (2nd ed.). Cambridge: Cambridge University Press.

Ballenger, J. A. M. (2015). The Densest Concentration on Earth? Quantifying Clovis-Mammoth Associations in the San Pedro Valley, Southeastern Arizona, U.S.A. In A. M. Smallwood & T. A. Jennings (Eds.), *Clovis: On the Edge of a New Understanding* (pp. 183–204). College Station: Texas A&M University Press.

Bamforth, D. B. (1986). Technological Efficiency and Tool Curation. *American Antiquity*, *51*, 38–50.

Bettinger, R. L. (1991). *Hunter-Gatherers: Archaeological and Evolutionary Theory.* New York: Plenum.

Binford, L. R. (1978). Analysis of Behavior and Site Structure: Learning from an Eskimo Hunting Stand. *American Antiquity*, *43*, 330–361.

Binford, L. R. (1980). Willow Smoke and Dogs' Tails: Hunter-Gatherer Settlement Systems and Archaeological Site Formation. *American Antiquity*, *45*, 4–20.

Black, S. L. (1986). *The Clemente and Hermina Hinojosa Site, 41 JW 8: A Toyah Horizon Campsite in Southern Texas. Special Report, no.18.* Center for Archaeological Research, University of Texas at San Antonio.

Bleed, P. (1986). The Optimal Design of Hunting Weapons: Maintainability or Reliability. *American Antiquity*, *51*, 737–747.

Bordes, F. (1967). *The Old Stone Age.* New York: McGraw-Hill.

Bordes, F., & Crabtree, D. (1969). The Corbiac Blade Technique and Other Experiments. *Tebiwa*, *12*, 1–21.

Bousman, C. B. (1998). Paleoenvironmental Change in Central Texas: The Palynological Evidence. *Plains Anthropologist*, *43*, 201–219.

Bradley, B. A., Collins, M. B., & Hemmings, A. (2010). *Clovis Technology.* Ann Arbor, Mich.: International Monographs in Prehistory.

Buchanan, B., Kilby, J. D., Huckell, B. B., O'Brien, M. J., & Collard, M. (2012). A Morphometric Assessment of the Intended Function of Cached Clovis Points. *PLoS One*, *7*(2), e30530.

Clark, G. A., & Riel-Salvatore, J. (2006). Observations on Systematics in Paleolithic Archaeology. In E. Hovers & S. L. Kuhn (Eds.), *Transitions before the Transition* (pp. 29–56). New York: Springer.

Collard, M., Shennan, S. J., & Tehrani, J. J. (2006). Branching, Blending, and the Evolution of Cultural Similarities and Differences among Human Populations. *Evolution and Human Behavior*, *27*, 169–184.

Collins, M. B. (1999). *Clovis Blade Technology*. Austin: University of Texas Press.

Collins, M. B. (2004). Archeology in Central Texas. In T. K. Perttula (Ed.), *The Prehistory of Texas* (pp. 101–126). College Station: Texas A&M University Press.

Collins, M. B. (2007). Discerning Clovis Subsistence from Stone Artifacts and Site Distributions on the Southern Plains Periphery. In R. B. Walker & B. N. Driskell (Eds.), *Foragers of the Terminal Pleistocene in North America* (pp. 59–87). Lincoln: University of Nebraska Press.

Cook, E. R., & Krusic, P. J. (2004). *The North American Drought Atlas*. Lamont-Doherty Earth Observatory. http://iridl.ldeo.columbia.edu/SOURCES/.LDEO/.TRL/.NADA2004/.pdsi-atlas.html

DeAngelis, J. A., & Lyman, R. L. (2016). Evaluation of the Early Paleo-Indian Zooarchaeological Record as Evidence of Diet Breadth. *Archaeological and Anthropological Sciences*. doi: 10.1007/s12520-016-0377-1.

Dering, P. (2008). Late Prehistoric Subsistence Economy on the Edwards Plateau. *Plains Anthropologist*, *53*, 59–77.

Dillehay, T. D. (1974). Late Quaternary Bison Population Changes on the Southern Plains. *Plains Anthropologist*, *19*, 180–196.

Eerkens, J. W., & Bettinger, R. L. (2001). Techniques for Assessing Standardization in Artifact Assemblages: Can We Scale Material Variability? *American Antiquity*, *66*, 493–504.

Eerkens, J. W., & Bettinger, R. L. (2008). Cultural Transmission and the Analysis of Stylistic and Functional Variation. In M. J. O'Brien (Ed.), *Transmission and Archaeology: Issues and Case-Studies* (pp. 21–38). Washington, D.C.: Society for American Archaeology.

Eren, M. I., and B. Buchanan. (2016). Clovis Technology. *eLS*. 1–9. doi: 10.1002/9780470015902.a0026512.

Eren, M. I., Greenspan, A., & Sampson, C. G. (2008). Are Upper Paleolithic Blade Cores More Productive Than Middle Paleolithic Discoidal Cores? A Replication Experiment. *Journal of Human Evolution*, *55*, 952–961.

Eren, M. I., Patten, R. J., O'Brien, M. J., & Meltzer, D. J. (2013). Refuting the Technological Cornerstone of the Ice-Age Atlantic Crossing Hypothesis. *Journal of Archaeological Science*, *40*, 2934–2941.

Ferring, C. R. (2001). The Archaeology and Paleoecology of the Aubrey Clovis Site (41DN479) Denton County, Texas. Denton: Center for Environmental Archaeology, Department of Geography, University of North Texas.

Frison, G. C., & Bradley, B. A. (1980). *Folsom Tools and Technology of the Hanson Site, Wyoming*. Albuquerque: University of New Mexico Press.

Goebel, T., & Buvit, I. (2011). *From the Yenisei to the Yukon: Interpreting Lithic Assemblage Variability in Late Pleistocene/Early Holocene Beringia*. College Station: Texas A&M Press.

Goodyear, A. C. (1989). A Hypothesis for the Use of Cryptocrystalline Raw Materials among Paleoindian Groups of North America. In C. J. Ellis & J. C. Lothrop (Eds.), *Eastern Paleoindian Lithic Resource Use* (pp. 1–9). Boulder, Colo.: Westview Press.

Green, L. M., & Hester, T. R. (1973). The Finis Frost Site: A Toyah Phase Occupation of San Saba County, Central Texas. *Bulletin of the Texas Archeological Society*, *44*, 69–88.

Grove, M. (2010). Logistical Mobility Reduces Subsistence Risk in Hunting Economies. *Journal of Archaeological Science, 37*, 1913–1921.

Halligan, J. J., Waters, M. R., Perrotti, A., Owens, I. J., Feinberg, J. M., Bourne, M. D., et al. (2016). Pre-Clovis Occupation 14,550 Years Ago at the Page-Ladson Site, Florida, and the Peopling of the Americas. *Science Advances, 2*, e1600375.

Haynes, C. V., Jr., & Huckell, B. B. (Eds.). (2007). *Murray Springs, A Clovis Site with Multiple Activity Areas in the San Pedro Valley, Arizona. Anthropological Papers, no. 71.* Tucson: University of Arizona.

Haynes, G. (2002). *The Early Settlement of North America: The Clovis Era.* Cambridge: Cambridge University Press.

Haynes, G. (2011). Extinctions in North America's Late Glacial Landscapes. *Quaternary International, 285*, 89–98.

Hester, T. R. (1975). Late Prehistoric Cultural Patterns along the Lower Rio Grande of Texas. *Bulletin of the Texas Archeological Society, 46*, 107–125.

Hester, T. R. (1980). *Digging into South Texas Prehistory: A Guide for Amateur Archaeologists.* San Antonio, TX: Corona.

Hester, T. R. (1995). The Prehistory of South Texas. *Bulletin of the Texas Archeological Society, 66*, 427–459.

Hester, T. R., & Parker, R. (1970). The Berclair Site: A Late Prehistoric Component in Goliad County, Southern Texas. *Bulletin of the Texas Archeological Society, 41*, 1–24.

Hester, T. R., & Shafer, H. J. (1975). An Initial Study of Blade Technology on the Central and Southern Texas Coast. *Plains Anthropologist, 20*, 175–185.

Hill, M. E. (2007). A Moveable Feast: Variation in Faunal Resource Use among Central and Western North American Paleoindian Sites. *American Antiquity, 72*, 417–438.

Huckell, B. B., & Kilby, J. D. (Eds.). (2014). *Clovis Caches: Recent Discoveries and New Research.* Albuquerque: University of New Mexico Press.

Jelks, E. B. (1962). *The Kyle Site: A Stratified Central Texas Aspect Site in Hill County, Texas. Archaeology Series, no. 5.* Austin: Department of Anthropology, University of Texas.

Jelks, E. B. (1993). Observations on the Distributions of Certain Arrow-Point Types in Texas and Adjoining Regions. *Lithic Technology, 18*, 9–15.

Jennings, T. A. (2015). Clovis Adaptations in the Great Plains. In A. M. Smallwood & T. A. Jennings (Eds.), *Clovis: On the Edge of a New Understanding* (pp. 277–296). College Station: Texas A&M University Press.

Jennings, T. A. (2016). The Impact of Stone Supply Stress on the Innovation of a Cultural Variant: The Relationship of Folsom and Midland. *PaleoAmerica, 2*, 116–123.

Jennings, T. A., & Waters, M. R. (2014). Pre-Clovis Lithic Technology at the Debra L. Friedkin Site, Texas: Comparisons to Clovis through Site-Level Behavior, Technological Trait-List, and Cladistic Analyses. *American Antiquity, 79*, 25–44.

Jennings, T. A., Pevny, C. D., & Dickens, W. A. (2010). A Biface and Blade Core Efficiency Experiment: Implications for Early Paleoindian Technological Organization. *Journal of Archaeological Science, 37*, 2155–2164.

Jennings, T. A., Smallwood, A. M., & Waters, M. R. (2015). Exploring Late Paleoindian and Early Archaic Unfluted Lanceolate Point Classification in the Southern Plains. *North American Archaeologist, 36*, 243–265.

Johnson, J. K. (1994). Prehistoric Exchange in the Southeast. In T. G. Baugh & E. E. Jonathon (Eds.), *Prehistoric Exchange Systems in North America* (pp. 99–125). New York: Plenum.

Kelley, J. C. (1947). The Lehmann Rock Shelter: A Stratified Site of the Toyah, Uvalde, and Round Rock Foci. *Bulletin of the Texas Archeological and Paleontological Society*, *18*, 115–128.

Kelly, R. L. (1995). *The Foraging Spectrum: Diversity in Hunter-Gatherer Lifeways*. Washington, D.C.: Smithsonian Institution Press.

Kelly, R. L., & Todd, L. C. (1988). Coming into the Country: Early Paleoindian Hunting and Mobility. *American Antiquity*, *53*, 231–244.

Kenmotsu, N. A., & Boyd, D. K. (Eds.). (2012). *The Toyah Phase of Central Texas: Late Prehistoric Economic and Social Processes*. College Station: Texas A&M University Press.

Kilby, J. D. (2008). *An Investigation of Clovis Caches: Content, Function, and Technological Organization*. Ph.D. dissertation, University of New Mexico, Albuquerque.

Kilby, J. D. (2015). A Regional Perspective on Clovis Blades and Caching Behavior. In A. M. Smallwood & T. A. Jennings (Eds.), *Clovis: On the Edge of a New Understanding* (pp. 145–159). College Station: Texas A&M University Press.

Kuhn, S. L. (1994). A Formal Approach to the Design and Assembly of Mobile Toolkits. *American Antiquity*, *59*, 426–442.

Kuhn, S. L. (1995). *Mousterian Lithic Technology: An Ecological Perspective*. Princeton, N.J.: Princeton University Press.

Lipo, C. P., O'Brien, M. J., Collard, M., & Shennan, S. (Eds.). (2006). *Mapping Our Ancestors: Phylogenetic Approaches in Anthropology and Prehistory*. New York: Aldine.

Lohse, J. C., Black, S. L., & Cholak, L. M. (2014a). Toward an Improved Archaic Radiocarbon Chronology for Central Texas. *Bulletin of the Texas Archeological Society*, *85*, 259–287.

Lohse, J. C., Hemmings, C. A., Collins, M. B., & Yelacic, D. M. (2014b). Putting the Specialization Back in Clovis: What Some Caches Reveal about Skill and the Organization of Production in the Terminal Pleistocene. In B. B. Huckell & J. D. Kilby (Eds.), *Clovis Caches: Recent Discoveries and New Research* (pp. 153–176). Albuquerque: University of New Mexico Press.

Lohse, J. C., Culleton, B. J., Black, S. L., & Kennett, D. J. (2014c). A Precise Chronology of Middle to Late Holocene Bison Exploitation in the Far Southern Great Plains. *Journal of Texas Archeology and History*, *1*, 94–126.

Louys, J., & Faith, J. T. (2015). Phylogenetic Topology Mapped onto Dietary Ecospace Reveals Multiple Pathways in the Evolution of the Herbivorous Niche in African Bovidae. *Journal of Zoological Systematics and Evolutionary Research*, *53*, 140–154.

Lycett, S. J. (2009). Understanding Ancient Hominin Dispersals Using Artefactual Data: A Phylogeographic Analysis of Acheulean Handaxes. *PLoS One*, *4*(10), e7404.

Lycett, S. J. (2015). Differing Patterns of Material Culture Intergroup Variation on the High Plains: A Quantitative Analysis of Parfleche Characteristics vs. Moccasin Decoration. *American Antiquity*, *80*, 714–731.

Lycett, S. J., von Cramon-Taubadel, N., & Foley, R. A. (2006). A Crossbeam Co-ordinate Caliper for the Morphometric Analysis of Lithic Nuclei: A Description, Test, and Empirical Examples of Application. *Journal of Archaeological Science*, *33*, 847–861.

Mandel, R. D. (2000). Introduction. In R. D. Mandel (Ed.), *Geoarchaeology in the Great Plains* (pp. 1–43). Norman: University of Oklahoma Press.

Manninen, M. A., & Knutsson, K. (2014). Lithic Raw Material Diversification as an Adaptive Strategy—Technology, Mobility, and Site Structure in Late Mesolithic Northernmost Europe. *Journal of Anthropological Archaeology*, *33*, 84–98.

Mauldin, R., Thompson, J., & Kemp, L. (2012). Reconsidering the Role of Bison in the Terminal Late Prehistoric (Toyah) Period in Texas. In N. A. Kenmotsu & D. K. Boyd (Eds.), *The Toyah Phase of Central Texas: Late Prehistoric Economic and Social Processes* (pp. 90–110). College Station: Texas A&M University Press.

Mauldin, R. P., Hard, R. J., Munoz, C. M., Rice, J. L. Z., Verostick, K., Potter, D. R., et al. (2013). Carbon and Nitrogen Stable Isotope Analysis of Hunter-Gatherers from the Coleman Site, a Late Prehistoric Cemetery in Central Texas. *Journal of Archaeological Science*, *40*, 1369–1381.

McGhee, G. R. (2011). *Convergent Evolution: Limited Forms Most Beautiful*. Cambridge, Mass.: MIT Press.

Meltzer, D. J. (2004). Modeling the Initial Colonization of the Americas: Issues of Scale, Demography, and Landscape Learning. In C. M. Barton, G. A. Clark, D. R. Yesner, & G. A. Pearson (Eds.), *The Settlement of the American Continent: A Multidisciplinary Approach to Human Biogeography* (pp. 123–137). Tucson: University of Arizona Press.

Meltzer, D. J., & Holliday, V. T. (2010). Would North American Paleoindians Have Noticed Younger Dryas Age Climate Changes? *Journal of World Prehistory*, *23*, 1–41.

Morgan, B. M., Eren, M. I., Khreisheh, N., Hill, G., & Bradley, B. A. (2015). Clovis Bipolar Lithic Reduction at Paleo Crossing, Ohio: A Reinterpretation Based on the Examination of Experimental Replications. In A. M. Smallwood & T. A. Jennings (Eds.), *Clovis: On the Edge of a New Understanding* (pp. 121–143). College Station: Texas A&M University Press.

O'Brien, M. J., & Lyman, R. L. R. (2003). *Cladistics and Archaeology*. Salt Lake City: University of Utah Press.

O'Brien, M. J., Boulanger, M. T., Buchanan, B., Collard, M., Lyman, R. L., & Darwent, J. (2014). Innovation and Cultural Transmission in the American Paleolithic: Phylogenetic Analysis of Eastern Paleoindian Projectile-Point Classes. *Journal of Anthropological Archaeology*, *34*, 100–119.

Parry, W., & Kelly, R. (1987). Expedient Core Technology and Sedentism. In J. Johnson & C. Morrow (Eds.), *The Organization of Core Technology* (pp. 285–304). Boulder, Colo.: Westview Press.

Perreault, C., & Brantingham, P. J. (2011). Mobility-Driven Cultural Transmission along the Forager-Collector Continuum. *Journal of Anthropological Archaeology*, *30*, 62–68.

Perttula, T. K. (2004). An Introduction to Texas Prehistoric Archeology. In T. K. Perttula (Ed.), *The Prehistory of Texas* (pp. 5–14). College Station: Texas A&M University Press.

Prasciunas, M. M. (2007). Bifacial Cores and Flake Production Efficiency: An Experimental Test of Technological Assumptions. *American Antiquity*, *72*, 334–348.

Rasic, J., & Andrefsky, W., Jr. (2001). Alaskan Blade Cores as Specialized Components of Mobile Toolkits: Assessing Design Parameters and Toolkit Organization through Debitage Analysis. In W. Andrefsky, Jr., (Ed.), *Lithic Debitage: Context, Form, and Meaning* (pp. 61–79). Salt Lake City: University of Utah Press.

Sanchez, G., Holliday, V. T., Gaines, E. P., Arroyo-Cabrales, J., Martinez-Taguena, N., Kowler, A., et al. (2014). Human (Clovis)-Gomphothere (*Cuvieronius* sp.) Association ~13,390 Calibrated BP in Sonora, Mexico. *Proceedings of the National Academy of Sciences of the United States of America*, *111*, 10972–10977.

Schiffer, M. B., & Skibo, J. M. (1987). Theory and Experiment in the Study of Technological Change. *Current Anthropology*, *28*, 595–622.

Schillinger, K., Mesoudi, A., & Lycett, S. J. (2016). Copying Error, Evolution, and Phylogenetic Signal in Artifactual Traditions: An Experimental Approach Using "Model Artifacts." *Journal of Archaeological Science*, *70*, 23–34.

Schneider, C. A., Rasband, W. S., & Eliceiri, K. W. (2012). NIH Image to ImageJ: 25 Years of Image Analysis. *Nature Methods*, *9*, 671–675.

Shennan, S. (2008). Evolution in Archaeology. *Annual Review of Anthropology*, *37*, 75–91.

Smallwood, A. M. (2012). Clovis Technology and Settlement in the American Southeast: Using Biface Analysis to Evaluate Dispersal Models. *American Antiquity*, *77*, 689–713.

Smallwood, A. M., & Jennings, T. A. (Eds.). (2015). *Clovis: On the Edge of a New Understanding*. College Station: Texas A&M University Press.

Straus, L. G., Meltzer, D. J., & Goebel, T. (2005). Ice Age Atlantis? Exploring the Solutrean-Clovis "Connection." *World Archaeology*, *37*, 507–532.

Suhm, D. A., & Jelks, E. B. (Eds.). (1962). *Handbook of Texas Archeology: Type Descriptions. Special Publication, no. 1*. Austin: Texas Archeological Society.

Surovell, T. A. (2009). *Toward a Behavioral Ecology of Lithic Technology: Cases from Paleoindian Archaeology*. Tucson: University of Arizona Press.

Swofford, D. (1998). *PAUP*: Phylogenetic Analysis Using Parsimony (*and Other Methods) (version 4)*. Sunderland, Mass.: Sinauer.

Tunnell, C. (1989). Versatility of a Late Prehistoric Flint Knapper: The Weaver-Ramage Chert Cache of the Texas Rolling Plains. In B. C. Roper (Ed.), *In the Light of Past Experience: Papers in Honor of Jack T. Hughes* (pp. 369–397). Publication, no. 5. Canyon, Texas: Panhandle Archeological Society.

Turner, E. S., Hester, T. R., & McReynolds, R. L. (2011). *Stone Artifacts of Texas Indians*. Lanham, Md.: Taylor.

VanPool, T. L., & Leonard, R. D. (2011). *Quantitative Analysis in Archaeology*. Oxford: Wiley-Blackwell.

Wang, W., Lycett, S. J., von Cramon-Taubadel, N., Jin, J. J. H., & Bae, C. J. (2012). Comparison of Handaxes from Bose Basin (China) and Western Acheulean Indicates Convergence of Form, not Cognitive Differences. *PLoS One*, *7*(4), e35804.

Waters, M. R., & Stafford, T. W. (2007). Redefining the Age of Clovis: Implications for the Peopling of the Americas. *Science*, *315*, 1122–1126.

Waters, M. R., Pevny, C. D., Carlson, D. L., Dickens, W. A., Smallwood, A. M., Minchak, S. A., et al. (2011). *A Clovis Workshop in Central Texas: Archaeological Investigations of Excavation Area 8 at the Gault Site*. College Station: Texas A&M University Press.

Will, M., Mackay, A., & Phillips, N. (2015). Implications of Nubian-like Core Reduction Systems in Southern Africa for the Identification of Early Modern Human Dispersals. *PLoS One*, *10*(6), e0131824.

13 The "Levallois-like" Technological System of the Western Stemmed Tradition: A Case of Convergent Evolution in Early North American Prehistory?

Loren G. Davis and Samuel C. Willis

In western North America, the Western Stemmed Tradition (WST) is an archaeological pattern that is contemporaneous with but technologically different from the Clovis Paleoindian tradition (Beck and Jones 2010; Bryan 1979, 1980, 1988, 1991; Davis and Schweger 2004; Davis, Willis, and Macfarlan 2012; Davis, Nyers, and Willis 2014; Goebel and Keene 2014; Grayson 2011; Jenkins et al. 2012, 2013, 2016; Madsen 2015) that is dated to 13,340–12,700 calendar years before present (cal BP) (Haynes 2005; Waters and Stafford 2007). WST lithic-reduction strategies differ from what is found in the Clovis tradition (Davis et al. 2012): a diverse use of tool-stone types, often from local sources of varying quality; a range of core-reduction strategies, including centripetal, "Levallois-like," unidirectional, and multidirectional, all of which are commonly employed for the production of linear macroflakes; the use of linear macroflakes as blanks for tool production, including the manufacture of stemmed and foliate projectile points; and a parallel but less prominent pattern of direct, multistage reduction of large bifacial preforms to finished biface forms.

Of these significant technological aspects, the presence of the Levallois-like pattern of core production remains enigmatic. Elements of core reduction similar in form to Levallois technology were first described from the Akmak site of southeastern Alaska (Anderson 1970; Dumond 1972). Soon after, Muto (1976) described an unusual pattern of lithic reduction from terminal Pleistocene- to early Holocene–age Windust and Cascade components, which he termed the *Cascade technique* and, alternatively, *Levallois-like*. Muto (1976: vi) defined the Levallois-like pattern in this general way:

> The Cascade Technique is a Levallois-like stone reduction system characteristic of early (ca. 10,000–4,500 BP) archaeological phases in the Lower Snake River Region, Southeastern Washington. This New World Levallois-like technique produced end products morphologically identical to products of the Old World Levallois technique. The two differ in one respect. In the Old World system different cores were made for flakes, blades, or points. In the New World system all products were produced from the same core through modification of the reduction trajectory.

Muto based his definition of the Cascade technique on an examination of the lithic-artifact assemblage from the Wexpusnime (45GA61) and Granite Point (45WT41) sites in Washington state and by making comparisons with Bordes' (1961) study of Levallois technology in France. The former site contains an early Holocene-age Cascade phase component, whereas the latter includes both late Pleistocene to early Holocene-aged Windust and Cascade components (Bense 1972; Leonhardy 1970; Leonhardy and Rice 1970; Leonhardy, Schrodel, Bense, and Beckerman 1971). Muto (1976: 2) saw a larger spatial distribution of the technique based on an "inspection of lithic collections from throughout the Columbia Plateau, Snake River Plain, and the northern Great Basin (that) indicates a very early lithic technological tradition that is analogous to the Levallois technique of the Old World." Muto provided no specifics on the exact sites or collections examined for this comparative analysis, however. He emphasized two-dimensional concepts of Levallois in his pattern matching of Lower Snake River Canyon Levallois-like technological products against European Paleolithic forms, deemphasizing geometric morphometric approaches (apart from gross size categories of flakes and tools). He occasionally mentioned convexity and edge/platform angles, signaling a deeper understanding of Levallois attributes, but these were not systematically studied nor used for identifying Levallois-like artifacts in his collections.

Still, Muto's Levallois-like examples bear similar morphometric and technological elements that are consistent with definitions of Levallois core reduction provided by Boëda (1994, 1995) and explained by Lycett, von Cramon-Taubadel, and Eren (2016: 24):

> In essence, cores were identified as "Levallois" if they were clearly bifacial in knapped form and possessed a plane of intersection produced by intersection of the two faces at the core's margin. They also possessed a disproportionately large negative flake scar, or scars, on their "Levallois" surface (≥50% of total surface area) and, in accordance with Boëda (1995), the axis of this flaking was broadly parallel to the plane of intersection. These cores also exhibited some remnant of the distal and lateral convexity possessed by the core prior to the removal of the final flake, such that removal of this final flake visibly truncated prior flake removals from the same surface (Van Peer 1992: 10).

Boëda (1988) identified two primary modes of Levallois flake-product manufacture: a *méthode lineal*, which generates one flake product from a prepared core surface, and a *méthode recurrent*, which can produce multiple Levallois flake blanks from a prepared core surface. Dibble (1989) arrived at a similar conclusion in his determination that Levallois core production can lead to the manufacture of multiple, often different kinds of flake products from a single core.

Beyond the interior Pacific Northwest, the Levallois-like technological pattern appears to be a regional aspect of late Pleistocene to early Holocene sites along the Northwest Coast. Carlson (1996) reports "Levalloisoid" cores and flakes from the period 1B (9,000–8,000 radiocarbon years B.P. [RYBP]) component of the Namu site, located in southern British Columbia, and Fedje, Mackie, McSporran, and Wilson (2011) provide a preliminary description of discoidal core technology from late Pleistocene–early Holocene Kingii components at Haida Gwaii in coastal British Columbia. These cores bear opposed convex faces, one with a steep angle (relative to the convergent margin) and another with a lower-angle convex surface. Although this core technology is described as having a "superficial appearance of Levallois centripetal cores" (Fedje et al. 2011: 327), the authors believe the cores were used to strike unidirectional linear macroflakes from the lowermost, more steeply convex face, in contrast to Levallois core production of flake blanks from the upper, less convex face (Boëda 1995).

Notably, several "bladelike" flake blanks and flake tools are also reported (Fedje et al. 2011), which contain relatively flat longitudinal profiles, high-angle striking platforms, and multiple centripetal scars on their dorsal faces. Because the discoidal cores are smaller than the "bladelike" flakes, and given the retention of centripetal dorsal-scar patterning on the flake products, it is possible that these cores were at the end of their use-lives and had been structured to support the production of linear macroflake blanks from the upper, less-convex surface. Further, many Kingii complex stemmed projectile points bear evidence of having been made on large flakes and are also larger than the cores.

Although the Levallois-like pattern was originally conceptualized as part of the post-8,000 RYBP Cascade phase of the Lower Snake River Canyon culture-history model, Leonhardy and Rice (1970) mentioned Levallois-like lithic technology as a part of the earlier Windust phase (8,000–11,000 RYBP) as well. Despite the fact that the Levallois-like pattern has been attributed to the WST—by way of its inclusion in the Windust phase—it has not yet been reported from clear stratigraphic and temporal association with the earliest archaeological components in the Pacific Northwest. However, excavations conducted since 2009 at the Cooper's Ferry site in western Idaho have discovered multiple examples of what appear to be Levallois-like technology in clear association with WST cultural components. Here we describe these discoveries.

The Levallois-like Technological Pattern at Cooper's Ferry, Idaho

Cooper's Ferry is located in the lower Salmon River canyon of western Idaho, within an alluvial terrace approximately 17 kilometers south of the town of Cottonwood. The lower Salmon River is a tributary of the Snake and subsequently the Columbia River basin, which drains the interior Pacific Northwest and the southern portion of British Columbia. The basin is extensively eroded into the Columbia Basalt Formation, which is one of the world's largest, and the youngest, continental basaltic-flow provinces (Baksi 1989; Barry et al. 2010; Tolan et al. 1989; figure 13.1). Archaeological excavations conducted at the site by Butler in the 1960s (Butler 1969), and later by Davis, beginning in 1997 and from 2009 to 2017 (Davis and Sisson 1998; Davis and Schweger 2004; Davis et al. 2014, 2017), revealed multiple WST cultural components in a stratified sequence of aeolian and alluvial deposits. Although Davis's investigation is ongoing, geoarchaeological and archaeological studies to date indicate that the earliest cultural occupation at the site may date to at least 11,410 ± 130 RYBP (13,129–13,372 cal BP; figure 13.2).

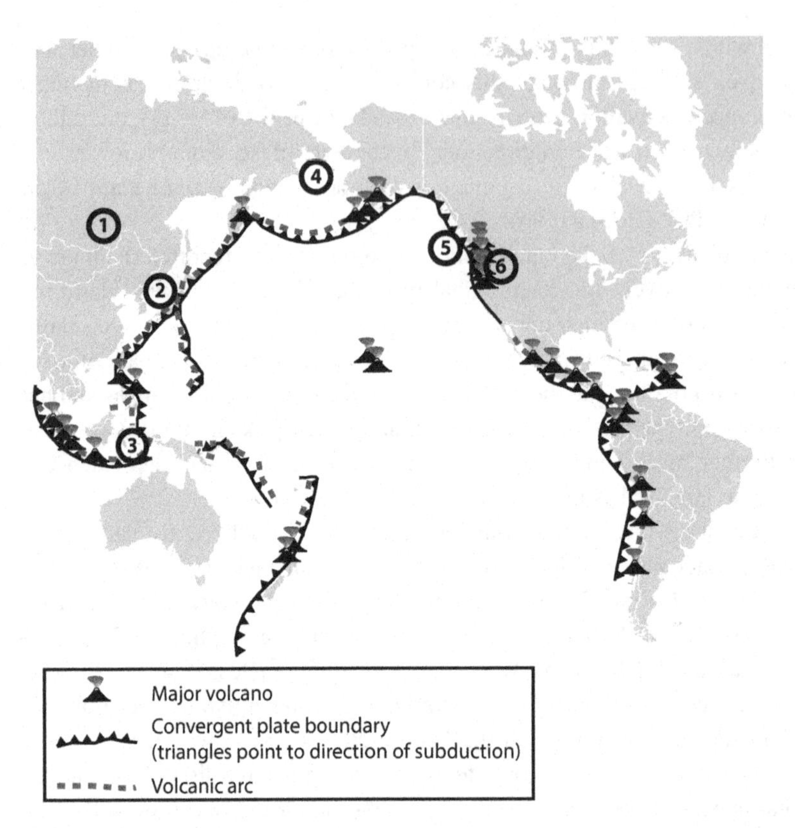

Figure 13.1
Convergent tectonic plate boundaries, volcanic arc terrains, and major volcanoes in the Pacific "Ring of Fire." Key regions and sites discussed in text: 1, Trans-Baikal; 2, Korean and Japanese peninsulas; 3, East Timor; 4, Beringia; 5, Northwest Coast; 6, Cooper's Ferry.

At Cooper's Ferry, artifacts attributed to the Levallois-like pattern—based on the aforementioned Levallois criteria (Boëda 1994, 1995; Lycett et al. 2016; Van Peer 1992)—include several cores and flake products found in WST components. Specifically, Levallois-like artifacts were excavated from lithostratigraphic units (LU) 3–6, overlapping with the stratigraphic distribution of numerous WST projectile points (figure 13.2). Radiocarbon dating of river-mussel shell and wood charcoal indicates a temporal span for Levallois-like technology of ca. 11,400–8030 RYBP. The presence of the Rock Creek paleosol in LU3, which has been dated elsewhere in the lower Salmon River canyon to ≥10,750 RYBP (Davis and Schweger 2004), provides a minimum age for the Levallois-like and WST artifacts found in LU3.

The Levallois-like lithic technology at Cooper's Ferry is currently represented by seven cores, eight flake blanks, and two unifaces made on flakes (table 13.1). These artifacts were clustered mainly in LU3 and LU6, with two additional cores and two flake

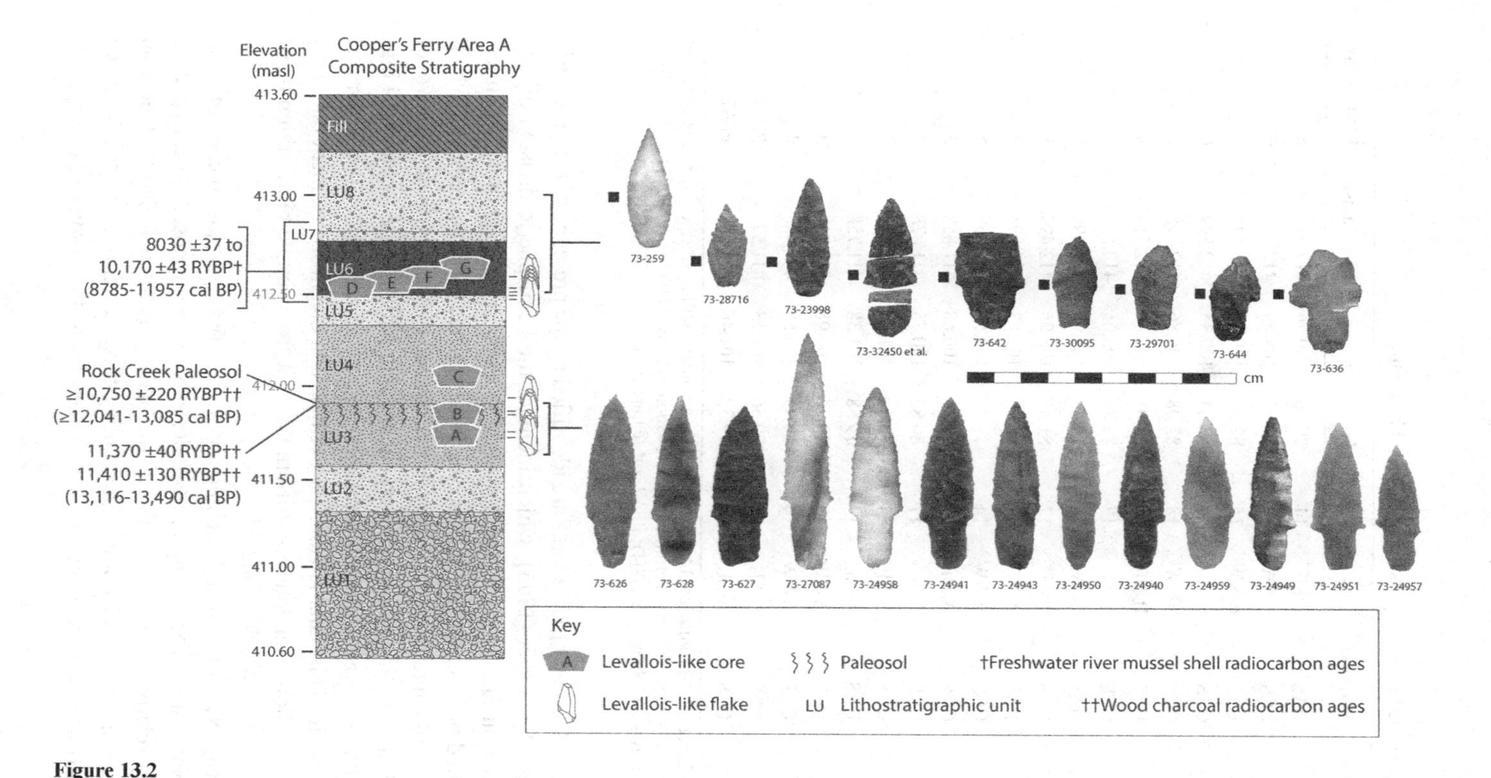

Figure 13.2
Composite stratigraphic profile from Area A at Cooper's Ferry showing the vertical distribution and stratigraphic association of excavated Levallois-like cores and flake products and selected Western Stemmed Tradition and foliate (Cascade) projectile points. Catalog numbers appear at bottom of projectile point images. Bases of Levallois-like cores sit at the elevation of their *in situ* discovery. Stratigraphic units and chronometric dating follow Davis and Schweger (2004) and Davis et al. (2014). Stratigraphic position of two Levallois-like flake-blank products found in pit feature A2 (PFA2) are not shown here but were associated with a cache that originated from the upper surface of LU3 (Davis et al. 2014). Calendrical calibration of radiocarbon ages performed with Calib (ver. 7.1; Stuiver et al. 1993).

Table 13.1
Levallois-like artifacts recovered from Cooper's Ferry organized by depth

Catalog #	Description	Material	Unit	Level	N	E	Elev. (masl)	LU/Feature
73–25444	Core	BAS	E-SE	14	79.46	109.76	412.619	6
RN-145	DEB	CCS	E-SW	9	79.33	108.49	412.598	6
73–30170	Core	BAS	F-NW	15	82.33	100.11	412.541	6
73–29993	UNI	CCS	J-NE	19	82.44	109.35	412.529	6
73–28255	DEB	CCS	I-NW	18	82.08	106.86	412.518	6
73–29133	Core	BAS	B-NE	20	81.01	103.35	412.513	6
73–28834	DEB	CCS	G-SE	17	81.2	103.05	412.505	5
73–31109	Core	BAS	E-SW	16	79.48	108.94	412.496	5
73–28911	DEB	CCS	G-SE	17	81.87	103.05	412.495	5
73–640	Core	BAS	A-NE	11	80.43	101.88	412.078	4
73–3996	DEB	CCS	A-SW	14	79.4	100.51	411.938	5
73–946	DEB	CCS	A-NE	13	80.21	100.68	411.858	3
73–4513	DEB	CCS	A-NW	17	80.82	100.75	411.828	3
73–634	Core	CCS	A-SE	14	79.6	101.16	411.818	3
73–637	UNI	BAS	A-NE	14	79.58	101.38	411.758	3
73–4358	DEB	CCS	A-SW	17	79.76	100.98	411.738	3
73–635	Core	CCS	A-SW	14	79.39	100.58	411.688	3
73–631	DEB	CCS	A-SE	21	79.56	101.13	411.59	PFA2
73–4634	DEB	CCS	A-SE	24	79.38	101.46	411.38	PFA2

Abbreviations: DEB, debitage; UNI, uniface; BAS, basalt; CCS, cryptocrystalline silicate; N, E, northing and easting positions (in meters from N100 E100 datum); masl, elevation in meters above sea level; LU, lithostratigraphic unit (following Davis and Schweger 2004); PFA2, pit feature A2 (Davis et al. 2014).

products coming from LU4 and LU5. Cores are identified by the presence of the geometric framework described by Boëda (1995), which includes the hemispheric division of two core faces (a production face and a preparatory face) separated by a continuous convergent margin; a domed production face that is clearly designated for flake removal, which includes the presence of lateral and distal convexities created by the centripetal removal of flakes in an angle that is oblique to the convergent margin; a high-angle faceted striking-platform area; and the removal of linear macroflake products in a trajectory that is parallel to the convergent margin. This latter attribute is a critical and diagnostic element that most clearly defines the Levallois technique and differentiates it from simple biface thinning or discoidal/centripetal core-reduction methods.

Lycett et al. (2016: 28) explain that detaching a Levallois flake "is akin to chopping the top off an egg; albeit a stone egg laying on its side, and having imposed a very specific margin morphology relative to that top." To use an analogy familiar to North American archaeologists, in a generally similar way that a fluting flake is removed from a prepared faceted striking platform along a trajectory that is parallel to the convergent margin of its biface, so too are linear macroflakes produced from the Cooper's Ferry Levallois-like cores. In figure 13.3, specimens 73–634 and 73–30170 show this quite clearly, as both retain a large dorsal flake scar that is oriented parallel to the convergent margin. Levallois-like macroflake products struck from these cores are expected to retain telltale elements of their associated cores. Figure 13.4 shows three representative Levallois-like macroflakes that bear prior centripetal flake-removal scars on their dorsal faces, relatively high-angle faceted striking platforms, flattish longitudinal profiles, a central and often sinuous dorsal ridge created by the convergence of centripetal flake removals prior to the macroflake's reduction from the core, and a cross-sectional area that approximates a prismatic form (Boëda, Hou, Forestier, Sarel, and Wang 2013).

Excavations conducted at Cooper's Ferry since 2009 have also recovered many examples of WST projectile points, the close study of which has identified key geometric attributes that reveal their mode of manufacture (Davis et al. 2012, 2014, 2017). For example, WST projectile points from Cooper's Ferry occasionally retain unworked ventral surfaces, slightly curved longitudinal profiles, and plano-convex cross sections (figure 13.5). These key attributes evidence the production of WST projectile points from linear macroflakes, similar in shape and volume to the Levallois-like linear flake blanks shown in figure 13.4, most probably through a process of limited percussive reduction and more-extensive pressure flaking (Davis et al. 2014). It is important to point out the significance of linking WST projectile-point manufacture at Cooper's Ferry to the Levallois-like linear flake blanks, given that these particular blank forms provide a thicker central volume as a result of initial preparation of dorsal-convexity shaping that takes place prior to its removal from the core. In contrast, it would be difficult if not impossible to create a WST projectile point with forms such as those in figure 13.5 from a thin bifacial reduction flake or a prismatic blade. Such thinner blanks would be problematic because they lack the cross-sectional area needed to shape the point, not to mention the fact that bifacial thinning flakes commonly bear more pronounced longitudinal curvature.

Origins of the Levallois-like Technological Pattern

Levallois-like assemblages are reported from multiple sites in western North America, generally associated with late Pleistocene– to early Holocene–age cultural components. Two hypotheses are advanced to explain this pattern. First, the Levallois-like pattern is similar in morphometry and concept to the Initial Upper Paleolithic (IUP), which includes "industries dating to between 35 ka and 50 ka, and that show features of Levallois technology in blade production. … This includes assemblages scattered from N. Africa to central Europe to northwest China" (Kuhn and Zwyns 2014: 2; chapter 8, this volume) and reflects direct cultural diffusion of IUP concepts to the New World during the late Pleistocene. Second, the Levallois-like pattern is similar to the IUP Levallois, because it reflects the

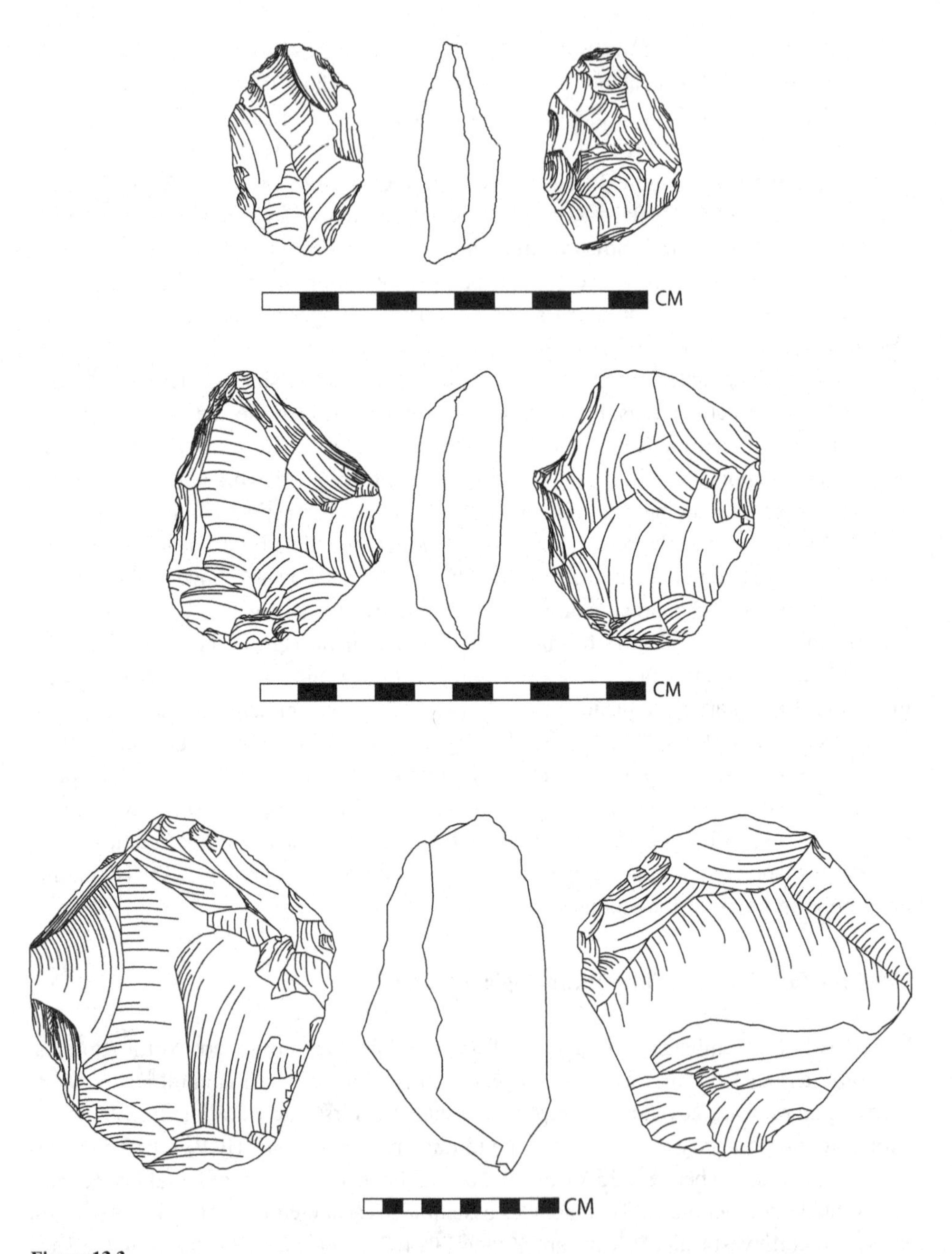

Figure 13.3
Representative Levallois-like cores 73–635 (top), 73–634 (middle), and 73–30170 (bottom) from Cooper's Ferry. Production faces (dorsal) of cores appear at left; high-angle, faceted striking platforms associated with dorsal macroflake removals are at the bottom. The dorsal face of core 73–635 retains the signs of a prepared Levallois-like core: multiple centripetal flake removals that maintain lateral and distal convexities and converge to produce a central sinuous dorsal ridge.

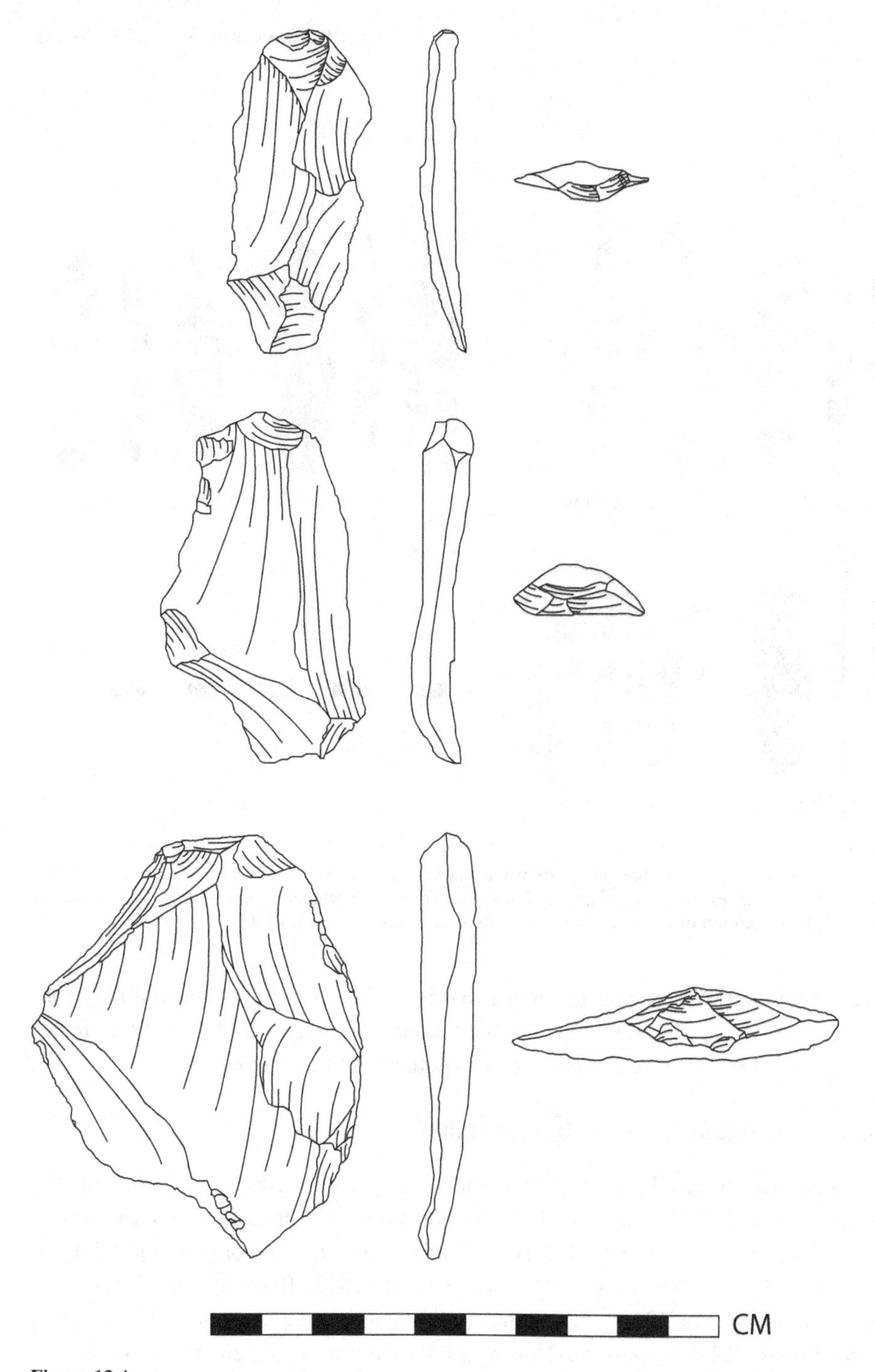

Figure 13.4
Representative Levallois-like flake products from Cooper's Ferry, including macroflakes 73–946 (top; note the faceted *chapeau de gendarme* platform [Debénath and Dibble 1994]) and 73–4513 (middle); and uniface 73–637 (bottom).

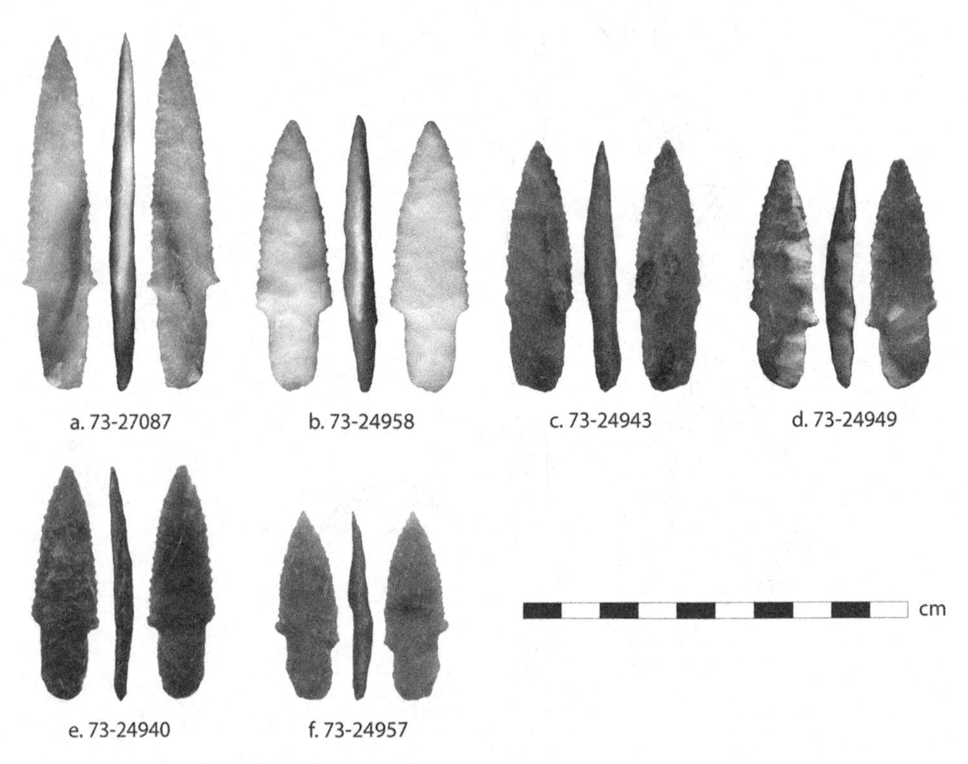

Figure 13.5
Western Stemmed Tradition projectile points from Cooper's Ferry that bear evidence of manufacture from linear macroflakes or blades. Note the pronounced plano-convex cross sectional form (a–f), slight longitudinal curvature (c–f), and retention of unworked original ventral flake-blank surfaces (d–f).

convergent evolution of a general technique that is applied to the manufacture of predetermined flake products from lower-quality but spatially ubiquitous tool-stone types found in western North America. We consider each hypothesis in turn below.

Hypothesis 1: Levallois-like as Cultural Diffusion

Archaeologists report that Levallois cores and blade technologies were present in the Trans-Baikal region of Siberia and Mongolia and northern China between ca. 40,000 RYBP and 20,000 RYBP at the Kara Bom, Chikhen-Agui, Shuigonggou, and Jinsitai Cave sites (Bar-Yosef and Wang 2012; Boëda et al. 2013; Brantingham, Krivoshapkin, Li, and Tserendagva 2001; Brantingham, Kerry, Krivoshapkin, and Kuzmin 2004; Kuhn and Zwyns 2014; Lycett and Norton 2010; Madsen et al. 2014; Zwyns 2012). Kuhn and Zwyns (2014) suggest that the emergence of the IUP in northeastern Asia, which includes a Levallois technological pattern (Derevianko and Markin 1995), resulted from an expansion of distinct technological and cultural systems from the Siberian Altai region, where it dates to ca. 43,000 RYBP at Kara Bom, and into Mongolia by ca. 25,000 RYBP (Bar-Yosef and Wang 2012; Brantingham et al. 2001; Derevianko and Petrin 1995; Kuhn and Zwyns 2014). However, Kuhn and Zwyns (2014) and Madsen et al. (2014) also note a paucity of sites and chronological controls to demonstrate this conclusively. Diagnostic Levallois artifacts of northeast Asia include flake cores, point cores, blade cores, narrow-faced cores, and broad-faced cores, as well as Levallois flakes, projectile points, and macroblades (Boëda et al. 2013; Brantingham et al. 2001; Brantingham et al. 2004). Levallois technologies disappeared from the Trans-Baikal region after the Last Glacial Maximum at ca. 18,000 RYBP and were replaced by other core-and-blade technologies, including prismatic macroblade and microlithic technologies (Bar-Yosef and Wang 2012).

Farther south along the western Pacific Rim, the presence of a Pleistocene-age Levallois technological pattern is less certain. For Japan, Abe (1976) presents a “Levallois-like” core from the Takseyama site that retains the telltale morphometry of a recurrent Levallois core (compare with Boëda 1988); however, Sato, Nishiaki, and Suzuki (1995: 497) state, “Levallois in Japan seems to be a myth,” and argue instead that non-Levallois discoidal-core technology is characteristic of regional Middle Paleolithic sites (>30,000 RYBP). Bae (2010) describes bidirectional blade cores from South Korea’s Sokchang-ni site, which has but a single radiocarbon date (ca. 24,000 RYBP; Bae 1992), that are similar to the flat-faced and broad-faced cores described by Brantingham et al. (2001) from the Trans-Baikal region. Lycett and Norton (2010: 60–61) argue that “Shuidonggou is the only site currently identified in China or Korea that has Levallois flakes and cores” and that “no evidence of the Levallois technology has been identified in Korea (Norton 2000).” O’Connor, Ono, and Clarkson (2011) report Levallois flakes from the lowest archaeological component (phase I) at Jerimalai Shelter in East Timor, with radiocarbon ages of 38,000–42,000 RYBP. O’Connor et al. (2011: S2) noted the flake assemblage reflects “the frequent use of faceting and radial flake removals from small nodules and larger flakes, with Levallois-like flakes, pointed blades with faceted platforms and other faceted flakes struck from the ventral surfaces of larger flakes all present in the assemblage.”

Overall, early western Beringian sites date younger than 20,000 RYBP and apparently lack Levallois core and blade technology; however, there is clear evidence for projectile-point manufacture from macroflakes or macroblades (Slobodin 2011). The earliest sites in eastern Beringia, associated with the Nenana Complex, date as early as 11,770 RYBP (Hamilton and Goebel 1999; Holmes 1996; Yesner, Holmes, and Crossen 1992) and also appear to lack Levallois-like technologies, although there is clear evidence of projectile points (e.g., Chindadn points) made from thin macroflakes or macroblades (Goebel 2004, 2011; Hoffecker and Elias 2007; Powers and Hoffecker 1989).

Although it is clear that Pleistocene foragers used flake-based manufacturing techniques on both sides of Beringia—an important technological pattern that links early Beringian peoples (Hoffecker and Elias 2007)—arguing that these particular patterns relate specifically to a Levallois technology is entirely another matter in the absence of representative cores and flake-blank products. That said, the emphasis on tool production from thin macroflakes seen among late Pleistocene sites on both sides of Beringia can be read to signal an evolutionary connection that requires further explanation. Although Levallois technology appeared in the Trans-Baikal region by ca. 40,000 RYBP, it disappeared by ca. 20,000 RYBP and has not been found farther east into central and eastern Beringia. Thus, the degree of separation, both temporally (>8,000 radiocarbon years) and spatially (over 7,500 kilometers from the Northwest Coast to the Trans-Baikal region), makes it difficult, in the context of currently available archaeological evidence, to support the hypothesis that the appearance of Levallois-like technology in northwestern North America resulted from direct cultural diffusion from northeast Asia. Of course, if the route of movement of Levallois-like technological ideas occurred along a Pacific coastal route, then its archaeological evidence would be held in sites now submerged on the continental shelf.

Hypothesis 2: Levallois-like as Evolutionary Convergence

Geological reconstructions of Late Wisconsin glacial ice sheets (Dyke, Moore, and Robertson 2003) indicate that the Copper River basin was deglaciated by 13,500 RYBP, providing a direct route from eastern Beringia to the Pacific Ocean. From that point, early foragers could have moved along the eastern Pacific margin through a combination of short boat transits and walking along nonglaciated terrains, in what Davis et al. (2012) call a Partial Amphibious Migration. Alternatively, foragers bearing a fully maritime adaptation could have traveled along the glaciated and nonglaciated margin of North America, in what Davis et al. (2012) call a Full Maritime Migration. If the first migrants to the New World followed a Pacific coastal route of entry south of the Cordilleran ice sheet, they would have found their first major inland route at the mouth of the Columbia River. Following it upstream, early foragers would have encountered a landscape dominated by lower-quality igneous, sedimentary, and metamorphic tool stones, with only occasional occurrences of higher-quality chert, quartz, and obsidian. Once they moved south and east of the Columbia River basin and into the Great Basin, they would have found high-quality silica-rich tool stones, including obsidian and, in areas east of the Rocky Mountains, extensive chert and flint deposits. Beyond the Columbia River basin, we see an emphasis on the reduction of large bifaces, manufactured through extensive percussive reduction of bifacial preforms (Beck and Jones 2010; Bradley 1993; Bradley, Collins, and Hemmings 2010; Collins and Lohse 2004), which may be related to the kind of scenario described here.

This general distribution of early lithic reduction patterns and tool-stone types in North America mirrors its macroscale geologic history (figure 13.6). Large formations of cherts, flints, and other silicate-rich knappable tool stones are more readily available east of the

Rocky Mountains in rocks overlying the extensively weathered North American Craton. Tool stones are widely available in sedimentary bedrock deposits in the Canadian and American Great Plains and in the American Southwest and Southeast. Such notable sources of siliceous tool stone include Spanish Diggings in Wyoming, Wyandotte chert outcrops in Indiana, Edwards chert deposits in central Texas, and Fort Payne chert sources in the Southeast. In the western North America Accretionary Belt (figure 13.6), chert tool stone is available in far more limited quantities. There, fine-grained microcrystalline volcanic and metamorphic rocks such as basalt and argillite are ubiquitous. Obsidian, which possesses superior physical properties for knapping, is much more restricted in area.

Although obsidian artifacts are found in archaeological sites far removed from their volcanic sources, the distances that foragers transported large quantities of obsidian appear to be relatively short. For example, glassy volcanic material accounts for less than 1 percent of the total tool stone at Cooper's Ferry, although sources of the material occur 180 kilometers away at Oregon's Dooley Mountain source and 220 kilometers away at Idaho's Timber Butte source. In contrast, Paisley Caves, located in Oregon's northern Great Basin region, has a lithic assemblage that is composed almost exclusively of obsidians from multiple sources located within 50 kilometers of the site (Jenkins et al. 2016).

The particular geologic context of the eastern Pacific Rim would have provided a fitness landscape wherein selective pressures might have favored development of technologies that effectively exploited lower-grade igneous, metamorphic, and sedimentary rocks. On this landscape, early foragers might have experimented with different means of knapping nodular microcrystalline and cryptocrystalline rocks, independently developing techniques for producing predetermined flake blanks to be manufactured into particular tool types. Thus, it is no coincidence that early lithic technologies from southeastern Alaska and coastal British Columbia sites are dominated by the production of macroblades and microblades, bifaces, unifaces, and cobble tools from basalts, argillites, and cherts in much smaller proportions (Ackerman 1996; Carlson 1996; Fedje et al. 1996; Matson 1996; Mitchell and Pokotylo 1996). This pattern of material use is seen at Cooper's Ferry, where five of the seven Levallois-like cores were made on basalt, and fine-grained volcanic tool stones were also used to make WST projectile points.

Thus it is not difficult to image that the Levallois-like technological pattern could have been independently invented in the Columbia River Plateau and Northwest Coast regions. In his description of cores from the early lithic assemblage at Namu, Carlson (1996: 90) remarked that "several are similar to the tortoise or Levallois type core prepared for the removal of a large single flake; they are part of a logical process of cobble reduction and are not necessarily related to Old World Levallois industries." Indeed, Otte (1995) pointed to evolutionary convergence as the reason why Levallois concepts appear so widely scattered and disconnected through space and time in Africa, Central Asia, China, and Australia, offering an explanation for its inevitable appearance:

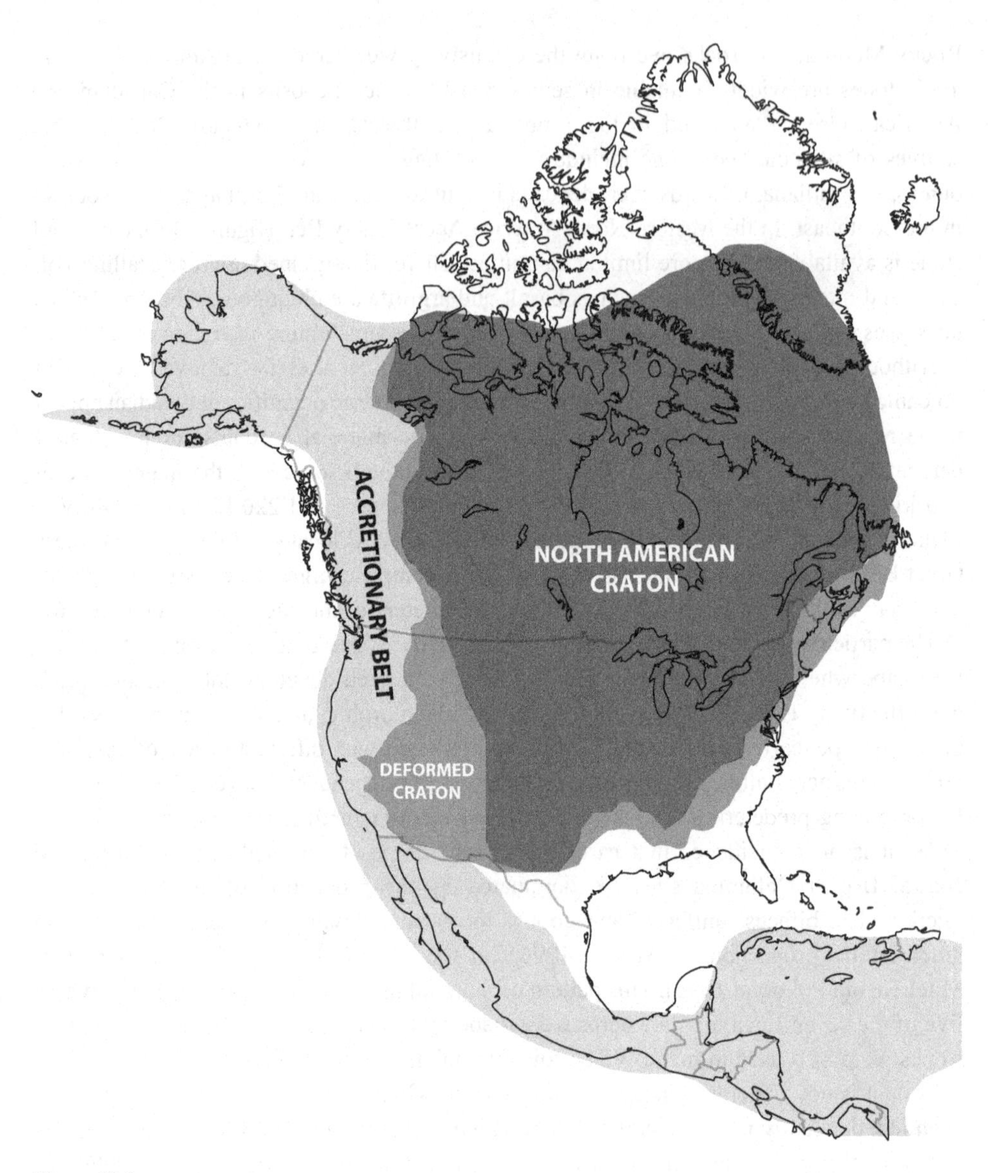

Figure 13.6
Macroscale geological map of North America showing locations of the North American Craton and the North American Accretionary Belt.

Seen from a global perspective, the Levallois method can appear independently in humanity's technical history … an indefinite number of times, provided that at least three factors are found in association: 1. Lithic materials with appropriate mechanical laws ("brittle" rocks), 2. Technical needs for a particular form fulfilling potential uses, 3. The conceptual capacities necessary for foreseeing the action, its coordination and its realization in stages. (pp. 117–118)

Because the archaeological record of Beringia lacks evidence for a direct introduction of Levallois lithic technology into North America's lower latitudes, and given the propensity of Levallois concepts to appear at different times and places around the world, the early presence of the Levallois-like technological pattern at Cooper's Ferry and elsewhere in western North America can be most parsimoniously explained as a result of evolutionary convergence.

Conclusion

The discovery of Levallois-like artifacts at Cooper's Ferry, Idaho, in clear stratigraphic association with WST cultural components provides proof of the pattern's existence in the Columbia River Plateau. Levallois-like cores and flake blanks found at Cooper's Ferry in association with the site's earliest WST components extend the timing of this technological pattern back before 11,410 RYBP. The continued presence of Levallois-like artifacts into the site's late WST foliate-bearing early Cascade components demonstrates the cultural continuity of this technological pattern, providing a key link between the regional Windust and Cascade phases.

The association of the Levallois-like pattern with early WST components at Cooper's Ferry has implications that extend far beyond western Idaho or the Columbia River Plateau. Although comparisons have been made between the Nenana Complex and the Clovis tradition to support an argument that a Clovis progenitor exists in interior Alaskan Pleistocene-age sites (Goebel 2004, 2011; Hoffecker 2011; Powers and Hoffecker 1989), a stronger case can be made that the Nenana Complex and the WST bear much closer technological similarities, given that both emphasize the manufacture of tools, including projectile points, from flake blanks. This line of argument mirrors that expressed by Hoffecker and Elias (2007), who saw strong similarities between the flake-based lithic technologies of Kamchatka's Ushki Layer 7 assemblage of stemmed points made on linear macroflakes and the Chindadn projectile point types of the Nenana Complex.

Indeed, by focusing on the shared aspects of flake-based lithic technology (which contrasts with the extensive bifacial reduction pattern of the Clovis tradition) as the primary basis for comparison instead of simply matching the presence of artifact type categories or the two-dimensional shape of projectile points, we can reasonably link Ushki Layer 7, the Nenana Complex, and Cooper's Ferry's early WST components on the basis of their shared patterns of lithic manufacture. The timing of these archaeological patterns is compelling as well, given that all three are associated with radiocarbon ages in excess of 11,000 RYBP and the Ushki Layer 7 component may date as early as 14,300 RYBP (Goebel, Waters, and Dikova 2003; Slobodin 2006: 13).

Whether Ushki Layer 7 or Nenana flake blanks were made from a Levallois-like core reduction is not readily apparent from the literature and may require additional study of artifact collections. Regardless of the specific method for flake-blank manufacture, the fact that Ushki Layer 7, the Nenana Complex, and Cooper's Ferry WST components share a lithic technology that is primarily flake based is a key fact that is worth highlighting in the context of this paper's central topic. Adding also the fact that fluted points clearly date younger in Alaska than in the mid-continental North America, and considering recent arguments that the Ice Free Corridor was not likely a viable route of migration early enough to facilitate the initial movement of humans south of the continental ice sheets (Pedersen et al. 2016), the strong similarities seen among the Ushki, Nenana, and WST lithic technological products and their modes of manufacture could be explained if Pleistocene-age foragers followed the Partial Amphibious Model route from eastern Beringia, down the Copper River drainage (thought to be deglaciated by ca. 16,000 cal BP [Dyke et al. 2003]) to the Pacific, and southward along the Northwest Coast beyond the Cordilleran Ice Sheet, as proposed by Davis et al. (2012).

Equally plausible is the prospect that Pleistocene foragers followed a coastal route along the southern margin of Beringia and the southern coast of Alaska, with some migrants traveling up the unglaciated Copper River basin into eastern Beringia while others continued southward. Although similarities observed between Nenana and WST lithic technologies may indicate a close evolutionary connection, the apparent absence of a Levallois-like core technology among interior Alaskan sites may signal that the advent of Levallois-like strategies occurred along the Pacific coastline.

At this time, because of the absence of evidence that earlier Levallois lithic technology from the Trans-Baikal region of Siberia persisted later in the Pleistocene associated with a spread of humans into Beringia, we conclude that the Levallois-like technological pattern of northwestern North America represents an independent invention of a lithic-technology strategy to support movement into a landscape dominated by fine-grained microcrystalline volcanic and metamorphic rocks such as basalt and argillite and thus a case of evolutionary convergence. Of course, although the Levallois-like pattern might ultimately postdate the initial migration of humans into North America and simply be a product of Pacific Northwest and Northwest Coast foragers settling into the geological context of their landscapes, this interpretation does not change the conclusion that evolutionary convergence is at work here. Additional discoveries of Levallois-like technologies in good contexts dated in excess of 12,500 RYBP (ca. 14,500 cal BP) could resolve this issue.

References

Abe, Y. (1976). Levallois-like Core from Yamagata Prefecture, Japan. *Journal of the Anthropological Society of Nippon, 84*, 246–251.

Ackerman, R. E. (1996). Ground Hog Bay, Site 2. In F. H. West (Ed.), *American Beginnings: The Prehistory and Paleoecology of Beringia* (pp. 456–460). Chicago: University of Chicago Press.

Anderson, S. R. (1970). Akmak: An Early Archaeological Assemblage from Onion Portage, Northwest Alaska. *Acta Arctica, 16*, 1–80.

Bae, K. (1992). Pleistocene Environment and Paleolithic Stone Industries of the Korean Peninsula. In C. M. Aikens & S. N. Rhee (Eds.), *Pacific Northeast Asia in Prehistory* (pp. 13–21). Pullman: Washington State University Press.

Bae, K. (2010). Origin and Patterns of the Upper Paleolithic Industries in the Korean Peninsula and Movement of Modern Humans in East Asia. *Quaternary International, 211*, 103–112.

Baksi, A. K. (1989). Reevaluation of the Timing and Duration of Extrusion of the Imnaha, Picture Gorge, and Grande Ronde Basalts, Columbia River Basalt Group. In S. P. Reidel & P. R. Hooper (Eds.), Geological Society of America Special Paper (Vol. 239). *Volcanism and Tectonism in the Columbia Flood—Basalt Province* (pp. 105–112). Boulder, Colo.

Barry, T. L., Self, S., Kelley, S. P., Reidel, S., Hooper, P., & Widdowson, M. (2010). New $^{40}Ar/^{39}Ar$ dating of the Grande Ronde Lavas, Columbia River Basalts, USA: Implications for Duration of Flood Basalt Eruption Episodes. *Lithos, 118*, 213–222.

Bar-Yosef, O., & Wang, Y. (2012). Paleolithic Archaeology in China. *Annual Review of Anthropology, 41*, 319–335.

Beck, C., & Jones, G. T. (2010). Clovis and Western Stemmed: Population Migration and the Meeting of Two Technologies in the Intermountain West. *American Antiquity, 75*, 81–116.

Bense, J. A. (1972). *The Cascade Phase: A Study of the Effect of the Altithermal on a Cultural System.* Ph.D. dissertation, Washington State University, Pullman.

Boëda, E. (1988). Le Concepts Laminaire: Repture et Filiation avec le Concept Levallois. In M. Otte (Ed.), *L'homme de Néanderthal* (pp. 41–49). Belgium: Etudes et Recherches Archéologiques de l'Université de Liège.

Boëda, E. (1994). *Le Concept Levallois: Variabilité des Méthodes.* Paris: Éditions Centre National de la Recherche Scientifique.

Boëda, E. (1995). Levallois: Volumetric Construction, Methods, a Technique. In H. L. Dibble & O. Bar-Yosef (Eds.), *The Definition and Interpretation of Levallois Technology* (pp. 41–68). Monographs in World Archaeology, no. 23. Madison, Wis.: Prehistory Press.

Boëda, E., Hou, Y. M., Forestier, H., Sarel, J., & Wang, H. M. (2013). Levallois and Non-Levallois Blade Production at Shuidonggou in Ningxia, North China. *Quaternary International, 295*, 191–203.

Bordes, F. (1961). *Typologie du Paléolithique Ancien et Moyen.* Paris: Éditions Centre National de la Recherche Scientifique.

Bradley, B. A. (1993). Paleo-Indian Flaked Stone Technology in the North American High Plains. In O. Soffer & N. D. Praslov (Eds.), *From Kostenki to Clovis* (pp. 251–262). New York: Plenum.

Bradley, B. A., Collins, M. B., & Hemmings, A. (2010). *Clovis Technology*. International Monographs in Prehistory, no. 17. Ann Arbor, Mich.

Brantingham, P. J., Krivoshapkin, A., Li, L., & Tserendagva, Y. (2001). Constraints on Levallois Core Technology: A Mathematical Model. *Journal of Archaeological Science, 28*, 747–761.

Brantingham, P. J., Kerry, K., Krivoshapkin, A., & Kuzmin, Y. (2004). Time-Space Dynamics in the Early Upper Paleolithic of Northeast Asia. In D. B. Madsen (Ed.), *Entering America: Northeast Asia and Beringia before the Last Glacial Maximum* (pp. 255–283). Salt Lake City: University of Utah Press.

Bryan, A. L. (1979). Smith Creek Cave. In D. R. Tuohy & D. L. Rendall (Eds.), *The Archaeology of Smith Creek Canyon, Eastern Nevada* (pp. 162–253). Anthropological Papers, no. 17. Carson City: Nevada State Museum.

Bryan, A. L. (1980). The Stemmed Point Tradition: An Early Technological Tradition in Western North America. In L. B. Harten, C. N. Warren, & D. R. Tuohy (Eds.), *Anthropological Papers in Memory of Earl H. Swanson* (pp. 77–107). Idaho State Museum of Natural History, Pocatello.

Bryan, A. L. (1988). The Relationship of the Stemmed Point and Fluted Point Traditions in the Great Basin. In J. A. Willig, C. M. Aikens, & J. L. Fagan (Eds.), *Early Human Occupation in Far Western North America: The Clovis-Archaic Interface* (pp. 53–74). Anthropological Papers, no. 21. Carson City: Nevada State Museum.

Bryan, A. L. (1991). The Fluted-Point Tradition in the Americas—One of Several Adaptations to Late Pleistocene American Environments. In R. Bonnichsen & K. L. Turnmire (Eds.), *Clovis Origins and Adaptations* (pp. 15–34). Corvallis: Center for the Study of the First Americans, Oregon State University.

Butler, B. R. (1969). The Earlier Cultural Remains at Cooper's Ferry. *Tebiwa, 12*, 35–50.

Carlson, R. L. (1996). Early Namu. In R. L. Carlson & L. Dalla Bona (Eds.), *Early Human Occupation in British Columbia* (pp. 83–102). Vancouver: University of British Columbia Press.

Collins, M. B., & Lohse, J. C. (2004). The Nature of Clovis Blades and Blade Cores. In D. B. Madsen (Ed.), *Entering America: Northeast Asia and Beringia before the Last Glacial Maximum* (pp. 159–186). Salt Lake City: University of Utah Press.

Davis, L. G., & Schweger, C. E. (2004). Geoarchaeological Context of Late Pleistocene and Early Holocene Occupation at the Cooper's Ferry Site, Western Idaho, USA. *Geoarchaeology, 19*, 685–697.

Davis, L. G., & Sisson, D. A. (1998). An Early Stemmed Point Cache from the Lower Salmon River Canyon of West-Central Idaho. *Current Research in the Pleistocene, 15*, 12–14.

Davis, L. G., Willis, S. C., & Macfarlan, S. J. (2012). Lithic Technology, Cultural Transmission, and the Nature of the Far Western Paleoarchaic—Paleoindian Co-Tradition. In D. Rhode (Ed.), *Meetings at the Margins: Prehistoric Cultural Interactions in the Intermountain West* (pp. 47–64). Salt Lake City: University of Utah Press.

Davis, L. G., Nyers, A. J., & Willis, S. C. (2014). Context, Provenance and Technology of a Western Stemmed Tradition Artifact Cache from the Cooper's Ferry Site, Idaho. *American Antiquity, 79*, 596–615.

Davis, L. G., Bean, D. W., & Nyers, A. J. (2017). Morphometric and Technological Attributes of Western Stemmed Tradition Projectile Points Revealed in a Second Artifact Cache Discovered at the Cooper's Ferry Site, Idaho. *American Antiquity, 82*, 536–557.

Debénath, A., & Dibble, H. L. (1994). *Handbook of Paleolithic Typology*. Vol. 1: Lower and Middle Paleolithic of Europe. Philadelphia: University Museum, University of Pennsylvania Press.

Derevianko, A. P., & Markin, S. V. (1995). The Mousterian of the Altai in the Context of the Middle Paleolithic Culture of Eurasia. In H. L. Dibble & O. Bar-Yosef (Eds.), *The Definition and Interpretation of Levallois Technology* (pp. 473–484). Madison, Wis.: Prehistory Press.

Derevianko, A. P., & Petrin, V. T. (1995). Levallois of Mongolia. In H. L. Dibble & O. Bar-Yosef (Eds.), *The Definition and Interpretation of Levallois Technology* (pp. 455–472). Madison, Wis.: Prehistory Press.

Dibble, H. L. (1989). The Implications of Stone Tool Types for the Presence of Language during the Lower and Middle Palaeolithic. In P. Mellars & C. Stringer (Eds.), *The Human Revolution: Behavioural and Biological Perspectives on the Origins of Modern Humans* (pp. 415–432). Edinburgh: Edinburgh University Press.

Dumond, D. E. (1972). Review of "Akmak: An Early Archaeological Assemblage from Onion Portage, Northwest Alaska" by D. D. Anderson. *American Anthropologist, 74*, 1504–1505.

Dyke, A. S., Moore, A., & Robertson, L. (2003). *Deglaciation of North America. Open File, no. 1574*. Ottawa: Geological Survey of Canada.

Fedje, D. W., Mackie, A. P., McSporran, J. B., & Wilson, B. (1996). Early Period Archaeology in Gwaii Haanas: Results of the 1993 Field Program. In R. L. Carlson & L. Dalla Bona (Eds.), *Early Human Occupation in British Columbia* (pp. 133–150). Vancouver: University of British Columbia Press.

Fedje, D. W., Mackie, Q., Smith, N., & Mclaren, D. (2011). Function, Visibility, and Interpretation of Archaeological Assemblages at the Pleistocene/Holocene Transition in Haida Gwaii. In T. Goebel & I. Buvit (Eds.), *From the Yenisei to the Yukon: Interpreting Lithic Assemblage Variability in Late Pleistocene/Early Holocene Beringia* (pp. 323–339). College Station: Texas A&M University Press.

Goebel, T. (2004). The Search for a Clovis Progenitor in Subarctic Siberia. In D. B. Madsen (Ed.), *Entering America: Northeast Asia and Beringia before the Last Glacial Maximum* (pp. 311–358). Salt Lake City: University of Utah Press.

Goebel, T. (2011). What is the Nenana Complex? Raw Material Procurement and Technological Organization at Walker Road, Central Alaska. In T. Goebel & I. Buvit (Eds.), *From the Yenisei to the Yukon: Interpreting Lithic Assemblage Variability in Late Pleistocene/Early Holocene Beringia* (pp. 199–214). College Station: Texas A&M University Press.

Goebel, T., & Keene, J. L. (2014). Are Great Basin Stemmed Points as Old as Clovis in the Intermountain West? A Review of the Geochronological Evidence. In *Archaeology in the Great Basin and Southwest: Papers in Honor of Don D. Fowler*, edited by N. Parezo & J. Janetski (pp. 35–60). Salt Lake City: University of Utah Press.

Goebel, T., Waters, M. R., & Dikova, M. M. (2003). The Archaeology of Ushki Lake, Kamchatka, and the Pleistocene Peopling of the Americas. *Science, 301*, 501–505.

Grayson, D. K. (2011). *The Great Basin: A Natural Prehistory*. Berkeley: University of California Press.

Hamilton, T. D., & Goebel, T. (1999). Late Pleistocene Peopling of Alaska. In R. Bonnichsen & K. L. Turnmire (Eds.), *Ice Age Peoples of North America: Environments, Origins and Adaptations of the First Americans* (pp. 156–199). Corvallis: Center for the Study of the First Americans, Oregon State University.

Haynes, C. V., Jr. (2005). Clovis, Pre-Clovis, Climate Change, and Extinction. In R. Bonnichsen, B. T. Lepper, D. Stanford, & M. R. Waters (Eds.), *Paleoamerican Origins: Beyond Clovis* (pp. 113–132). College Station: Center for the Study of the First Americans, Texas A&M University.

Hoffecker, J. F. (2011). Assemblage Variability in Beringia: The Mesa Factor. In T. Goebel & I. Buvit (Eds.), *From the Yenisei to the Yukon: Interpreting Lithic Assemblage Variability in Late Pleistocene/Early Holocene Beringia* (pp. 165–178). College Station: Texas A&M University Press.

Hoffecker, J. F., & Elias, S. A. (2007). *Human Ecology of Beringia*. New York: Columbia University Press.

Holmes, C. E. (1996). Broken Mammoth. In F. H. West (Ed.), *American Beginnings: The Prehistory and Palaeoecology of Beringia* (pp. 312–318). Chicago: University of Chicago Press.

Jenkins, D. L., Davis, L. G., Stafford, T. W., Jr., Campos, P. F., Hockett, B., Jones, G. T., et al. (2012). Clovis Age Western Stemmed Projectile Points and Human Coprolites at the Paisley Caves. *Science, 337*, 223–228.

Jenkins, D. L., Davis, L. G., Stafford, T. W., Jr., Campos, P. F., Connolly, T. J., Scott Cummings, L., et al. (2013). Geochronology, Archaeological Context, and DNA at the Paisley Caves. In K. E. Graf, C. V. Ketron, & M. R. Waters (Eds.), *Paleoamerican Odyssey* (pp. 485–510). College Station: Texas A&M University Press.

Jenkins, D. L., Davis, L. G., Stafford, T. W., Jr., Connolly, T. J., Jones, G. T., Rondeau, M., et al. (2016). Younger Dryas Archaeology and Human Experience at the Paisley Caves in the Northern Great Basin. In M. Kornfeld & B. Huckell (Eds.), *Stones, Bones and Profiles* (pp. 127–206). Boulder: University Press of Colorado.

Kuhn, S., & Zwyns, N. (2014). Rethinking the Initial Upper Paleolithic. *Quaternary International, 347*, 29–38.

Leonhardy, F. C. (1970). *Artifact Assemblages and Archaeological Units at Granite Point Locality I (45WT41), Southeastern Washington*. Ph.D. dissertation, Washington State University, Pullman.

Leonhardy, F. C., & Rice, D. G. (1970). A Proposed Culture Typology for the Lower Snake River Region, Southeastern Washington. *Northwest Anthropological Research Notes, 4*, 1–29.

Leonhardy, F. C., Schrodel, G. C., Bense, J. A., & Beckerman, S. (1971). Wexpusnime (45GA61): Preliminary Report. Reports of Investigations, no. 49. Pullman: Washington State University Laboratory of Anthropology.

Lycett, S., & Norton, C. (2010). A Demographic Model for Paleolithic Technological Evolution: The Case of East Asia and the Movius Line. *Quaternary International, 211*, 55–65.

Lycett, S. J., von Cramon-Taubadel, N., & Eren, M. I. (2016). Levallois: Potential Implications for Learning and Cultural Transmission Capacities. *Lithic Technology, 41*, 19–38.

Madsen, D. B. (2015). A Framework for the Initial Occupation of the Americas. *PaleoAmerica, 1*, 217–250.

Madsen, D. B., Oviatt, C. G., Zhu, Y., Brantingham, P. J., Elston, R. G., Chen, F., et al. (2014). The Early Appearance of Shuidonggou Core-and-Blade Technology in North China: Implications for the Spread of Anatomically Modern Humans in Northeast Asia? *Quaternary International, 347*, 21–28.

Matson, R. G. (1996). The Old Cordilleran Component at the Glenrose Cannery Site. In R. L. Carlson & L. Dalla Bona (Eds.), *Early Human Occupation in British Columbia* (pp. 111–122). Vancouver: University of British Columbia Press.

Mitchell, D., & Pokotylo, D. L. (1996). Early Period Components at the Milliken Site. In R. L. Carlson & L. Dalla Bona (Eds.), *Early Human Occupation in British Columbia* (pp. 65–82). Vancouver: University of British Columbia Press.

Muto, G. R. (1976). *The Cascade Technique: An Examination of a Levallois-like Reduction System in Early Snake River Prehistory*. Ph.D. dissertation, Washington State University, Pullman.

Norton, C. J. (2000). The Current State of Korean Paleoanthropology. *Journal of Human Evolution, 38*, 803–825.

O'Connor, S., Ono, R., & Clarkson, C. (2011). Pelagic Fishing at 42,000 Years before the Present and the Maritime Skills of Modern Humans. *Science, 334*, 1117–1121.

Otte, M. (1995). The Nature of Levallois. In H. L. Dibble & O. Bar-Yosef (Eds.), *The Definition and Interpretation of Levallois Technology* (pp. 117–124). Madison, Wis.: Prehistory Press.

Pedersen, M. W., Ruter, A., Schweger, C., Friebe, H., Staff, R. A., Kjeldsen, K. K., et al. (2016). Postglacial Viability and Colonization in North America's Ice-Free Corridor. *Nature, 537*, 45–50.

Powers, W. R., & Hoffecker, J. F. (1989). Late Pleistocene Settlement in the Nenana Valley, Central Alaska. *American Antiquity, 54*, 263–287.

Sato, H., Nishiaki, Y., & Suzuki, M. (1995). Lithic Technology of the Japanese Middle Paleolithic Levallois in Japan? In H. L. Dibble & O. Bar-Yosef (Eds.), *The Definition and Interpretation of Levallois Technology* (pp. 485–500). Madison, Wis.: Prehistory Press.

Slobodin, S. (2006). The Paleolithic of Western Beringia: A Summary of Research. In D. Dumond & R. E. Bland (Eds.), *Archaeology in Northeast Asia on the Pathway to the Bering Strait* (pp. 9–23). Anthropological Papers, no. 65. Eugene: University of Oregon.

Slobodin, S. (2011). Late Pleistocene and Early Holocene Cultures of Beringia: The General and the Specific. In T. Goebel & I. Buvit (Eds.), *From the Yenisei to the Yukon: Interpreting Lithic Assemblage Variability in Late Pleistocene/Early Holocene Beringia* (pp. 91–118). College Station: Texas A&M University Press.

Stuiver, M., Reimer, P. J., & Reimer, R. (1993). CALIB Radiocarbon Calibration (ver. 7.1). *Radiocarbon, 35*, 215–230.

Tolan, T. L., Reidel, S. P., Beeson, M. H., Anderson, J. L., Fecht, K. R., & Swanson, D. A. (1989). Revisions to the Estimates of the Areal Extent and Volume of the Columbia River Basalt Group. In S. P. Reidel & P. R. Hooper (Eds.), *Volcanism and Tectonism in the Columbia River Flood–Basalt Province* (pp. 1–20). Special Paper, no. 239. Boulder, Colo.: Geological Society of America.

Van Peer, P. (1992). *The Levallois Reduction Strategy*. Madison, Wis.: Prehistory Press.

Waters, M. R., & Stafford, T. W. (2007). Redefining the Age of Clovis: Implications for the Peopling of the Americas. *Science*, *315*, 1122–1126.

Yesner, D. R., Holmes, C. E., & Crossen, K. J. (1992). Archaeology and Paleoecology of the Broken Mammoth Site, Central Tanana Valley, Interior Alaska. *Current Research in the Pleistocene*, *9*, 53–57.

Zwyns, N. (2012). *Laminar Technology and the Onset of the Upper Paleolithic in the Altai, Siberia*. Leiden, Netherlands: Leiden University Press.

14 Assessing the Likelihood of Convergence among North American Projectile-Point Types

Briggs Buchanan, Metin I. Eren, and Michael J. O'Brien

Convergence is regarded as an important phenomenon in evolutionary biology and has been thoroughly documented across numerous genera (McGhee 2011; chapter 2, this volume). It likely assumes a similar importance in cultural evolution, although it has not been as well researched in anthropology and archaeology (for exceptions, see Adler et al. 2014; Boulanger and Eren 2015; Eren, Patten, O'Brien, and Meltzer 2013; Eren, Patten, O'Brien, and Meltzer 2014; Lycett 2009, 2011; O'Brien, Boulanger, Buchanan et al. 2014; O'Brien et al. 2014b, 2014c; Straus, Meltzer, and Goebel 2005; Wang, Lycett, von Cramon-Taubadel, Jin, and Bae 2012; Will, Mackay, and Phillips 2015). There are numerous reasons for this (see chapter 1, this volume), one being that critics of evolutionary models of culture have long assumed that blending obscures evidence of evolutionary processes (Dewar 1995; Kroeber 1948; Moore 1994, 2001; Terrell 1988, 2001; Terrell, Hunt, and Gosden 1997; Terrell, Kelly, and Rainbird 2001). However, numerous studies have now shown that blending is not more prevalent in culture than biology and that it does not necessarily destroy phylogenetic signals in evolutionary sequences of culture (e.g., Lipo, O'Brien, Collard, and Shennan 2006; O'Brien and Lyman 2003). The conclusion that follows from this is that cultural lineages are subject to the same processes as biological organisms. A further caveat to this is that the evolutionary processes (adaptation and constraints) that produce convergence in distantly related lineages of organisms can similarly produce convergence in distantly related cultural lineages.

If we focus on material culture and those objects that are produced with a relatively intractable reductive technology—one in which only a limited set of possible shapes can be taken away to form the final product (chapter 4, this volume)—we can posit that convergence will occur more often when compared to other forms of culture. In particular, stone tools should have a high likelihood of convergence, as not only is tool production a reductive technology, but only a limited suite of techniques has ever been used to reduce stone throughout the more two million years that it has been worked by hominins (Patten 2009; Whittaker 1994). Here we assess the likelihood of convergence in a particular class of stone tools, namely North American bifacially flaked points.

Bifacial points are an important component of the North American archaeological record, as they have been documented for nearly all time periods and regions of the continent. Bifacial points were used with a variety of delivery systems, including a spear, an atlatl, and a bow. Within this vast record of points used with different delivery systems is a wide range of variation in functional and neutral (stylistic) characters that reflect local adaptation, shared history, and a variety of historically contingent stochastic processes (Eerkens and Lipo 2005; Eren, Buchanan, and O'Brien 2015; Lycett 2015; Lycett and von Cramon-Taubadel 2015; O'Brien 2008; Shennan 2002). Researchers have classified these points into hundreds of different types using combinations of characters (e.g., Justice 1987, 2002a, 2002b; Justice and Kudlaty 1999; Turner and Hester 1999). As archeologists have long known (Lyman, O'Brien, and Dunnell 1997; O'Brien and Lyman 1999), types that have specimens with limited distributions in space and time serve as spatial and temporal markers and can help archaeologists assign an age to an assemblage or site when the materials necessary to carry out other forms of dating are unavailable (e.g., a lack of suitable organic material for radiometric dating).

However, the information gleaned from stone points can be misleading if cases of convergence go unrecognized (chapters 11 and 12, this volume). For example, unidentified convergence of two distantly related point types separated in space and time would incorrectly expand the spatiotemporal limits of the type. Recognizing cases of convergence among points will help to reduce errors when developing point lineages, as identifying probable cases of convergence can be used to refute hypotheses of historical connections among points. Thus, having an idea of the likelihood of convergence among types is an important first step for these kinds of analyses.

We carried out a pilot assessment of convergence among North American point types by recording characters in a large sample of types recorded across the continent. To identify probable cases of convergence, we isolated pairs of point types that shared a common set of characters. Next, we assessed the spatial and temporal distance separating the two types. We then ranked potentially convergent types by spatiotemporal distance. Two potentially convergent types with significant temporal separation but overlapping spatial distributions were identified as possible cases of functional convergence. We defined this as functional convergence because the regional environment remains more similar within a region over time than it does between regions. On the other hand, cases of potential convergence that were separated by significant distances in space and time were considered to be more likely the result of neutral convergence, as adaptation is less likely to have played a similar role in regions separated in space and time. Finally, we compared the number of potentially functional and neutral point convergences. Our working assumption was that functional convergence should be more likely than neutral convergence, as points were used as weapon tips and therefore tied directly to the adaptations of the hunters using them.

Methods

We examined point types described in four well-known volumes compiled by Noel Justice on point types in the United States for the Midcontinent and East (Justice 1987), the Midwest (Justice and Kudlaty 1999), the Southwest (Justice 2002a), and California and the Great Basin (Justice 2002b). These volumes do not include the Great Plains or the Pacific Northwest regions, but as shown in figure 14.1, the distribution of types spans all 48 contiguous states because several point types have distributions that extend beyond the regions covered in the volumes. For each of the 210 types described in the four volumes, we recorded the type name, the state or states in which the type is found, the type's suspected age range, and a set of 10 characters that are described below.

Types and the descriptions that accompany them are potentially problematic, because they are extensionally derived (Dunnell 1986; O'Brien and Lyman 2000, 2002). This means that the types are based on observed characteristics of the specimens in the set of points that are being scrutinized as opposed to outlining these criteria a priori and then assigning specimens to them. O'Brien and Lyman (2002) underscored two potentially negative outcomes of extensional units. First, they often include both descriptive and diagnostic criteria, and in some cases only descriptive criteria. Problems arise with types

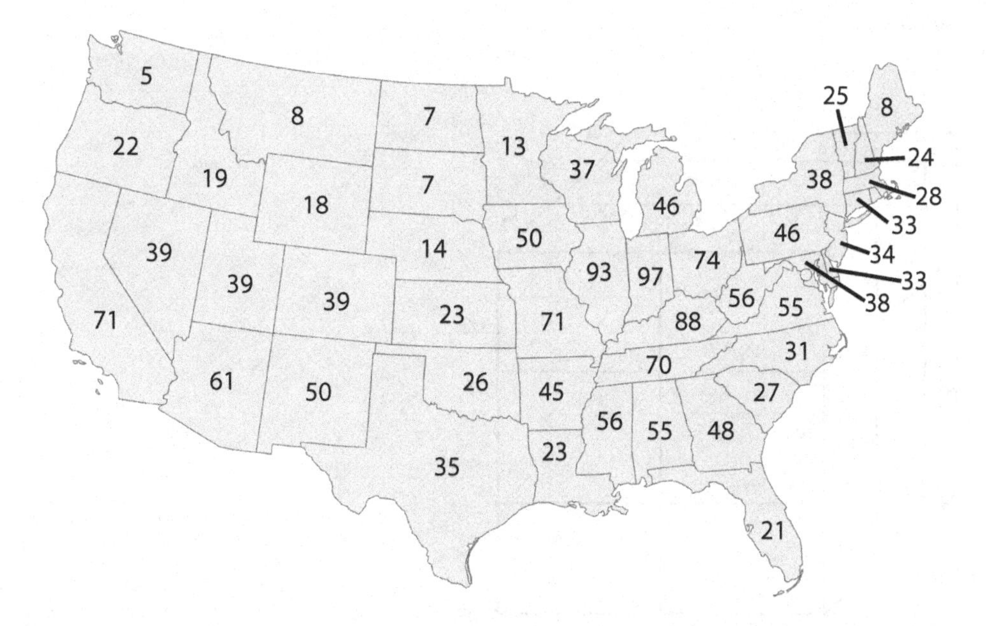

Figure 14.1
Map of the 48 contiguous U.S. states showing the number of point types recorded for each state, with multistate types recorded for each state in which they occur.

defined in this way because descriptive characteristics can be vague and overlap among types, making it difficult to assign particular specimens to a type. Second, because extensional definitions are derived from inspection of actual specimens, they tend to reinforce the notion that types are in some sense real (or emic) categories rather than analytical categories. We recognize these issues and here use types only in an exploratory manner. We also avoid some of the problems inherent in traditional types by employing paradigmatic classification, in which the units—classes—are defined by the intersection of character states of each mutually exclusive character (Dunnell 1971; Eren, Chao, Hwang, and Colwell 2012; Eren et al. 2016; Meltzer 1981; O'Brien and Lyman 2003; O'Brien et al. 2014a). Class definitions are then defined by the unique combination of character states defined by the classification. Classes formed by this procedure possess important properties, including the equivalency of character states, the unambiguous nature of character states, and the ability for the analyst to compare a class to all other classes in the same classification (Dunnell 1971). These properties make paradigmatic classification ideal for assessing archaeological convergence, because types with unique combinations of characters—the types with the potential to be convergent—can be easily distinguished. Note also that whereas all states of each character can in theory combine with all states of every other character (figure 14.2), they rarely will. In other words, there are filled segments of morphospace as well as empty segments (O'Brien et al. 2016; chapters 2 and 5, this volume).

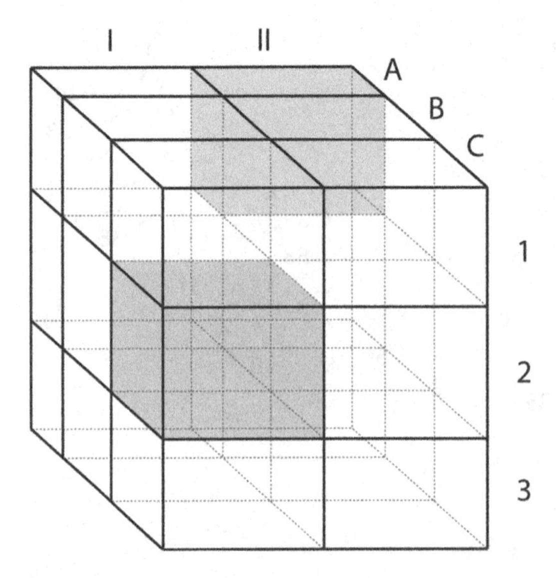

Figure 14.2
A simple three-dimensional paradigmatic classification system showing the intersection of the character states of each character to create 18 taxa (2 × 3 × 3). The upper box represents hypothetical class II, A, 1, and the lower box represents hypothetical class I, C, 2.

We used ten characters to generate the classification (examples of traditional point types are listed in parentheses):

Character I. Presence or Absence of a Channel Flake

A channel flake is a flake removed from the proximal base toward the tip:

1. Channel flake present (Clovis, Folsom)
2. Channel flake absent (Milnesand, Plainview)

Character II. Overall Point Shape

1. Triangular (Cottonwood Triangular)
2. Oval (no examples)
3. Lanceolate (Agate Basin)
4. Parallel-sided (Stilwell)
5. Fluted (Clovis)
6. Triangular/lanceolate (Silver Lake)
7. Lanceolate/parallel (Tuolumne)
8. Curved (Stockton Curve)
9. Truncated (Livermore)
10. Triangular/truncated (Eva II)
11. Lanceolate/oval (Nodena)

Character III. Base Shape

1. Arc/round (San José)
2. Normal curve (Buchanan-Eared)
3. Triangular (Delta Side-Notched)
4. Folsomoid (Folsom)
5. Flat (Milnesand)
6. Convex (Channel Island Barbed)
7. Flat/convex (Mayacmas Corner-Notched)
8. Flat/concave (Pandale)
9. Bifurcated (Awatovi Side-Notched)

Character IV. Stem Shape

0. Absent (Milnesand)
1. Square/rectangular (Scottsbluff)
2. Rounded (Stockton Notched Leaf)

3. Expanding (San José)
4. Contracting (Mojave)
5. Pointed (no examples)
6. Side-Notched (Snaketown Side-Notched)
7. Corner-Notched (Rose Spring Corner-Notched)
8. Basal-Notched (Norden Basal-Notched)
9. Basal/Corner-Notched (Gatecliff Split-Stem)
10. Side/Corner-Notched (Contra Costa Notched)
11. Square/Rounded (Stockton Parallel-Stemmed)
12. Basal/Side-Notched (Sierra Side-Notched)

Character V. Presence or Absence of a Basal Notch

1. Basal notch present (LeCroy Bifurcated Stem, Rice-Lobed)
2. Basal notch absent (Agate Basin)

Character VI. Presence or Absence of a Side Notch

1. Side notch present (Snaketown Side-Notched, Northern Side-Notched)
2. Side notch absent (Scottsbluff)

Character VII. Presence or Absence of a Corner Notch

1. Corner notch present (Rattlesnake Corner-Notched, Rose Spring Corner-Notched)
2. Corner notch absent (Scottsbluff)

Character VIII. Presence or Absence of Serration

1. Serration present (Big Valley Stemmed, Gunther Barbed)
2. Serration absent (Scottsbluff)

Character IX. Location of Maximum Width

1. Proximal quarter (Big Sandy)
2. Second-most proximal quarter (Hi-Lo)
3. Third-most proximal quarter (Agate Basin)
4. Distal quarter (no examples)

Character X. Tang-tip Shape

1. Pointed (Rattlesnake Corner-Notched)
2. Round (Scottsbluff)

The characters in the paradigmatic classification specified above produce 329,472 possible classes, 157 of which account for our sample of 210 point types.

To evaluate the likelihood of convergence, we isolated pairs of point types in the paradigmatic classification that were in the same class—that is, two point types with the same character states for all 10 characters. We then examined the temporal and spatial range of each pair of point types. We took a conservative approach in that point types had to be in the same class *and* come from different regions or date to different time ranges to be considered convergent. We considered type pairs from the same region and time period to be possible cases of "misclassification" rather than convergence.

We then counted the number of possible cases of convergence with similar or overlapping spatial distributions but with different time ranges. We considered these cases more likely to be examples of functional convergence. Again, we are defining functional convergence as convergence due to adaptation to similar environments. Although the environmental conditions within a given region change over time, we assume that the environment within a region at any given time over the past 14,000 years is more similar to the environment in the same region at another point time than it is between regions. Thus, similar types within the same region, but separated in time, are likely to be cases of functional convergence. We also counted the number of cases more likely to be neutral convergence—cases of convergence where similar point types are separated in space and time, which suggests that adaptation played an insignificant role in the convergence of characters. We then compared the number of functional and neutral type convergences.

Results

Of the 157 classes, 121 contained only single types. Of the remaining 36 classes, 27 contained two types, four contained three types, four contained four types, and one contained seven types. We view the incidences of multiple types in the same class as possible cases of convergence. Of the 84 pairwise combinations of types, 16 occur in the same region and date to the same time period. As we indicated above, these cases have a higher likelihood of being typological mistakes rather than probable cases of convergence. The remaining 68 pairwise combinations contain types that come from different regions and/or date to different time periods. Table 14.1 lists the 68 pairwise combinations of possible type convergence. Of those, eight are located in the same region or overlap in their regions and have date ranges that are separated by at least 1,000 years in time (table 14.2). These eight pairwise combinations of 15 point types (one type is represented twice) represent the most likely cases of functional convergence in our sample. Cases of possible neutral convergence are represented by 33 pairwise combinations of point types separated in space

Table 14.1
Cases of possible point-type convergence based on shared classes and different spatiotemporal distributions

Point type 1	Point type 2	Point type class
Bull Creek	Canaliño Triangular	0 1 1 0 0 0 0 0 1 1
Big Sandy	Brewerton Eared-Notched	0 1 1 6 0 1 0 0 1 1
Big Sandy	Gatlin Side-Notched	0 1 1 6 0 1 0 0 1 1
Big Sandy	Panoche Side-Notched	0 1 1 6 0 1 0 0 1 1
Big Sandy	Redding Side-Notched	0 1 1 6 0 1 0 0 1 1
Big Sandy	Ridge Ruin Side-Notched	0 1 1 6 0 1 0 0 1 1
Big Sandy	White Mountain Side-Notched	0 1 1 6 0 1 0 0 1 1
Brewerton Eared-Notched	Gatlin Side-Notched	0 1 1 6 0 1 0 0 1 1
Brewerton Eared-Notched	Panoche Side-Notched	0 1 1 6 0 1 0 0 1 1
Brewerton Eared-Notched	Redding Side-Notched	0 1 1 6 0 1 0 0 1 1
Brewerton Eared-Notched	Ridge Ruin Side-Notched	0 1 1 6 0 1 0 0 1 1
Brewerton Eared-Notched	White Mountain Side-Notched	0 1 1 6 0 1 0 0 1 1
Gatlin Side-Notched	Panoche Side-Notched	0 1 1 6 0 1 0 0 1 1
Gatlin Side-Notched	Redding Side-Notched	0 1 1 6 0 1 0 0 1 1
Gatlin Side-Notched	White Mountain Side-Notched	0 1 1 6 0 1 0 0 1 1
Panoche Side-Notched	Redding Side-Notched	0 1 1 6 0 1 0 0 1 1
Panoche Side-Notched	Ridge Ruin Side-Notched	0 1 1 6 0 1 0 0 1 1
Panoche Side-Notched	White Mountain Side-Notched	0 1 1 6 0 1 0 0 1 1
Redding Side-Notched	Ridge Ruin Side-Notched	0 1 1 6 0 1 0 0 1 1
Redding Side-Notched	White Mountain Side-Notched	0 1 1 6 0 1 0 0 1 1
Ridge Ruin Side-Notched	White Mountain Side-Notched	0 1 1 6 0 1 0 0 1 1
Hardaway Side-Notched	Point of Pines Side-Notched	0 1 2 6 0 1 0 0 1 1
Hamilton Incurvate	Fort Ancient	0 1 5 0 0 0 0 0 1 1
Hamilton Incurvate	Snaketown Triangular Straight Base	0 1 5 0 0 0 0 0 1 1
Madison	Snaketown Triangular Straight Base	0 1 5 0 0 0 0 0 1 1
Fort Ancient	Snaketown Triangular Straight Base	0 1 5 0 0 0 0 1 1 1
Borax Wide Stemmed	Genesee	0 1 5 1 0 0 0 0 1 1
Borax Wide Stemmed	Sykes	0 1 5 1 0 0 0 0 1 1
Borax Wide Stemmed	White Springs	0 1 5 1 0 0 0 0 1 1
Genesee	White Springs	0 1 5 1 0 0 0 0 1 1
Buck Creek Barbed	Wade	0 1 5 1 0 0 1 0 1 1
Bakers Creek	Perkiomen Broad	0 1 5 3 0 0 0 0 2 1
Bear River Side-Notched	Citrus Side-Notched	0 1 5 6 0 1 0 0 1 1
Bear River Side-Notched	Sudden Side-Notched	0 1 5 6 0 1 0 0 1 1
Citrus Side-Notched	Sudden Side-Notched	0 1 5 6 0 1 0 0 1 1
Rattlesnake/Clear Lake Corner-Notched	Vosburg Corner-Notched	0 1 5 7 0 0 1 0 1 1
Cache Creek Corner-Notched	Palmer Corner-Notched	0 1 5 7 0 0 1 1 1 1

Table 14.1 (continued)

Point type 1	Point type 2	Point type class
Morrow Mountain	Sierra Contracting Stem	0 1 6 4 0 0 0 0 1 1
Round Valley Corner-Notched	Tularosa Corner-Notched	0 1 6 7 0 0 1 1 1 1
Brewerton Side-Notched	Trimble Side-Notched	0 1 7 6 0 1 0 0 1 1
Dolores Expanding Stem	Mayacmas Corner-Notched	0 1 7 7 0 0 1 0 1 1
Elko Corner-Notched	Mayacmas Corner-Notched	0 1 7 7 0 0 1 0 1 1
Cottonwood Triangular	Levanna	0 1 8 0 0 0 0 0 1 1
Jalama Side-Notched	Northern Side-Notched	0 1 8 6 0 1 0 0 1 1
Jalama Side-Notched	Pueblo Side-Notched	0 1 8 6 0 1 0 0 1 1
Jalama Side-Notched	Snaketown Side-Notched	0 1 8 6 0 1 0 0 1 1
Northern Side-Notched	Pueblo Side-Notched	0 1 8 6 0 1 0 0 1 1
Northern Side-Notched	Snaketown Side-Notched	0 1 8 6 0 1 0 0 1 1
Morris	St. Albans Side-Notched	0 1 9 8 1 1 0 1 1 1
Gatecliff Split Stem	Pinto	0 1 9 9 1 0 1 1 1 1
Guadalupe	Martis Corner-Notched	0 10 7 7 0 0 1 1 1 1
Cortaro	Price Ranch Triangular	0 3 1 0 0 0 0 0 1 1
Black Rock Concave	Mendocino Concave Base	0 3 1 0 0 0 0 0 2 1
Black Rock Concave	Nebo Hill Lanceolate	0 3 1 0 0 0 0 0 2 1
Black Rock Concave	Quad	0 3 1 0 0 0 0 0 2 1
Mendocino Concave Base	Nebo Hill Lanceolate	0 3 1 0 0 0 0 0 2 1
Mendocino Concave Base	Quad	0 3 1 0 0 0 0 0 2 1
Nebo Hill Lanceolate	Quad	0 3 1 0 0 0 0 0 2 1
Beaver Lake	Hi-Lo	0 3 1 0 0 0 0 0 2 2
Milnesand	Wadlow	0 3 5 0 0 0 0 0 2 2
Excelsior	Malaga Cove Leaf	0 3 6 0 0 0 0 1 2 1
Adena Stemmed	Need Stemmed Lanceolate	0 3 6 2 0 1 0 0 2 1
Copena Triangular	Pomranky	0 6 5 0 0 0 0 0 1 1
Kin Kletso Side-Notched	Matanzas Side-Notched	0 6 5 6 0 1 0 0 1 1
Kramer	Little Bear Creek	0 6 7 1 0 0 0 0 2 1
Contra Costa Notched	San Pedro	0 6 7 10 0 1 1 1 1 1
Black Mesa Narrow Neck	Diablo Canyon Side-Notched	0 6 7 6 0 1 0 0 1 1
Humboldt	Lake Erie Bifurcated Base	0 6 9 8 1 0 0 0 2 1

Table 14.2
Cases of possible functional convergence among point types based on shared classes from similar regions with different temporal ranges

Type 1	Type 2	Point type class	Number of years separating type time ranges
Genesee	White Springs	0 1 5 1 0 0 0 0 1 1	1,020
Black Rock Concave	Mendocino Concave Base	0 3 1 0 0 0 0 0 2 1	2,000
Citrus Side-Notched	Sudden Side-Notched	0 1 5 6 0 1 0 0 1 1	3,200
Northern Side-Notched	Pueblo Side-Notched	0 1 8 6 0 1 0 0 1 1	4,150
Northern Side-Notched	Snaketown Side-Notched	0 1 8 6 0 1 0 0 1 1	4,200
Big Sandy	Brewerton Eared-Notched	0 1 1 6 0 1 0 0 1 1	4,920
Nebo Hill Lanceolate	Quad	0 3 1 0 0 0 0 0 2 1	6,300
Morris	St. Albans Side-Notched	0 1 9 8 1 1 0 1 1 1	7,400

and time—that is, they have nonoverlapping regional distributions and have time ranges that are separated by at least 1,000 years.

Discussion

Interestingly, the comparison of functional versus neutral convergence showed that cases of possible neutral convergence were almost twice as likely as functional convergence. To reiterate, we defined cases of possible functional convergence as two point types with the same characteristics coming from overlapping regions or the same region but separated in time by at least 1,000 years. We considered possible neutral convergence to be cases where two point types were significantly separated by space and time. Our underlying, and untested, assumption was that functional convergence would be more likely, since points are used as weapon tips and are tied directly to the adaptations of the hunters using them. Our finding suggests the contrary, that neutral convergence is more likely. Regardless of kind, our results show that we should expect convergence to have occurred relatively frequently, in about 14 types for every 100 encountered. Clearly, this finding has significant implications for analyses relying on point types, in particular for analyses that construct and interpret point-type lineages.

References

Adler, D. S., Wilkinson, K. N., Blockley, S., Mark, D. F., Pinhasi, R., Schmidt-Magee, B. A., et al. (2014). Early Levallois Technology and the Lower to Middle Paleolithic Transition in the Southern Caucasus. *Science*, *345*, 1609–1613.

Boulanger, M. T., & Eren, M. I. (2015). On the Inferred Age and Origin of Lithic Bi-points on the Eastern Seaboard and Their Relevance to the Pleistocene Peopling of North America. *American Antiquity*, *80*, 134–145.

Dewar, R. E. (1995). Of Nets and Trees: Untangling the Reticulate and Dendritic in Madagascar's Prehistory. *World Archaeology*, *26*, 301–318.

Dunnell, R. C. (1971). *Systematics in Prehistory*. New York: Free Press.

Dunnell, R. C. (1986). Methodological Issues in Americanist Artifact Classification. In M. B. Schiffer (Ed.), *Advances in Archaeological Method and Theory* (Vol. 9, pp. 149–207). New York: Academic Press.

Eerkens, J. W., & Lipo, C. P. (2005). Cultural Transmission, Copying Errors, and the Generation of Variation in Material Culture and the Archaeological Record. *Journal of Anthropological Archaeology*, *24*, 316–334.

Eren, M. I., Chao, A., Hwang, W., & Colwell, R. (2012). Estimating the Richness of a Population When the Maximum Number of Classes Is Fixed: A Nonparametric Solution to an Archaeological Problem. *PLoS One*, *7*(5), e34179.

Eren, M. I., Patten, R. J., O'Brien, M. J., & Meltzer, D. J. (2013). Refuting the Technological Cornerstone of the Ice-Age Atlantic Crossing Hypothesis. *Journal of Archaeological Science*, *40*, 2934–2941.

Eren, M. I., Patten, R. J., O'Brien, M. J., & Meltzer, D. J. (2014). More on the Rumor of "Intentional Overshot Flaking" and the Purported Ice-Age Atlantic Crossing. *Lithic Technology*, *39*, 55–65.

Eren, M. I., Buchanan, B., & O'Brien, M. J. (2015). Social Learning and Technological Evolution during the Clovis Colonization of the New World. *Journal of Human Evolution*, *80*, 159–170.

Eren, M. I., Chao, A., Chiu, C., Colwell, R. K., Buchanan, B., Boulanger, M. T., et al. (2016). Statistical Analysis of Paradigmatic Class Richness Supports Greater Paleoindian Projectile-Point Diversity in the Southeast. *American Antiquity*, *81*, 174–192.

Justice, N. D. (1987). *Stone Age Spear and Arrow Points of the Midcontinental and Eastern United States*. Bloomington: Indiana University Press.

Justice, N. D. (2002a). *Stone Age Spear and Arrow Points of the Southwestern United States*. Bloomington: Indiana University Press.

Justice, N. D. (2002b). *Stone Age Spear and Arrow Points of California and the Great Basin*. Bloomington: Indiana University Press.

Justice, N. D., & Kudlaty, S. K. (1999). *Field Guide to Projectile Points of the Midwest*. Bloomington: Indiana University Press.

Kroeber, A. L. (1948). *Anthropology: Race, Language, Culture, Psychology, Prehistory*. New York: Harcourt, Brace.

Lipo, C. P., O'Brien, M. J., Collard, M., & Shennan, S. (Eds.). (2006). *Mapping Our Ancestors: Phylogenetic Approaches in Anthropology and Prehistory*. New York: Aldine.

Lycett, S. J. (2009). Are Victoria West Cores "Proto-Levallois"? A Phylogenetic Assessment. *Journal of Human Evolution*, *56*, 175–191.

Lycett, S. J. (2011). "Most Beautiful and Most Wonderful": Those Endless Stone Tool Forms. *Journal of Evolutionary Psychology (Budapest)*, *9*, 143–171.

Lycett, S. J. (2015). Cultural Evolutionary Approaches to Artifact Variation over Time and Space: Basis, Progress, and Prospects. *Journal of Archaeological Science*, *56*, 21–31.

Lycett, S. J., & von Cramon-Taubadel, N. (2015). Toward a "Quantitative Genetic" Approach to Lithic Variation. *Journal of Archaeological Method and Theory*, *22*, 646–675.

Lyman, R. L., O'Brien, M. J., & Dunnell, R. C. (1997). *The Rise and Fall of Culture History*. New York: Plenum.

McGhee, G. R. (2011). *Convergent Evolution: Limited Forms Most Beautiful*. Cambridge, Mass.: MIT Press.

Meltzer, D. J. (1981). A Study of Style and Function in a Class of Stone Tools. *Journal of Field Archaeology*, *8*, 313–326.

Moore, J. H. (1994). Putting Anthropology Back Together Again: The Ethnogenetic Critique of Cladistic Theory. *American Anthropologist*, *96*, 370–396.

Moore, J. H. (2001). Ethnogenetic Patterns in Native North America. In J. E. Terrell (Ed.), *Archaeology, Language and History: Essays on Culture and Ethnicity* (pp. 30–56). Westport, Conn.: Bergin and Garvey.

O'Brien, M. J. (Ed.). (2008). *Cultural Transmission and Archaeology: Issues and Case Studies*. Washington, D.C.: Society for American Archaeology Press.

O'Brien, M. J., & Lyman, R. L. (1999). *Seriation, Stratigraphy, and Index Fossils: The Backbone of Archaeological Dating*. Plenum, New York: Kluwer Academic.

O'Brien, M. J., & Lyman, R. L. (2000). *Applying Evolutionary Archaeology: A Systematic Approach*. Plenum, New York: Kluwer Academic.

O'Brien, M. J., & Lyman, R. L. (2002). The Epistemological Nature of Archaeological Units. *Anthropological Theory*, *2*, 37–56.

O'Brien, M. J., & Lyman, R. L. R. (2003). *Cladistics and Archaeology*. Salt Lake City: University of Utah Press.

O'Brien, M. J., Boulanger, M. T., Buchanan, B., Collard, M., Lyman, R. L., & Darwent, J. (2014a). Innovation and Cultural Transmission in the American Paleolithic: Phylogenetic Analysis of Eastern Paleoindian Projectile-Point Classes. *Journal of Anthropological Archaeology*, *34*, 100–119.

O'Brien, M. J., Boulanger, M. T., Collard, M., Buchanan, B., Tarle, L., Straus, L., et al. (2014b). On Thin Ice: Problems with Stanford and Bradley's Proposed Solutrean-Clovis Colonization. *Antiquity*, *88*, 606–613.

O'Brien, M. J., Boulanger, M. T., Collard, M., Buchanan, B., Tarle, L., Straus, L., et al. (2014c). Solutreanism. *Antiquity*, *88*, 622–624.

O'Brien, M. J., Boulanger, M. T., Buchanan, B., Bentley, R. A., Lyman, R. L., Lipo, C. P., et al. (2016). Design Space and Cultural Transmission: Case Studies from Paleoindian Eastern North America. *Journal of Archaeological Method and Theory*, *23*, 692–740.

Patten, R. J. (2009). *Old Tools, New Eyes* (2nd ed.). Lakewood, Colo.: Stone Dagger Publications.

Shennan, S. (2002). *Genes, Memes, and Human History: Darwinian Archaeology and Cultural Evolution*. London: Thames and Hudson.

Straus, L. G., Meltzer, D. J., & Goebel, T. (2005). Ice Age Atlantis? Exploring the Solutrean-Clovis "Connection." *World Archaeology*, *37*, 507–532.

Terrell, J. E. (1988). History as a Family Tree, History as a Tangled Bank. *Antiquity*, *62*, 642–657.

Terrell, J. E. (2001). Introduction. In J. E. Terrell (Ed.), *Archaeology, Language, and History: Essays on Culture and Ethnicity* (pp. 1–10). Westport, Colo.: Bergin and Garvey.

Terrell, J. E., Hunt, T. L., & Gosden, C. (1997). The Dimensions of Social Life in the Pacific: Human Diversity and the Myth of the Primitive Isolate. *Current Anthropology*, *38*, 155–195.

Terrell, J. E., Kelly, K. M., & Rainbird, P. (2001). Foregone Conclusions? In Search of "Papuans" and "Austronesians." *Current Anthropology*, *42*, 97–124.

Turner, E. S., & Hester, T. R. (1999). *A Field Guide to Stone Artifacts of Texas Indians*. Houston: Gulf Publishing.

Wang, W., Lycett, S. J., von Cramon-Taubadel, N., Jin, J. J. H., & Bae, C. J. (2012). Comparison of Handaxes from Bose Basin (China) and the Western Acheulean Indicates Convergence of Form, not Cognitive Differences. *PLoS One*, *7*(4), e35804.

Whittaker, J. C. (1994). *Flintknapping: Making and Understanding Stone Tools*. Austin: University of Texas Press.

Will, M., Mackay, A., & Phillips, N. (2015). Implications of Nubian-like Core Reduction Systems in Southern Africa for the Identification of Early Modern Human Dispersals. *PLoS One*, *10*(6), e0131824.

Contributors

R. Alexander Bentley, University of Tennessee

Briggs Buchanan, University of Tulsa

Marcelo Cardillo, Consejo Nacional de Investigaciones Científicas y Técnicas, Buenos Aires

Mathieu Charbonneau, Central European University

Judith Charlin, Consejo Nacional de Investigaciones Científicas y Técnicas, Buenos Aires

Chris Clarkson, University of Queensland

Loren G. Davis, Oregon State University

Metin I. Eren, Kent State University

Peter Hiscock, University of Sydney

Thomas A. Jennings, University of West Georgia

Steven L. Kuhn, University of Arizona

Daniel E. Lieberman, Harvard University

George R. McGhee, Rutgers University

Alex Mackay, University of Wollongong

Michael J. O'Brien, Texas A&M University–San Antonio

Charlotte D. Pevny, SEARCH, Inc.

Ceri Shipton, Cambridge University

Ashley M. Smallwood, University of West Georgia

Heather L. Smith, Eastern New Mexico University

Jayne Wilkins, University of Cape Town

Samuel C. Willis, Far Western Anthropological Research Group

Nicolas Zwyns, University of California, Davis

Index

Printed in the United States
by Baker & Taylor Publisher Services